Web Marketing for the Music Business

Tom Hutchison

NEW YORK AND LONDON

First published 2013
by Focal Press
70 Blanchard Rd Suite 402
Burlington, MA 01803

Simultaneously published in the UK
by Focal Press
2 Park Square, Milton Park, Abingdon, Oxon OX14 4RN

Focal Press is an imprint of the Taylor & Francis Group, an informa business

Notices
Knowledge and best practice in this field are constantly changing. As new research and experience broaden our understanding, changes in research methods, professional practices, or medical treatment may become necessary.

Practitioners and researchers must always rely on their own experience and knowledge in evaluating and using any information, methods, compounds, or experiments described herein. In using such information or methods they should be mindful of their own safety and the safety of others, including parties for whom they have a professional responsibility.

Library of Congress Cataloging in Publication Data
Hutchison, Thomas W. (Thomas William) Web marketing for the music business / Tom Hutchison.
p. cm. Includes index.
1. Music trade. 2. Internet marketing. I. Title.
ML3790.H986
2013781.34068'8--dc23 2012031320

ISBN: 978-0-240-82370-6 (pbk)
ISBN: 978-0-240-82385-0 (ebk)

Typeset in Giovanni
by HWA Text and Data Management, London

Web Marketing for the Music Business

Learn to create a powerful online presence that captures your audience by exposing them to the sights and sounds of your band or music project and allowing them to easily become paying fans. *Web Marketing for the Music Business* includes basics and advice on web site creation:

- Setting up your web site and web site design
- Selecting your domain name and host
- Using HTML, Java, widgets, Flash, and RSS to charge up your web site
- Using search engine optimization (SEO) methods for the best search engine rankings
- Maximizing social media sites like Facebook, YouTube, and Twitter for easy sharing by fans
- Monitoring site traffic and using analytic tools
- Adding audio and video to your site
- Choosing and using commercial download services
- Creating and managing an online store
- Finding your market online
- Creating a mobile web site and mobile media campaign

Market your band using sites like Facebook, SonicBids, and ReverbNation, where fan interaction is key, and fan-generated content can be encouraged. Learn techniques to coordinate your offline and online promotions for maximum impact.

Drawing on his own experience and the knowledge of industry experts, author Tom Hutchison brings you solid marketing advice. The companion web site for the book, www.focalpress.com/cw/Hutchison, gives you more on the ever-changing world of online promotion.

This is the perfect book for do-it-yourself musicians, managers, and labels who want to maximize sales and exposure or industry professionals seeking information on new media.

Tom Hutchison was a professor of marketing in the Department of Recording Industry at MTSU, but was on leave to serve as the director of the School of Business and Management at Husson University. He worked with a wide range of popular artists including Faith Hill, The Dixie Chicks, The Roots, and Beck. Tom also conducted market research projects for Sony, MCA/Universal, DreamWorks, and Warner Music Group. Tom passed away suddenly on Memorial Day, 2012, just after finishing the manuscript for this new edition.

Contents

Figures

List of Figures

Tables

Foreword

My friend and colleague Tom Hutchison and I knew in the fall of 2005 that the transitioning business model for the music industry required that we add to our curriculum to prepare our students for success after graduation. We were both professors in the largest music business degree program in the U.S. near Nashville at Middle Tennessee State University, and we knew we needed to develop a course that teaches how to effectively incorporate new technologies into music marketing. We had already co-written with Amy Macy a leading textbook called *Record Label Marketing,* Tom had written three other books, and I had written *Artist Management for the Music Business*. But we really needed a specific book to offer students in our new course at MTSU as well as a written navigational guide for all students and professionals in the music business. We began offering the course in the fall of 2006, and Tom was able to complete his book manuscript late the next year with the first edition in print in 2008. It was and still is the leading book on the subject. In recent years he became a national expert in online education, winning awards for Outstanding Achievement in Instructional Technology and Distinguished Educator in Distance Learning, as well as developing online curricula for three universities.

Tom Hutchison has a lot of letters after his name and a résumé packed with big-time experience in the music business. He has been a leader in academia as an author, a researcher, and through his work in countless classrooms. Said another way, he has been one of the top experts in the nation on technology and the music business, and he was always my go-to person with my questions about adopting new technologies into my work in both physical and virtual classrooms. In this way, both you and I are lucky because we have *Web Marketing for the Music Business* as our guide in this new era of music marketing.

This book is important to you because you simply can't be in the "business" part of the music industry without understanding how to connect music with willing buyers. And frankly, you'll either be left behind or never get the chance to be commercially successful marketing music unless you keep this book close to you in your work. Tom says in this edition of the book, "… the old-school marketing techniques are giving way to a new type of marketing that incorporates the latest in communication and entertainment technology, and that is keeping up with the rapid pace of innovation. But old habits die hard." This brief, last sentence goes to the heart of what many elder statesmen of the music business are often heard saying, "I don't understand it, but we need to hire some people who do."

Clearly, a key reason this book is a must-have is because one of the primary hiring criteria music business companies seek is people with an understanding or experience using new technology and social media in a music business environment. Businesses that contact me for references for interns or job applicants most often ask whether my former students have training in music business marketing technology and social media. It often is the difference whether a graduate is hired or whether they're still waiting for a call or an email for an interview. Spending time with this book will give you the advantage many others won't have.

Finally, let's put the math to this quickly changing industry and the technologies driving it. SoundScan's Trudy Lartz said in October 2012 the sale of CDs is *down* 14% over 2011 and the sale of digital album downloads is *up* 16% over 2011. SoundScan data continues to describe the years-long decline of physical sales and the rapid increase of digital music. It's not music that's on the decline anymore – it's the way it is delivered to us and then how we consume it. This second edition of *Web Marketing for the Music Business* represents the last installment for the vision Dr. Thomas Hutchison had in 2005 to teach how these new technologies and social media should be used as key components of successful music marketing. And as these sales data continue to demonstrate the way the music business is changing, this – his final book – becomes even more important to all of us.

Tom died suddenly on Memorial Day this year. And in a way, *Web Marketing for the Music Business* became a memorial to celebrate his career of teaching and inspiring those thousands of students who came to him with their dreams of careers in the music business. The industry will miss his work on its behalf. And to his friends and colleagues? He's one of those irreplaceable treasures of this life.

As we use this book as part of Tom's legacy, I extend my wish to you for success in today's and tomorrow's music business.

Paul Allen
Author and professor of music business at MTSU,
Belmont University, and Cumberland University
Author of *Artist Management for the Music Business*
Nashville, Tennessee 2012

Chapter 1
Development of Music Marketing on the Internet

BACKGROUND ON THE INTERNET

The Internet is described as a global network connecting millions of computers. Unlike online services, which are centrally controlled, the Internet is decentralized by design. Each Internet computer, called a host, is independent. It got its start as several universities and government organizations saw the benefit in connecting mainframe computers together to share information. The military saw its use for communication in case the country was ever under attack. Little did they realize that in 2007, a blockbuster Bruce Willis movie would feature the country under attack of just such an infrastructure.

Electronic mail was introduced in the early 1970s along with communication protocol still in use today: Transmission Control Protocol/Internet Protocol (TCP/IP), and File Transfer Protocol (FTP). The domain name system was introduced in 1985 with the extensions of .com, .org, .net, .gov, .mil and .edu. Then in 1989, Tim Berners-Lee of CERN (European Laboratory for Particle Physics) developed a new technique for distributing information on the Internet he called the World Wide Web. Based on hypertext, the web permits the user to connect from one document to another at different sites on the Internet via hyperlinks (specially programmed words, phrases, buttons, or graphics). Unlike other Internet protocols, such as FTP and email, the Web is accessible through a graphical user interface. This brought the development of web browsers that could decipher coding in documents and display them according to a set of instructions called hypertext markup language. Meanwhile, the first efforts to index the contents of the Internet were introduced in the early 90s.

In 1995, several online services (America Online, CompuServe and Prodigy) began providing their software to computer manufacturers to include on new computer hard drives.[1] This propelled Internet use as new computer owners took advantage of the free trial period offered by several of these services. By 1996, 43.2 million (44%) U.S. households owned a personal computer, and 14 million of them were online.

In 1999, college student Shawn Fanning introduced Napster, a program that allowed users to swap music files over the Internet. This began the downfall of the recorded music industry as it had existed until that time. By 2001, the courts had ruled that Napster was in violation of copyright laws and it was forced to shut down its file-sharing operations. Other new file sharing services sprang up to fill the void, including Aimster, Grokster, Kazaa, Limewire and BitTorrent. By

January of 2002, 58.5% of the U.S. population (164.14 million people) were using the Internet. By 2003, it was estimated that illegal file-sharing of music files had grown to about 2.6 billion files per month.

WEB 2.0

In 2001, the "dot-com" bubble started to burst as traditional companies began to pull back on their Internet advertising budgets. Massive layoffs followed as content providers sought less expensive ways to provide content for Internet users. This brought about the development of Web 2.0, a system in which users are the content providers and the Web 2.0 companies just provide the platform. The term Web 2.0 was coined by media writer and analyst Tim O'Reilly, the founder and CEO of O'Reilly Media. Web 2.0 is basically the current generation of popular web applications that are democratic in nature, that put the user in control, and that rely on the aggregate wisdom of the masses to provide content and feedback. One of the early examples is Wikipedia, launched in early 2001 as "harnessing the wisdom of crowds to build an online encyclopedia" (*Information Week*); it went live in 2005.

MySpace was founded in 2003, which allowed casual users to build a web page for themselves and build social networks online. Flickr launched in early 2004, giving users the opportunity to store and share photos online. Yahoo bought Flickr in 2005. Other Web 2.0 innovations include Craigslist, Ebay, Digg, Facebook, YouTube, Second Life, etc. In 2006, *Time Magazine* awarded its person of the year award to "you—the creators on Web 2.0." *Time* describes Web 2.0 as "... a tool for bringing together the small contributions of millions of people and making them matter." By 2006, there were also more than 92 million web sites online.

Another innovation that has breathed life into Web 2.0 is Google AdSense. Introduced in 2005, AdSense provides a way for web site hobbyists to monetize their labor-of-love web pages. The millions of non-commercial web sites that were set up and run for personal interest now had a means to make money from their visitors by running ads. Many of these sites were simply advice, how-to, or other informational sites that did not necessarily sell products or charge money for the information they provide. So, with little setup and at no cost, these mom-and-pop web sites were able to sign up on Google to participate in AdSense and

> The concept of "Web 2.0" began with a conference brainstorming session between O'Reilly and MediaLive International. Dale Dougherty, web pioneer and O'Reilly VP, noted that far from having "crashed", the web was more important than ever, with exciting new applications and sites popping up with surprising regularity. What's more, the companies that had survived the collapse seemed to have some things in common. Could it be that the dot-com collapse marked some kind of turning point for the web, such that a call to action such as "Web 2.0" might make sense? We agreed that it did, and so the Web 2.0 Conference was born.
>
> **Tim O'Reilly, http://www.oreillynet.com/**

Table 1.1 Example Web 2.0 sites

Digg: offers a way to suggest stories for the votes of a wider readership.
Craigslist: online classified ads for free.
Wikipedia: online encyclopedia with users contributing content.
Flickr: post and share digital photos online.
Slide.com: create slide show and share or embed into your social networking page.
YouTube: Upload videos and share or embed into your social networking page.
MySpace: an online community with more than 70 million members, most notably teenagers and bands.
Meebo: lets you access your buddy list and IM all you want from its Web site.
Second Life: a virtual world where users can create an alternate life and interact with others.
Facebook: the world's largest social networking site with over one billion users.
Twitter.com: A text messaging service that lets people send notes to groups.
Pinterest: a social networking "pinboard" that allows users to add images of favorite products and services much like a shared online scrapbook.
Bebo: online community allowing you to share photos and blogs, draw on your own and other members' White Boards and find school and college friends
Amazon marketplace: ecommerce for small vendors.
LastFM: create your personal Internet radio station by rating songs.
Pandora: creates a radio station based on collaborative filtering based on your musical preferences.
Eyspot: Upload your video and use its tools to edit it and publish it on other sites.
LinkedIn: social networking site for career advancement.
NetFlix: uses collaborative filtering to allow its members to rate movies and receive recommendations.
Blogger.com: A site owned by Google that makes it easy to start and maintain a blog.
Skype: audio and video communication via the Internet.
eBay: online auction where eBay members sell items to other members.
Dodgeball: a service that lets people alert their friends to where they are via cell phone messages.
Google Earth: create mashups of aerial photographs overlaid with photos and descriptions.

start allowing Google to run topic-related ads on these sites. Site owners are paid a small sum each time a visitor "clicks through" one of the ads to visit the advertiser's web site. The advertisers participate through the other side of the equation, Google AdWords.

MUSIC AND THE INTERNET

When the Internet was in its infancy in the late 1980s, only a few audiophiles and tech-savvy musicians were online, communicating through the few rudimentary online services such as CompuServe and General Electric's Genie. The author of this book was one of those early online music pioneers, as a systems operator on CompuServe's MIDI forum. Modem baud rates were such back then that even small text files took a while to download, and the online services were charging by the minute. One of the most progressive music exchanges occurring at the time was in the form of MIDI data files,[2] small enough to move through the dial-up systems. The MIDI forum had three sections, (1) the message section, where members could post messages for each other in the exchange of ideas, (2) the libraries section that hosted the MIDI library of songs, and (3) the conference room where members held live chats, sometimes featuring a well-known artist or producer. As musicians started to exchange ideas via MIDI files over the Internet, the first online musical collaborations were created. A musician would start a song, record the data in MIDI, upload the file to the forum or send it to a particular forum member, and then allow them to modify the file, thus modifying the composition. The forum was also a place for the exchange of ideas, and some members published articles about music and the music business. The forums were not consumer-oriented at first. Then in the early 1990s, Geffen Records decided to have an online presence for fans and opened up a Geffen forum on CompuServe.

By 1994, Geffen Records was trying new ways to reach consumers through the Internet with their debut of an Aerosmith song online through their CompuServe forum (EW.com, 1994). For one week, they offered an outtake of the song "Head First" from the band's "Get a Grip" album. Early Internet adopters were able to download the song, which took up to several hours; the band waived their royalties while CompuServe suspended their usual $9.60 hourly charge. Geffen's Jim Griffin commented at the time "We're not saying this is how you'll

Geffen and CompuServe

"Geffen provides CompuServe and Internet users access to graphics, sound clips, tour schedules and artist biographies. Geffen artists interact with users with on-line messages and conferences. Geffen regularly runs contests offering prizes to promote on-line use. Geffen employees surf the Internet, CompuServe and other on-line services in the course of their work, including the pursuit of new talent for the label."

ComputerWorld Honors, 1995
http://www.cwhonors.org/search/his_4a_detail.asp?id=2213

get your music in the future. But we did want to try it out." Shortly after that, Geffen was one of the first labels to create its own web site. One way to look at old web pages from history is through the "Wayback Machine" at http://www.archive.org/web/web.php. Some of the original Geffen Records pages and assets are archived at http://web.archive.org/web/*/geffen.com/*

In 1995, Polygram Records introduced a prototype record label web site for consumers at the 1995 National Association of Recording Merchandisers (NARM) convention in San Diego. The label touted to skeptical music retailers that the site was not designed to siphon off retail business but to guide consumers to record stores that carried the products featured on the site. Soon after that, scores of record labels and artists began setting up web sites and procuring their domain names, while tiptoeing around the issue of cutting into retail store sales.

Music writer Peter Spellman wrote in 1995:

> "Online distribution–actually delivering that album order not through the mail but as digital bits down telephone or cable wires–is still pretty much in the "wouldn't it be neat?" stages...But the fact remains that the entire history of recorded music will soon be available for instant downloading. The technology to do this already exists, though availability to the public is still a few years away."
>
> **Peter Spellman, *The Musician's Internet*, 2001**

THE MP3 FORMAT

By 1997, the first viable audio compression format, MP3, was being introduced to consumers through several computer audio players such as the popular WinAmp.[3] Consumers were able to take audio files from their CDs or master tapes, compress the files and email them to other Internet users. In 1998, the web site MP3.com launched after noticing an abundance of web traffic using the search term MP3 to look for free music files. The site offered a way for people to post their favorite (or original) music files to share with others and was, at that time, strictly an advertiser-supported project. "At its peak, MP3.com delivered over 4 million MP3 formatted audio files per day to over 800,000 unique users on a customer base of 25 million registered users" (Wikipedia). In 2000, MP3.com launched a new service MyMP3.com that allowed registered users to upload a copy of music they already owned and then stream it to the computer of their choice. The company saw this as a legitimate way to honor copyright protection, but the Recording Industry Association of America (RIAA) and the labels did not agree, and sued MP3.com. The company settled with the labels and the service was discontinued, but by then Napster and other illegal peer-to-peer (P2P) file sharing networks were in full swing.

Also in 1997, the Rio was introduced as the first mass market MP3 player. The company, Diamond, was immediately sued by the RIAA who claimed the device allowed for copyright infringement and storage of illegal copies of songs. The RIAA claimed the Rio was made for illegal pirating. The courts disagreed and the

Figure 1.1
The Diamond Rio MP3 player (Courtesy of Diamond Multimedia).

Rio hit the market. In upholding this decision, the appeals court said the Audio Home Recording Act, which the RIAA had cited in its suit, prohibits devices that make copies from digital music recordings. The court said the Rio does not make copies but simply stores files it gets from computers (Salon, 06/16/1999). But the original Rio players were limited: storage capacity topped out at 64MB but most had only 32MB, and data transfer via the parallel or USB 1.0 port was very slow. Several companies created audio software to allow for MP3 encoding and playback facilities on the computer, both for the MacIntosh and the PC platforms.

The original players were more of a novelty than a practicality: "for a mere $200, [you can] listen to 60 minutes of CD-quality music in the MP3 audio format" (Salon, 10/28/1998). Users had to return to their computers and reload the player to listen to more songs or use removable media such as SmartMedia cards to hold additional music. By 1999, portable MP3 players from a number of manufacturers were popular, and other software developers, such as Liquid Audio and a2b, scrambled to develop audio compression formats that were more secure and would appeal to piracy-conscious record labels. In the fall of 1999, Napster was in full swing and MP3 players were popular with college students. The RIAA began contacting universities and targeting college students, who were the most likely copyright violators. Access to high speed Internet on campuses was fueling massive peer-to-peer music file sharing. The RIAA sent notices to over 300 universities warning them that students were hosting illegal MP3 files on campus servers. Janelle Brown, writer for Salon.com, wrote in November, 1999 "there are probably millions of illegal MP3 files and music traders online -- pity the poor fool whose job it is to track them all down" (Brown, 1999a).

By 2001, Brown was writing that the promise of a digital revolution that takes power away from the major conglomerates and puts it in the hands of the creative producers was all but dead. "Innovation is being sledge-hammered out of existence by legal threats and buyouts" (Brown, 06/01/2001). Vivendi Universal purchased MP3.com, BMG bought Myplay.com, the RIAA had defeated Napster and gone after several other services. Consumer frustration was growing because the major labels were not providing a legal, commercial alternative to illegal downloading at a time when the consumer was more than ready to convert from CDs to digital tracks stored on a computer system. "Napster had single-handedly turned millions of consumers on to the world of MP3s" (Brown, 06/01/2001). Many of the startups that avoided legal action were having a difficult time with profitability because the labels were unwilling to license music for legal downloads. The major labels were creating online distribution services of their own, but failed to capture the commercial market with their cobbled-together group of startups such as MusicNet and Duet. They had yet to hit on a formula for success in making a dent in the illegal P2P file trading. Until Steve Jobs of Apple convinced them he had the answer.

ITUNES STORY

In January of 2001, Apple introduced the iTunes jukebox software at Macworld. Unlike the predecessors, iTunes was simple, with most of the screen dedicated to a browser for finding music (www.ipodobserver.com), but the initial release was only for Mac users, with the promise of a Windows version to follow. iTunes included support for creating mixes, burning CDs and for downloading to popular MP3 players, including the Rio. Unfortunately, the current crop of MP3 players did not offer a very Mac user-friendly interface. Steve Jobs ordered the development of an MP3 player that would work seamlessly with the iTunes system. Part two of that plan was to create an online music downloading store. Thomas Hormby writes "Apple was not the first to create such a store, but it was the first not to fail spectacularly. The most notable pre-Apple music store was Pressplay, which was a joint venture between major record labels" (Hormby, 2007). Unlike the other legal downloading services that used a cacophony of policy restrictions (limited number of downloads and streams per month), the iTunes store was simple: 99 cents per download, up to three copies. It was an instant success, quickly capturing 76% of the hardware market and 82% of the legal music download market (Silverstein, 2006).

In 2007 the record labels sought a price restructuring of iTunes but were unable to persuade Jobs, who believed the current pricing strategy of 99 cents was crucial for the continued success of the service. According to an article in *Businessweek*, "Universal Music Group chief Doug Morris and his fellow music executives realized they had ceded too much control to Jobs" (*Businessweek*, 2007). Morris refused to re-sign a multi-year contract with iTunes and instead opted for a month-to-month agreement while nurturing alternative retailing options, including a new service called Total Music. One of the big question marks about this new business model is that it doesn't offer anything that hasn't already been tried—and failed. The plan does call for hardware makers and cell-

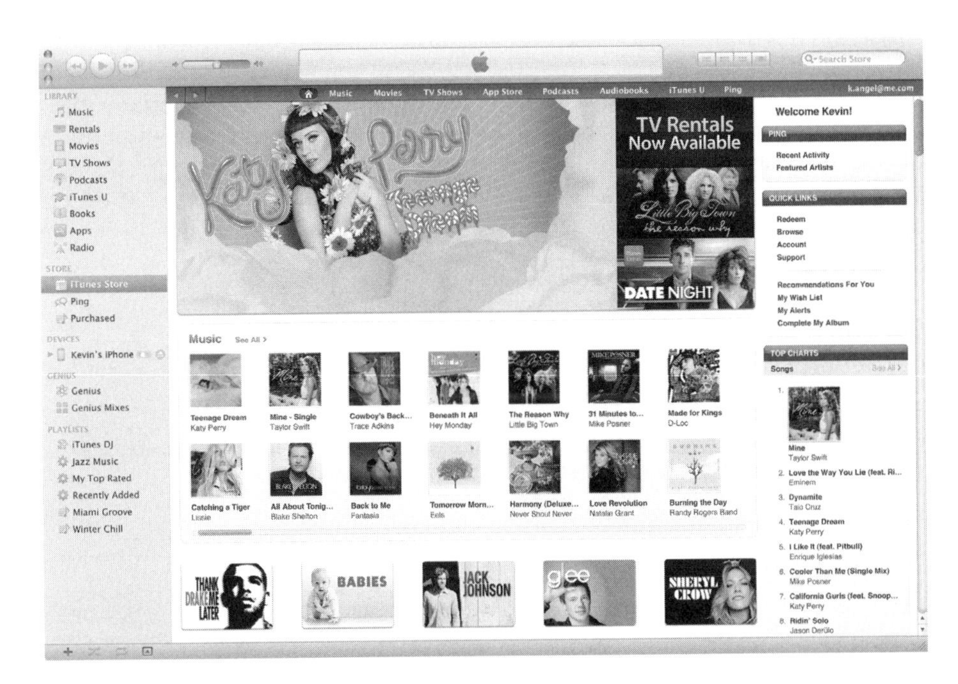

Figure 1.2
iTunes computer interface (courtesy of Apple, Inc.)

Figure 1.3
Apple mobile devices and iCloud (Courtesy of Apple, Inc.)

phone makers to absorb the cost of a monthly subscription service—something consumers themselves have so far rejected. Consumers don't seem to want to "rent" music by the month; they prefer to own the music. Consumers are concerned about losing access to their music library if they leave the service or if the service increases their fees. The existing subscription services have only a few million customers whereas iTunes has hundreds of millions of users.

This new model calls for hardware makers to add the $5 monthly subscription cost (for 18 months) to the price of the hardware, increasing the retail price by $90, so ultimately, the consumer is paying for the subscription service. Meanwhile, Amazon has launched an MP3 retail store and is in the process of licensing music from the major labels. Wal-Mart and Best Buy have also made an entrance into the music download market. None of these is giving the record labels a new pricing strategy that they were originally seeking from Apple. And in spring 2008, iTunes moved past Wal-Mart to become the leading music retailer in the U.S.

THE CURRENT ONLINE ENVIRONMENT

The development of social networking sites dedicated to discovering and sharing new music brings hope that the millions of unsigned musicians who have recently acquired the technology to create high quality recordings will now be able to achieve a moderate level of success, creating a middle class of professional entertainers that has been missing in the previous decades. A "record deal" with a major record label may no longer be necessary to enjoy the success of having a large fan base to purchase recorded music, concert tickets and artist memorabilia. Several chapters in this book are dedicated to outlining how to make that happen.

Glossary

Collaborative filtering Software applications that offer recommendations to consumers based upon that consumer's pattern of behavior and consumption and drawn from analyses of the consumption behavior of a large aggregate of consumers.

CompuServe An online information service that provides access to the Internet, email, instant messaging and an integrated contact list. Founded in 1969 as a timesharing service, CompuServe is one of the oldest online services, being the first to offer email in 1979 and online chat a year later.

Electronic mail (email) a store and forward method of composing, sending, storing, and receiving messages over electronic communication systems.

File Transfer Protocol (FTP) the protocol for exchanging files over the Internet. FTP is most commonly used to download a file from a server using the Internet or to upload a file to a server.

Genie An online information and bulletin board service that closed its doors at the end of 1999. Genie (General Electric Network for Information Exchange) was set up as a joint venture between GE and Ameritech in 1985. Its roundtable discussions, chat lines, games and Internet access attracted a niche of science fiction aficionados as well as horror and fantasy writers.

Google AdSense A contextual ad network that gives web publishers the opportunity to serve advertisements on their web site in return for commissions on a cost-per-click (CPC) basis.

Google AdWords cost-per-click (CPC) advertising. You pay only when users click on your ad. It has features that allow you to control your costs by setting a daily budget for what you are willing to spend per day. AdWords sponsored listings are also being shown on Google's partner sites.

Hyperlinks commonly referred to as a link, is a text string (i.e., sequence of characters) or an image in an electronic document that serves as a user-activated switch that causes another, predetermined location in the same page or document or an entirely different page to display on the screen.

Internet a global network connecting millions of computers. Unlike online services, which are centrally controlled, the Internet is decentralized by design.

MIDI MIDI (Musical Instrument Digital Interface) is a protocol designed for recording and playing back music on digital synthesizers.

MP3 (MPEG-1 Audio Layer-3) is a standard technology and format for compressing a sound sequence into a very small file (about one-twelfth the size of the original file[4]) while preserving the original level of sound quality when it is played.

Napster a controversial application that allowed people to share music over the Internet without having to purchase their own copy on CD. After downloading Napster, a user could get access to music recorded in the MP3 format from other users who are online at the same time.

Online service An organization that provides access to the Internet as well as proprietary content. Before the Internet became widely used by the general public, all online services were self-contained organizations known for their unique mix of databases and resources.

P2P On the Internet, peer-to-peer (referred to as P2P) is a type of transient Internet network that allows a group of computer users with the same networking program to connect with each other and directly access files from one another's hard drives. Kazaa and Limewire are examples of this kind of peer-to-peer software.

Web 2.0 the popular term for advanced Internet technology and applications including blogs, wikis, RSS and social bookmarking. The expression was originally coined by O'Reilly Media and MediaLive International in 2004, following a conference dealing with next-generation Web concepts and issues.

Notes

1 These companies, as well as GEnie and other smaller ones, developed dial-up systems as early as the late 1960s, but 1995 saw the first widely used WWW access and bundling.

2 MIDI files are the digital equivalent of a player piano scroll or music box disc. Bits of information instruct musical notes on a specialized musical instrument to turn on and off at particular points in the song. For MIDI, these are simple, small files that activate and direct the more complicated musical sound modules to create complex sounds that emulate a host of musical instruments.

3 WinPlay3 was the first MP3 player for PC's. It came out in 1995. WinAmp was second, in 1997.

4 Depending upon the selected bit rate.

Bibliography

Boehlert, Eric (2000). Rio's Pyrrhic victory. http://www.salon.com/2000/09/19/rio_fallout/

Brown, Janelle. (1998). Blame it on Rio. Netheads love the MP3 digital-music format. Why does the music industry hate it so much? Salon.com. http://archive.salon.com/21st/feature/1998/10/28feature.html (10/28/1998).

Brown, Janelle. (1999a). MP3 crackdown. As the recording industry "educates" universities about digital music piracy, students feel the heat. Salon.com. http://www.salon.com/tech/log/1999/11/17/riaa/index.html?source=search&aim=/tech/log. (11/17/1999).

Brown, Janelle. (1999b). MP3: Here, there, everywhere. The latest digital music players let you play MP3s on your home stereo, in your car or on the run -- but are they any good? Salon.com. http://www.salon.com/tech/review/1999/12/14/mp3/. (12/14/1999).

Businessweek (2007). Universal Music Takes on iTunes. *Business Week*. www.businessweek.com/print/magazine/content/07_43/b4055048.htm?chan=gl

ComputerWorld Honors: MIDI/Music Forums of CompuServe Award and Case Study. http://www.cwhonors.org/search/his_4a_detail.asp?id=2214

ComputerWorld Honors: Areosmith/Geffen Award and Case Study. http://www.cwhonors.org/search/his_4a_detail.asp?id=2213

Factmonster. Internet timeline. http://www.factmonster.com/ipka/A0193167.html

Go2Web20. Online tools and applications. http://www.go2web20.net/

Grossman, Lev. (2006). Time's Person of the Year: You. http://www.time.com/time/magazine/article/0,9171,1569514,00.html

Guglielmo, Connie (04/04/2008). Apple moves past Wal-Mart as leading music retailer in U.S.. *The Tennessean* newspaper, section e, pg. 3.

Hormby, Thomas (2007). History of iTunes and iPod. The iPod Observer. http://www.ipodobserver.com/story/31394 (May 10th, 2007)

Information Week. A brief history of Web 2.0. http://www.informationweek.com/1113/IDweb20_timeline.html

O'Reilly, Tim (2005). Whot is Web 2.0. http://oreilly.com/web2/archive/what-is-web-20.html

Silverstein, Jonathan (2006). iTunes: 1 billion served. ABC News. http://abcnews.go.com/print?id=1653881. (02/23/2006).

Spellman, Peter (2001). *The Musician's Internet: Online Strategies for Success In the Music Industry*, Berklee Press Publications, Boston, MA.

Webopedia. Definitions. www.webopedia.com

Chapter 2
State of the Market

TRENDS IN INTERNET USE

This chapter looks at the state of the market and current trends in Internet use. The discussion includes a description of the market, an overview of the state of e-commerce, and a look at specific trends in music sales online, including digital downloads.

Who's Using the Internet?

According to the Pew Internet Project's research in 2010, 79% of U.S. adults use the Internet, up slightly from 75% in 2007. Among the millennium generation, Internet use is at 95%, up from 93% in 2008 and from 92% in 2007. In 2011, Pew reported that two-thirds of U.S. adults had high-speed access at home, up from 47% in 2006. Point-topic.com puts U.S. broadband penetration at 77.5% of households in 2011.

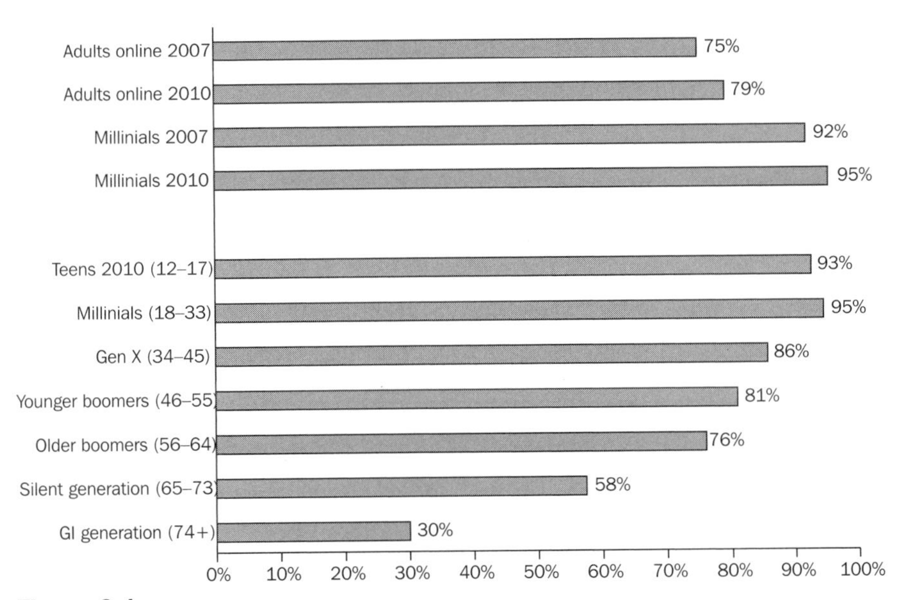

Figure 2.1
Internet usage among age groups (source: Pew Internet Project)

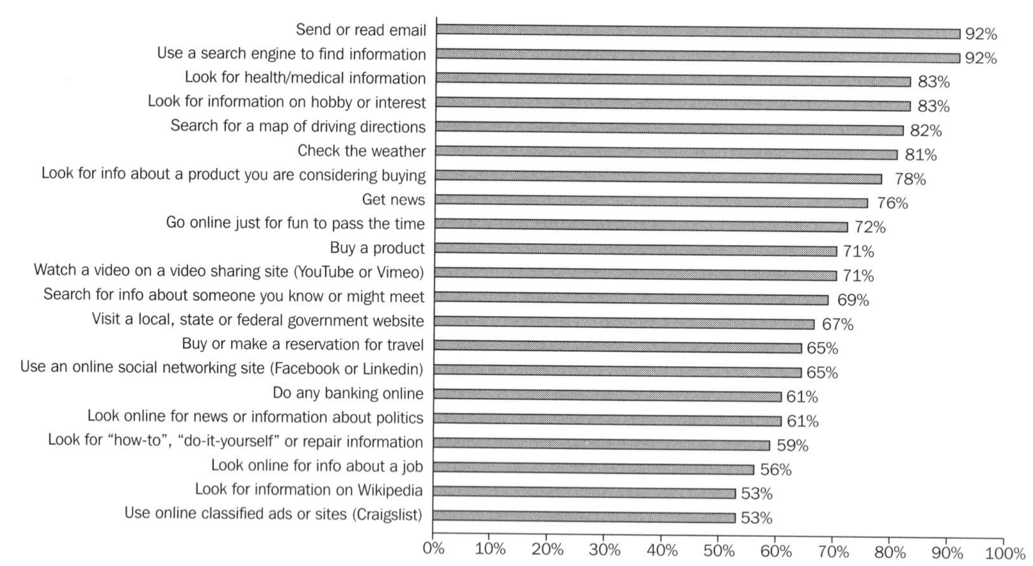

Figure 2.2
Percentage of Internet users engaged in online activities by age group

Ninety-three percent of teens use the Internet, yet it is the millinials who are more likely to use it for email, social interaction and to watch videos. Teens are the most predominant users for sending instant messages, playing online games and reading blogs. According to the *PEW 2010 Generations and Technology* study, "when teens do use email, they tend to use it more in formal situations or when communicating with adults [rather] than to communicate with friends" (Zickuhr, 2010). This is an important trend to remember when planning a marketing campaign aimed at teenagers.

Among adults, email and search engine use were the most popular of all online activities, with 92% of online users, according to a 2011 Pew research study (Purcell, 2011). Sixty-three percent of *all adults* have made an online purchase (not just online users). Fifty-one percent have listened to music online, up from 34% in 2006. And 27% have downloaded music files. An equal number (27%) state they have shared files from their computer with others.

Trends in e-Commerce

Many products are sold online as retailers shift some marketing efforts from retail stores or mail-order catalogs to online stores. Online commerce is just an extension of the mail-order business that has been around since the days of the first Sears-Roebuck catalog. In the 1960s and 1970s, credit cards and toll-free numbers helped to expand the mail-order industry. In the mid-1990s, as *online* became a household word and Internet use skyrocketed, consumers were still somewhat reluctant to purchase products online for a number of reasons: credit card security, lack of trust in the online retailers, inability to see the product before ordering, and clunky, complex storefronts and shopping carts.

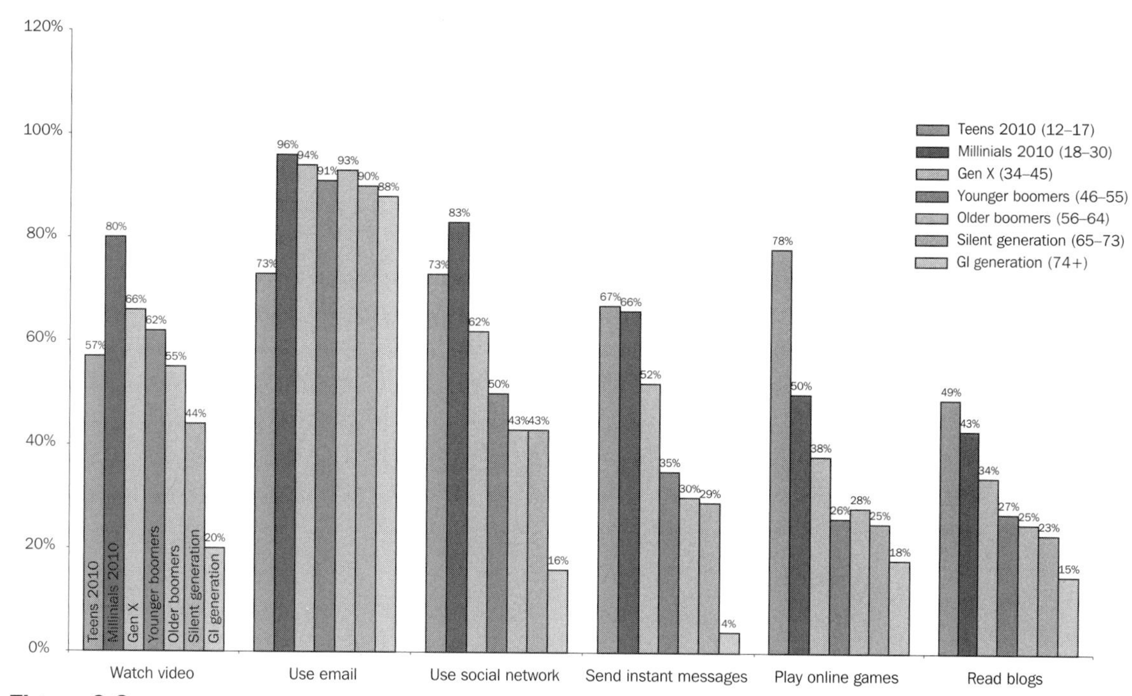

Figure 2.3
Activities by age group 2010

In 1992, CompuServe offered its users the chance to buy products online. In 1994, Netscape introduced a browser capable of using encryption technology, called secure socket layers (commonly referred to as SSL), to transmit financial information for commercial transactions online. Then as shopping interfaces improved, more established stores began an online presence, alternative payment methods evolved, and consumers became more comfortable with making purchases online. In 1995, two online retail giants, Amazon and eBay, were introduced.

In January 2008, Nielsen Online reported that globally, 875 million consumers had shopped online, which is more than 85% of web users. This was up by 40% from 2006. South Korea had the highest percentage of online shoppers, with 99% of Internet users, followed by the United Kingdom, Germany, and Japan, all with 97%. In the United States, 94% of Internet users purchase products online according to the Nielsen report.

U.S. retail e-commerce sales reached $165.4 billion in 2010, up 14.8% from $144.1 billion in 2009 and $142 billion in 2008. This now makes up 4.3 percent of total 2010 retail sales according to the U.S. Census Bureau, (2011). Forrester measured it at $176 billion in 2010. E-commerce sales are expected to rise in the U.S. to $279 billion by 2015 (Forrester, 2011). The top-selling online retailers for 2010 were Amazon, Wal-Mart, eBay, Best Buy, JC Penney, Kohls, Target, Macys, and Sears.

Table 2.1 shows the most popular products to purchase online.

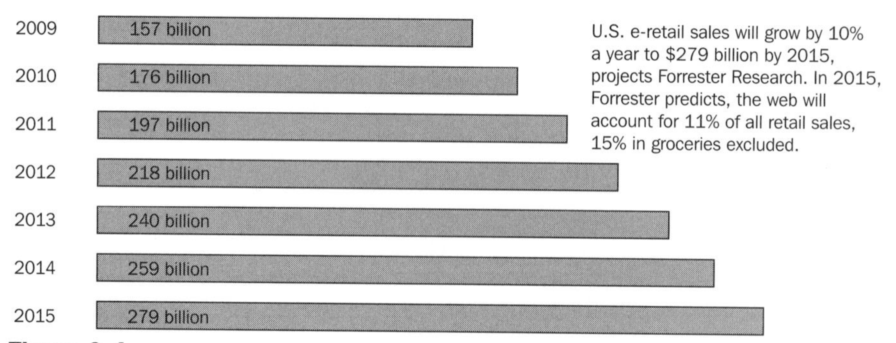

Figure 2.4
U.S. e-retail sales 2009–2015

Table 2.1 Most popular online purchases 2010

Software, books, music, videos and flowers	26%
Travel	21%
Computer hardware, consumer electronics and office supplies	16%
Apparel, footwear, jewelry and linens/home decorations	13%
Health, beauty, and food and beverages	8%
Toys, video games and sporting goods	7%

Source: EstoreFrontGuide

Online shoppers have become savvy. For the 2010 holiday shopping season, 58% of smartphone owners used the Internet and phone applications such as RedLaser to compare prices (CreditSense, 2011). Online consumers also rely on customer product reviews, with 57% stating in 2010 that they rely on customer reviews for their purchase decisions (Nielsen, 2010). According to Nielsen, shoppers tend to stick with shopping on sites they are familiar with, and 60% said they buy mostly from the same site. This information is important to consider for a small business as they decide how and where to sell their music online. Sixty percent of online shoppers used a credit card for purchases in 2007, whereas 25% had used PayPal. By 2011, PayPal's market share was up to 18% of sales (Williams, 2011a).

Mobile ecommerce transactions are on the rise. Juniper Research stated in 2011 that over 100 million people globally were using mobile money services (transactions, balance check, transfer of funds, and so forth). This is expected to rise to 200 million by 2013, with much of the growth coming from Africa and the Middle East (Williams, 2011b). Mobile commerce will be addressed in Chapters 14 and 15.

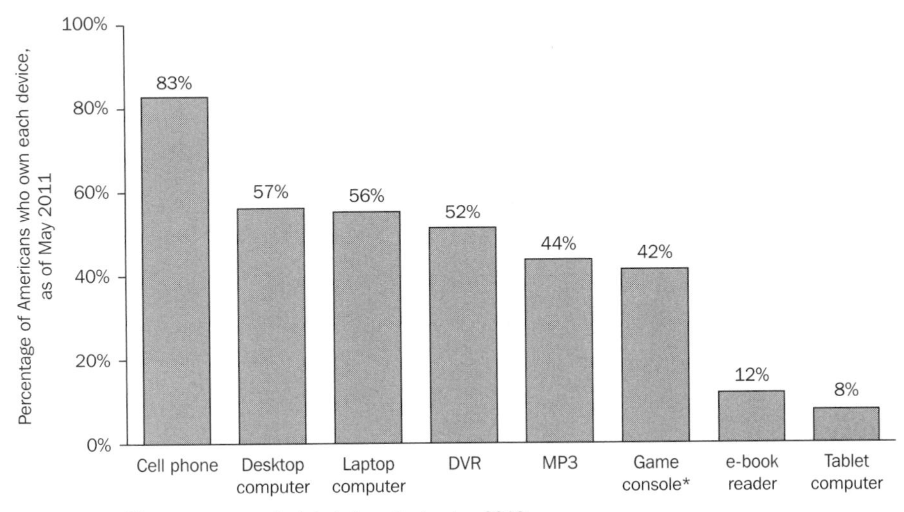

Figure 2.5
Most popular gadgets (source: Pew Center)

Music Sales Trends

Music is a dream product for e-commerce—it is one of only a handful of products that can be sold, distributed, and delivered all via the Internet. Of course, this is a double-edged sword. The ease with which music can be transferred over the Internet has also led to the continuing crisis of illegal file sharing. In an effort to monitor recorded music sales and determine trends and patterns, the industry in general and SoundScan in particular have come up with a way to measure digital album sales and compare them with music sales in previous years. When SoundScan first started tracking digital download sales, the unit of measurement for downloads was the single track, or in cases where the customer purchased the entire album, the unit of measurement was an album. But this did not give an accurate reflection of how music sales volume had changed, because most customers who download buy songs à la carte instead of in album form. In an attempt to more accurately compare previous years with the current sales trend, SoundScan came up with a unit of measurement called *track equivalent albums* (TEA), which means that 10 track downloads are counted as a single album. Thus, the total of all the downloaded singles is divided by ten and the resulting figure is added to album downloads and physical album units to give a total picture of "album" sales. Here is an example of how this works from Billboard. biz.

> When albums are tallied using the formula of 10 digital track downloads equaling one album, the 582 million digital track downloads last year translates into 58.2 million albums, giving overall albums a total of 646.4 million units. The overall 2006 total of 646.4 million is a drop of 1.2% from 2005's overall album sales of 654.1 million.
>
> **Ed Christman, *Billboard*, January 4, 2007**

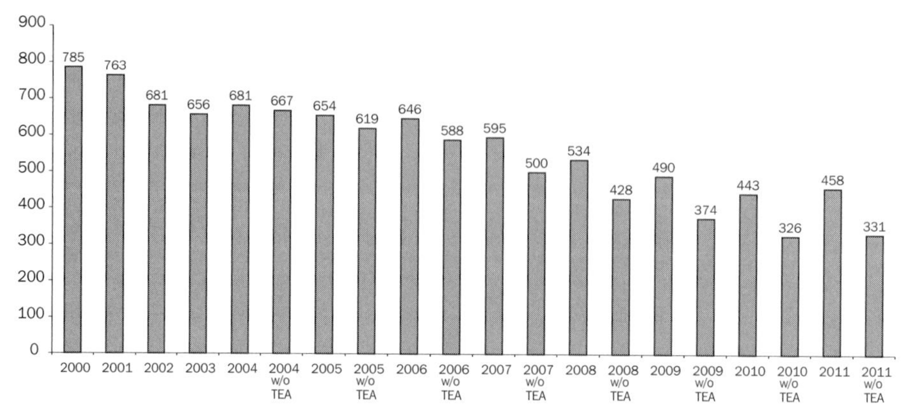

Figure 2.6
U.S. album sales in millions (source: SoundScan)

Having established the TEA as a new unit of measurement, industry trends show the following: U.S. album sales have continued to slide every year from a high in 2000 of 785 million units to 443.4 million units in 2010 (including TEA). (Without the addition of the TEA, album units in 2010 were 326.2 million.) This represents a drop of 9.5% from the 2009 overall album sales of 489.8 million.

By the close of 2010, total album sales were down 12.7% from the previous year and down 10% for the holiday season. With track-equivalent albums factored in, the slide was 9.5% from 2009. The one bright spot has been digital *album* sales, which rose 13% in 2010. Meanwhile, retail stores continued to reduce the amount of shelf space they devoted to recorded music as online sales and digital downloads continued to erode at the physical unit market share. Sales of physical albums fell by nearly 19%, down to 240 million physical units. Digital album sales reached 86.3 million in 2010, a 13% increase over 2006. Digital album sales accounted for 26% of total album sales (without TEA included). Labels have been making an effort to convert singles buyers into digital album buyers, sometimes by not making à-la-carte tracks available and sometimes by deep discounting the album compared to the individual tracks. When pricing albums, the DIY musician should consider a similar pricing strategy.

In the United States, digital music accounted for 46% of all *unit* purchases in 2010, up from 40% in 2009 and 32% in 2008. More than 1.172 billion digital tracks were sold in 2010 (SoundScan) up from 1.16 billion in 2009 (a mere 1%). When all music formats are included (music videos, singles, albums, digital, etc.), the U.S. industry fell 2.4% in 2010. Year-to-date digital tracks for November 2011 were up 10% over the same time in the previous year as the industry overall showed a slight uptick for 2011.

U.S. Historical Trends

In 2004, catalog products accounted for 46% of all digital music sales compared with 35% for physical products. This was at a time when the replacement cycle for CDs was at an end. In the late 1980s and early 1990s, labels enjoyed record profits as consumers sought to replace their vinyl and cassette music collections with the improved digital CD format. As a result, catalog sales ran as high as 50% of units sold. Although the conversion from CD to MP3 or other portable format has generated some interest in catalog sales, consumers are not repurchasing music they already own on CD. Instead they are rounding out their catalog collection by cherry-picking songs they want in their library. The digital downloading format has, however, allowed major record labels to reissue recordings that previously had been deleted from the active catalog because, now, the cost of making them available is low.

> Digital stores offer consumers a far greater virtual shelf space than the largest traditional brick and mortar stores. This means that a broader range of repertoire, including specialist, vintage or hard-to-find recordings is now available to fans.
>
> **IFPI Digital Music Report (2008)**

The switch from CDs to digital downloading varies when separated out by genres. According to Nielsen SoundScan, digital album sales for 2010 were as follows: the genres of alternative, rock, soundtracks, and electronic showed an active downloading market, with fans of these genres more likely to select to download an album rather than buy the CD. Fans of country, R&B, Latin, and gospel are more likely to buy CDs than download albums. The DIY musician may want to take this under consideration when ordering the production of physical product, however sales at live shows continue to be dominated by physical product. This may change as mobile devices become more sophisticated and smartphones are more widespread.

GLOBAL DIGITAL MUSIC SALES

The International Federation of the Phonographic Industry's (IFPI) Digital Music Reports for 2010 outlines some trends in global online music sales, including the following highlights:

- More than 400 licensed digital music services are available worldwide.
- Globally, digital music sales were up 9.2% in 2009 (IFPI). In 2010, digital sales continued to grow to account for about 29% of global sales, up from 15% in 2007.
- The increase in the value of the digital music market from 2004 to 2010 was 1000%.
- In 2010, the global digital music sector was worth an estimated US$ 4.6 billion, up six per cent on 2009.
- iTunes has sold more than 10 billion downloads since it was established in 2003.
- Digital album sales increased more sharply than singles in 2010, and accounted for 17.5 per cent of all album sales in the UK and 26.5 per cent in the U.S.

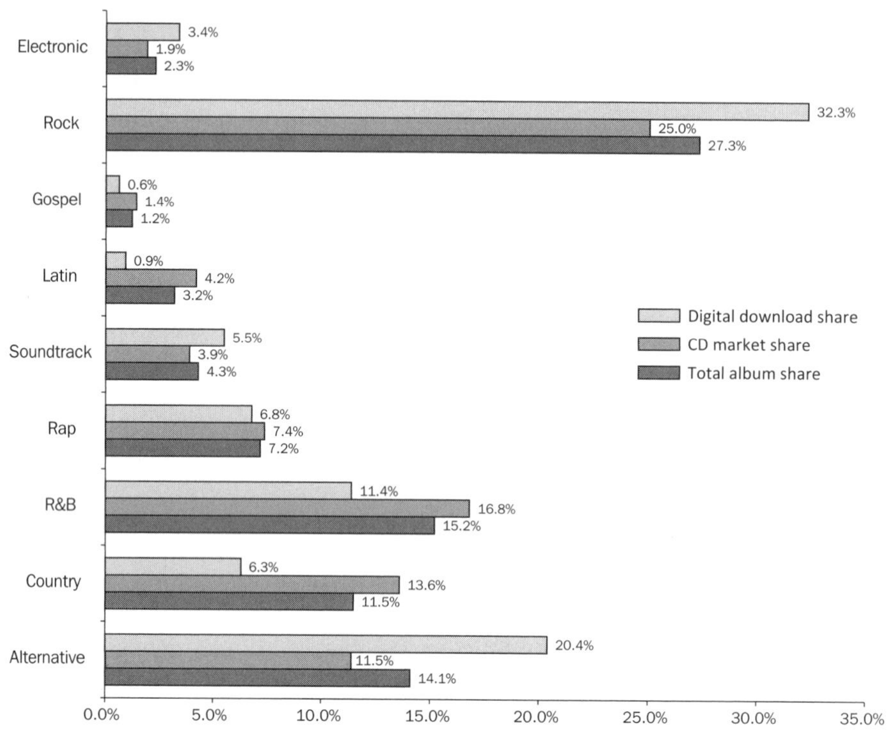

Figure 2.7
U.S. album sales by format for CD and digital (source: SoundScan)

- A new business model for subscription-based services provides a two-tiered service for customers: a free, advertiser-supported version, and a paid, "ad free" subscription version and is now known as the "freemium" model.

Sales of music downloads have continued to grow, although that growth is slowing from the rapid pace of just a few years earlier. In 2007, digital sales in the United States accounted for 30% of all music sold according to the IFPI (2008), but only 23% according to SoundScan. In 2010, SoundScan reported digital *album* sales of over 26% of albums sold in the U.S. (NARM, 2012).

The number of licensed tracks available for sale online increased from 1 million in 2003 to more than 6 million in 2007. By 2010, that was up to 13 million licensed tracks according to the IFPI. Digital sales, especially album sales, have been stronger for catalog titles than their CD counterpart. In 2008, *The Wall Street Journal* commented:

> [D]igital-album sales have consistently been more weighted to catalog and "deep catalog" items (generally speaking, releases more than 18 months old) than sales of physical CDs, and catalog and deep-catalog sales have shown stronger growth.
>
> **Fry (2008)**

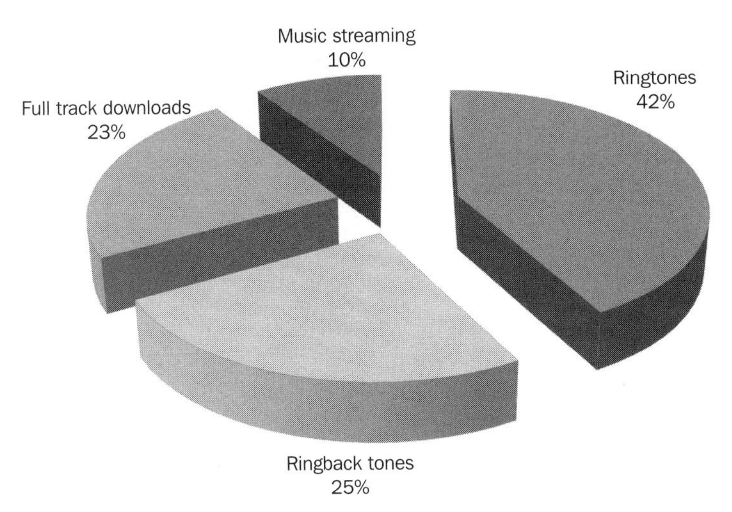

Figure 2.8
Mobile music revenue breakdown, global market worth US$14.4 billion in 2010 (source: Informa Telecoms and Media, 2011)

The South Korean market continues to lead the way in music shopping trends. In 2009, digital sales accounted for over 55.5% rising to 61.7% in 2010. According to *Music and Copyright*, the South Korean government has made a push for it to be the "most wired country in the world" (*Music & Copyright*, 2011).

Some countries are showing stronger sales of music via mobile devices than in the U.S. In Japan, digital music sales via mobile devices have topped 90% of all digital sales since 2007. Brazil reported 80% of digital sales via mobile devices in 2009; China reported that mobile sales comprised 90% of digital music sales in 2010 (Lai, 2011).

The sale of digital music via mobile devices was a growing field for several years but suffered a surprising slowdown in 2010, with a 28% drop in the U.S. due to the sharp decline in ringtone sales (Peoples, 2011). Declines for 2010 were also reported in France, Italy, Japan and Spain (Buskirk, 2011), now that the ringtone market has subsided.

Business Models

À-la-carte downloads have been the dominant business model for the first seven years of the digital music business, with iTunes leading the way. In the United States, iTunes surpassed Amazon and Target in 2007 to become the third largest music retailer, behind Wal-Mart and Best Buy. By early 2008, iTunes had moved to first place among music retailers. Lack of interoperability between services and devices hampered the development of the digital music market at first.[1] Labels were resistant to license digital downloads for fear it would increase piracy. The industry overcame this concern by offering digital rights management (DRM)

Table 2.2 Business models for monetizing recorded music

Physical product	Sale of physical CDs in physical retail stores WalMart, Best Buy	Sale of physical CDs online CD Baby, Amazon
Digital sales	Sale of digital tracks iTunes, then Amazon	Sale of digital tracks stored on server Amazon, then iTunes
Advertising-based	Advertiser-based user-influenced radio Pandora, Last FM	Advertiser-based on-demand listening Spotify
Subscription-based	Subscription-based user-influenced radio Pandora, Slacker	Subscription-based on-demand streaming Rhapsody, Spotify

downloads. In 2007, iTunes and other digital retailers began to abandon DRM, first on a trial basis and then a larger rollout, to address consumer complaints of the lack of compatibility between hardware devices.

The current business models for digital music income include: (1) the sales of digital tracks downloaded to consumer devices, (2) the sale of digital tracks stored on a central server (cloud computing) and available on demand, (3) advertising-based user-influenced radio streaming, (4) advertising-based on-demand listening, and (5) subscription-based on-demand streaming.

Labels, artists, and services have been experimenting with offering free, advertiser-supported music, either for streaming or downloading. Subscription-based services struggled early on in the digital realm, accounting for only 5% of global digital revenue (although these services grew 63% from mid-2005 to mid-2007). Now they are beginning to gain some ground with the rollout of mobile-device-based services and "freemium" tiered services. Freemium services offer a free, advertiser-supported limited functionality version of streaming "radio." These services range from user-controlled on-demand playlists (such as Spotify) to user-influenced playlists (such as Pandora). The freemium business models appear to be more popular with consumers than either the free advertising-based model or the paid subscription-only model.

Unlike its predecessor SpiralFrog, which offered only the advertising-based on-demand model, Spotify, which made its debut in the U.S. in 2011, has based its business model on providing entry-level advertiser-supported on-demand music, with efforts to upsell to the subscription version.

Headway has been made in streaming radio, providing ad-supported free versions and subscription services through services such as Pandora and LastFM. In January 2008, CBS-owned LastFM announced it had licensing agreements with all major labels and 150,000 indies to provide full-track streaming music to fans. Pandora is a playlist-based streaming radio service with a tiered system, (much like Spotify), and playlists are determined by collaborative filtering, a process that uses statistical analyses to identify and categorize music.

TRENDS IN MUSIC DELIVERY AND MARKETING

The trends in the industry continue with music sales moving away from the CD and toward a variety of web-based music services. The attraction for digital downloads for consumers is in the à la carte or cherry picking that consumers prefer over purchasing an entire album. However, this is now changing: record labels are starting to bundle digital songs so they cannot be purchased separately for newly-released albums.

For 2010, paid song downloads in the U.S. jumped to 1.172 billion units, up from 1.159 billion in 2009 and from a mere 581.9 million in 2006. Nine individual songs exceeded 2 million in download sales for 2007 with that number jumping to 37 in 2010. Five songs broke 4 million in sales.

For 2010, downloads:

- The top-selling digital artist based on digital tracks from July 2004 to the end of 2010: Taylor Swift with 34.269 million tracks sold.
- The most downloaded track: "California Gurls" by Katy Perry, with 4.398 million tracks.
- The most downloaded album: *Recovery* by Eminem, with 852,000 units.

In the first edition of this book, a list of 14 trends in digital music for 2008, based upon the Stuart Dredge article in *Tech Digest* offered notable developments in the coming years. A glance back at those trends reveals that many of them have been adopted and/or deserve further discussion:

1. *Advertiser-funded music.* Spotify and Pandora, MOG, Rdio and We7 are gaining ground, although Dredge's initial example SpiralFrog is no longer in business. Imeem has also vanished after being purchased by MySpace (Wray, 2010). The notion of giving away a track in exchange for exposure to an advertising message has been replaced with advertiser-supported streaming services—the same model commercial radio has used for decades. Spotify uses a unique method for delivering music to its subscribers; instead of relying on servers, Spotify uses a peer-to-peer network of subscribers that works like BitTorrent (Krietz & Niemelä, 2010). Less than 10% comes from their servers. MOG and Rdio both rely on integration with Facebook to drive revenue.
2. *Music search engines (collaborative filtering).* Collaborative filtering (CF) is the process of using large amounts of data to determine patterns of relevance. Wikipedia describes it as "a method of making automatic predictions (filtering) about the interests of a user by collecting preferences or taste information from many users (collaborating). The underlying assumption of the CF approach is that those who agreed in the past tend to agree again in the future." Programs like Apple's Genius use data from perhaps millions of users to determine and make recommendations of similarly liked songs. Programs like Pandora, with its music genome project, and LastFM depend on collaborative filtering to determine what songs to play next for each listener. Amazon makes extensive use of collaborative filtering to recommend specific products to each unique visitor.

3. *Record labels taking on iTunes.* Whether driven by labels or not, Amazon, and other retailers have made only slight inroads into the digital music market share. Despite the failed attempts of Doug Morris of Universal Music Group a few years ago to break Apple's stronghold on retailing digital downloads, Apple still dominates the market with over 63% of the share. With the launch of their cloud player and by dropping prices, Amazon has moved up to 10–12% of the market (Wilson, 2011; Keizer, 2010), but at the expense of other retailers more so than Apple. In 2011, WalMart shuttered their MP3 store (Albanesius, 2011).
4. *DRM-free music.* Ubiquitous. All digital music providers have dropped much of their DRM music, so this is no longer an issue among consumers.
5. *Music identification goes interactive.* Shazam is a smart-phone-based music identification application that is available for most mobile phones. It uses an audio fingerprint to match a song it detects with its database. The user holds the microphone near the music source and the system will return the results in less than a minute. Shazam also connects users to online retailers to purchase songs identified by its users. Gracenote is a program originally designed to supply track information to users who ripped their personal copies of popular CDs. A connection to Gracenote through iTunes will download the track information to be included for song playlists. Gracenote has now extended their services to include music recognition and music recommendation (gracenote.com).
6. *The rise of mobile music.* Smartphones allow for music purchases to be downloaded directly to the device. Pandora, Spotify and others offer apps for consumers to listen to streaming audio, rather than download music (see item #5 under new trends). In April 2011, Sprint introduced Sprint Music Plus, a partnership with RealNetworks to offer mobile music downloads and streaming. Verizon teamed up with Rdio to offer a cloud-based music library of 8.5 million songs (Ankeny, 2011).
7. *Bands giving away their music for free.* Distribution via the Internet has allowed bands to share sample tracks with fans without incurring much expense. In 2008, Nine Inch Nails followed Radiohead's lead to offer some tracks for free from its "Ghosts I-V" album. But both those bands already had a fan base in the millions (Netherby, 2008), a big factor in their success with music giveaways. In 2011, several bands experimented with giving away free song downloads to fans with smartphones at concerts via Bluetooth (Ehrlich, 2011). In 2011, the music blogs were alive with conversation debating the merits of giving away music with no agreement on the horizon.
8. *The Net Neutrality policy is challenged.* The FCC adopted rules in 2010 to protect net neutrality, but protects only lawful Internet traffic (Kang, 2010). The issue of neutrality is still to be addressed by legislators and the courts.
9. *Music-based widgets.* Widgets have made it easy for web users to transfer content from one web service to another in a simple, yet professional way. Instead of cutting and pasting, copying, or writing code, widgets provide the web user with a seamless, automated system for posting that interesting YouTube video on Facebook, dubbed "link and sell." In his music blog,

Chris Bracco states "the 'web widget' is an extremely effective application musicians use to make social networking easier" (Bracco, 2009). His article lists some uses of widgets, including: (1) uploading music for sale, (2) linking visitors to other social network sites featuring the artist, (3) syndication of information, and (4) collecting names for a mailing list.

10. *Social networking goes musical.* Social network sites have long realized the powerful role music plays in social interaction. Earlier entrants, such as imeem and iLike focused on music as the bond that brings users together. The next step has been how to monetize that opportunity. iTunes created Ping, allowing users to share their music tastes with friends. Others include iLike and BuzzNet. Facebook launched Facebook Music in fall 2011. Several streaming music providers (Spotify, Rdio, and MOG were ready with seamless integration. The goal for these services is to convert users to the subscription level.

11. *Ticket sales go mobile.* Mobile ticket sales are becoming popular for concerts, clubs, sporting events and other live entertainment events, accounting for 10% of ticket sales in 2011 (eTicketNews, 2011). (More on this in Chapter 15.)

To the trends discussed above, the following new trends are added from various sources and observation:

1. *Cloud computing.* Cloud computing is defined as hosted services for storing and delivering personal content over the Internet. In other words, software or your content that would normally be posted on your computer is, instead, hosted on a server and accessed by your computer via the Internet. What makes this a new trend is the fact that Amazon, iTunes and others are now offering consumers the option to store their music collections in the cloud so they can be accessed from remote locations. Ironically enough, back in 1999, My.MP3.com was an earlier player in the cloud game, offering consumers the option to store their music "in the cloud." They were sued by the record labels and dropped the idea.

2. *Digital mobility at home and on the go.* The combination of systems such as Apple's AirPlay, Airfoil and others has made it possible for various computer units in the home to communicate with one another, sending content back and forth. The introduction of the tablet format for computers had further hastened the way some consumers use computer-based services, especially media consumption. Movies and music can now be streamed from one device to another in the home and via Wi-Fi connections outside the home.

3. *Marketing with QR codes and Microsoft tags.* Quick response codes and Microsoft tags use an advanced version of barcode technology to implant the codes into print media. Smartphone applications allow the phone to "read" the barcode and respond as if it was a hot link to the web page. Smartphone users are instantly directed to the web page address encoded in the tag. (See Chapter 14.) The tags are available at no cost to anyone with a web site to promote. Search under "QR tag generators" or visit "Microsoft Tag" online. Tag providers also offer analytics of tag usage by consumers.

4. *Nearfield communications (NFC)*. NFC allows for simplified transactions, data exchange, and wireless connections between two devices in close proximity to each other, usually by no more than a few centimeters (NFC, 2011). It is anticipated that the technology will be used for mobile transactions, in addition to "sharing and pairing." It has been compared with the EZ Pass devices used at highway toll booths, except that it is two-way (Nosowitz, 2011).
5. *Internet Service Providers (ISP) and mobile service providers teaming up with on-demand or downloadable music services*. In 2010, several partnerships were formed between ISP providers and branded music services such as Spotify. In other instances, ISPs created their own branded music services offering a combination of streaming on demand and downloads. Mobile service providers such as Melon in South Korea, Vodaphone in Europe and Slacker in the U.S. began offering subscription-based music streaming services.

CURRENT ONLINE ENVIRONMENT FOR ENTREPRENEURS

Chris Anderson, in his book *The Long Tail*, describes how new Internet applications, including *collaborative filtering*, are moving the market away from the "head" consisting of the top-selling artists, writers, movies, and so forth and moving members of the market "down the tail" by allowing them to discover new, less popular products easily that would have, under previous circumstances, gone unnoticed in the marketplace.

The "leveling" effect that technology has created is resulting in lowering the barriers of entry for many small businesses in many areas. Costs associated with advertising, distribution and retailing have been reduced. Geographic barriers are melting. Access to business knowledge and contacts is more readily available. David Ingram from Demand Media states:

> The Internet has all but leveled the playing field for small-business marketers competing against established businesses. With advanced video and graphics editing software, small-business owners can create professional

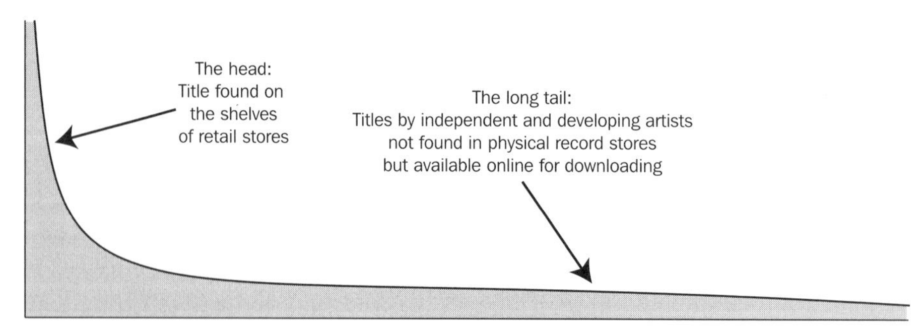

Figure 2.9
The head and the long tail

> marketing materials that reach thousands or millions of viewers online. Entrepreneurs can take advantage of cost-efficient web marketing tools such as Google AdWords to spread targeted marketing messages to a broad audience or a select niche. Several small businesses can share expensive advertising space online through banner- and traffic-swaps.
>
> **David Ingram, The Impact of Information Technology on Small Business**

The development of social networking sites dedicated to discovering and sharing new music brings hope that the millions of unsigned musicians who have recently acquired the technology to create high-quality recordings will now be able to achieve a moderate level of success, creating a middle class of professional entertainers that has been missing in the previous decades. A "record deal" with a major record label may no longer be necessary to enjoy the success of having a large fan base to purchase recorded music, concert tickets, and artist memorabilia. Several chapters in this book are dedicated to outlining how to make this happen.

In other words, the self-promoted artist and those on small independent labels now have several outlets through which they can promote and sell music to a wider market than in the past, basically lowering the barrier of entry into the music business. Not only are these small enterprises able to promote and distribute music at a fraction of the previous cost, but the newer, sophisticated techniques for music discovery via search engines, collaborative-filtering software, and music networking sites make it easier for music fans to discover and find some of these lesser-known artists that appeal to them. The merits of the long tail theory have been challenged but the concept remains: the barriers to entry have been leveled. Small businesses now have access to many of the same tools used by large media corporations and can make inroads into their market share.

Technology has impacted the music business in three ways:

1. The production of music. Desktop recording has enabled musicians to create a commercially viable recording.
2. Promotion and distribution. The Web and social networking have provided a means for the DIY musician to promote their recordings without the large expenses associated with major labels.
3. Consumers can find new music. Many new features, such as iTunes' Genius, have enabled consumers to seek and discover music they otherwise would have no knowledge about.

CONCLUSION

The Internet continues to siphon off recorded music sales and is accounting for a larger percentage of total retail sales in many sectors. Record labels are finding new, creative ways to sell music as the role of the CD as the primary source of revenue for recorded music diminishes. New business models are being developed that generate revenue from advertising on music-related sites, advertiser-sponsored music downloads and streaming, and using music as a promotional tool to sell other products such as concert tickets or portable music devices. Record labels have instituted something called a "360-degree model," where they are responsible for and share in all revenue streams in an artist's career. Then the music can be used as a tool to generate revenue elsewhere that is shared between the artist and the label. The DIY artists and small indie labels have always operated under the "360" model. As these models develop and as record labels scramble to remain relevant, recorded music is beginning to generate profits in new and unusual ways.

Glossary

Collaborative filtering Software applications that offer recommendations to consumers based on that consumer's pattern of behavior and consumption and drawn from analyses of the consumption behavior of a large aggregate of consumers.

Digital rights management (DRM) Copyright protection software incorporated into music files so that they cannot be shared freely with peers or copied to more than the maximum number of specific portable devices.

Exclusives CDs or special versions of albums available only in that chain's stores.

Internet service providers (ISP) An ISP (Internet service provider) is a company that provides individuals and other companies access to the Internet and other related services.

Loss leader Products are sold at below wholesale cost as an incentive to bring customers into the store so that they may purchase additional products with adequate markup to cover the losses of the sale product.

Music-based widgets A portable chunk of code that an end user can install and execute within any separate HTML-based web page without requiring additional configuration. End users use them to install add-ons such as iLike and Snocap to social networking pages.

Sideloading The act of loading content on to a mobile communication device through a port in the device rather than downloading through the wireless system.

Track equivalent albums (TEA) Ten track downloads are counted as a single album. All the downloaded singles are divided by ten and the resulting figure is added to album downloads and physical album units to give a total picture of "album" sales.

Note

1 Initially, labels sought to protect their music by including copy protection on music provided to legal downloading services so that the files could not be copied and shared with other consumers. As a result, different standards and platforms were developed that prevented the seamless transfer from one type of platform (such as ipod) to another (Zune). This process is referred to as digital rights management.

Bibliography

Albanesius, Chloe (2011). Walmart shutting down MP3 store. *PCMag.com*. http://www.pcmag.com/article2/0,2817,2390854,00.asp

Ankeny, Jason (2011). Mobile music remixed: Operators tune in new features and services. FierceMobileContent. http://www.fiercemobilecontent.com/special-reports/mobile-music-remixed-operators-tune-new-features-and-services

Bracco, C. (2009). A widgets primer and useful ones for musicians. http://tightmix.wordpress.com/2009/03/02/a-widgets-primer-useful-ones-for-musicians/

Buskirk, Eliot Van (2011). Mobile carriers are doing it wrong, music-wise. evolver.FM. http://evolver.fm/2011/06/13/mobile-carriers-are-doing-it-wrong-music-wise/

Christman, Ed (2007, January 4). Nielsen SoundScan releases year-end sales data. *Billboard.* http://www.billboard.biz/bbbiz/content_display/industry/e3iXZLO0IdrWuAOeIRwz3vtYA%3D%3D

CreditSense (2011). 2010 Online e-commerce sales skyrocket during holiday season. CreditSense. http://www.creditnet.com/blog/credit-news/2010-online-e-commerce-sales-skyrocket-during-holiday-season.

Dredge, Stuart (2008, January 4). 30 trends in digital music: the full list. *Tech Digest,* http://techdigest.tv/2008/01/30_trends_in_di_6.html.

DMW Daily (2008, January 3). U.S. album sales down 9.5%, digital sales up 45% in 2007, www.dmwmedia.com/news/2008/01/03/u.s.-album-sales-down-9.5%25,-digital-sales-45%25-2007.

E-Commerce as a revenue stream. http://newmedia.medill.northwestern.edu/courses/nmpspring01/brown/Revstream/history.htm.

Ehrlich, Brenna (2011). Band plans to give fans free music via bluetooth during concert. Mashable.com. http://mashable.com/2011/08/08/data-romance-bluetooth/

ETicketNews (2011). Report: mobile ticketing accounts for 10 percent of total ticket sales, mobile commerce in 2011—will see a boom in the near future. http://www.euticketnews.com/201103231155/report-mobile-ticketing-accounts-for-10-percent-of-total-ticket-sales-mobile-commerce-in-2011-will-see-a-boom-in-the-near-future.html

Forrester (2011). U.S. online retail forecast, 2010 to 2015. http://www.forrester.com/rb/Research/us_online_retail_forecast,_2010_to_2015/q/id/58596/t/2

Fry, Jason (2008, January 27). Beyond the album: 2007 brings new signs that the era of a beloved musical form is ending. *Wall Street Journal.*

IFPI Digital Music Report (2008). www.ifpi.org.

Ingram, David (n.d.). The impact of information technology on small business. Demand media. http://smallbusiness.chron.com/impact-information-technology-small-business-5255.html

Kang, Cecilia (2010). FCC passes first net neutrality rules. *The Washington Post.* http://voices.washingtonpost.com/posttech/2010/12/fcc.html?wpisrc=nl_polalert

Keizer, Gregg (2010). Apple controls 70% of U.S. music download biz. *ComputerWorld.* http://www.computerworld.com/s/article/9177395/Apple_controls_70_of_U.S._music_download_biz

Kreitz, Gunnar and Fredrik Niemela, (2010) Spotify – large scale, low latency, P2P music-on-demand streaming, *Proceedings of IEEE P2P.* Royal Institute of Technology, and Spotify, Stockholm, Sweden. http://www.csc.kth.se/~gkreitz/spotify-p2p10/

Lai, Herman (2011). China's online music sales up by 14 percent. MicGadget. http://micgadget.com/11733/chinas-online-music-sales-up-by-14-percent/

Los Angeles Times (2011). Pandora posts loss as revenue increases 117%. August 25, 2011. http://latimesblogs.latimes.com/entertainmentnewsbuzz/2011/08/pandoras-earnings-beat-expectations-drives-shares-up.html

Music-on-Demand Streaming. White Paper. http://www.csc.kth.se/~gkreitz/spotify-p2p10/spotify-p2p10.pdf

Music & Copyright (2011). South Korea continues to develop as a model for future recorded-music markets. *Music & Copyright.* http://musicandcopyright.wordpress.com/2011/03/10/south-korea-continues-to-develop-as-a-model-for-future-recorded-music-markets/

National Association of Recording Merchandisers (NARM) (2012). The Nielsen Company & Billboard's 2011 Music Industry Report. News release January 5, 2012. http://narm.com/PDF/NielsenMusic2011YEUpdate.pdf

NFC Forum (2011). What is NFC? http://www.nfc-forum.org/aboutnfc/.

Netherby, Jennifer (2008) . More bands embrace the option of giving away music. Reuters News Service. http://www.reuters.com/article/2008/03/15/us-free-idUSN1543936020080315

Nielsen (2010). Global trends in online shopping: A Nielsen global consumer report, June 2010. http://hk.nielsen.com/documents/Q12010OnlineShoppingTrendsReport.pdf

Nosowitz, Dan (2011, March 1). Everything you need to know about near field communication. *Popular Science Magazine*.

Peoples, Glenn (2011). Mobile sales plummet, subscription revenues slip in RIAA 2010 year-end stats. *Billboard*. http://www.printthis.clickability.com/pt/cpt?expire=&title=Mobile+Sales+Plummet%2C+Subscription+Revenues+Slip+in+RIAA+2010+Year-End+Stats&urlID=451884727&action=cpt&partnerID=81056&fb=Y&url=http%3A%2F%2Fwww.billboard.biz%2Fbbbiz%2Findustry%2Fdigital-and-mobile%2Fmobile-sales-plummet-subscription-revenues-1005160832.story

Pew Internet and American Life Project. www.pewinternet.org.

Point-Topic.com. World broadband penetration rankings by household. http://www.websiteoptimization.com/bw/1107/

Purcell, Kristin (2011). Search and email still top the list of most popular online activities. Pew Internet and American Life Project.

Williams, P. (2011a). PayPal's market share increases to 18% as more people use ecommerce, HostWay Global Web Solutions. http://www.hostway.co.uk/news/ecommerce/paypals-market-share-increases-to-18—as-more-people-use-ecommerce-800400807.html

Williams, P. (2011b). Ecommerce could rise after report says mobile money users set to double by 2013. HostWay Global Web Solutions. http://www.hostway.co.uk/news/ecommerce/ecommerce-could-rise-after-report-says-mobile-money-users-set-to-double-by-2013-800394957.html

Wilson, Renee (2011). Amazon drops song prices to take market share from iTunes. *Digital Journal*. http://digitaljournal.com/article/306183

Wray, S. (2010). We7 shows ad-funded model can work for online music. *The Guardian*. April 28, 2010. http://www.guardian.co.uk/media/2010/apr/28/we7-online-music-service/print

U.S. Census Data (2011). Estats.

Zickuhr, Kathryn (2010). Generations 2010. Pew Internet and American Life Project.

Chapter 3

Overall Music Marketing Strategy

Marketing an artist or any recorded music product involves a marketing plan with several elements; the resources available for marketing and the particular marketing goals of the project dictate the relative importance of each element. For instance, a local band with limited resources should probably focus more on developing a market in the geographic area where the artist performs, whereas an international star should focus more on the mass media and wide distribution. The components currently being used to market artists include the following:

- Publicity
- Advertising
- Radio promotion
- Retail promotion
- Music videos
- Grassroots marketing
- Internet marketing
- Tour support
- Special markets and products

For the do-it-yourself (DIY) artist or small indie label, priority is given to those marketing elements that are less expensive but involve more time: tour support, publicity, grassroots marketing, and Internet marketing.

PUBLICITY

Publicity consists of getting exposure for an artist in the mass media that is not in the form of advertising but is editorial content. In other words, publicists are responsible for getting news and feature coverage for an artist as well as appearances on television. These tasks include, but are not limited to, the following:

- Press releases
- News stories
- Feature stories
- Magazine covers
- Photos
- Television appearances
- Interviews

The one exception is radio airplay. The publicist is not responsible for getting songs played on the radio. That job would fall to the radio promoter (described in a later section of this chapter).

For publicity tools, publicists rely on press releases, biographies (*bios*), *tear sheets* (copies of previous articles), a *discography*, publicity photos, and publicity shots. It is the publicist who writes most of the copy for these items and prepares the *press kit*, although other branches of a record label may actually use the press kit in their marketing functions. Publicists are also responsible for media training the artist, for setting up television appearances and press interviews, sending out press releases and holding press conferences. They target a variety of media vehicles, including:

- Late night TV shows (Letterman, Leno, Conan O'Brien, *Saturday Night Live*, etc.)
- Daytime TV shows (Ellen DeGeneres, *The View*, etc.)
- Morning news shows (*Good Morning America*, etc.)
- Local newspapers (weeklies and dailies)
- National entertainment and music magazines
- Trade publications such as *Billboard* and *Radio and Records*
- Online e-zines and blogs
- Local TV shows (in support of concert touring)
- Cable TV shows (including music television channels, but for news and feature items, not airplay)

Small labels and DIY artists focus more on local media, especially in the markets where the artist performs and has a following. The marketing person for the artist will reach out to local daily and weekly newspapers, and local radio and TV stations for coverage of performances in the local market. Coverage is more likely if the marketer targets the music or entertainment writer or editor for each local publication. By offering up a press kit, promotional copies of recordings, concert tickets, photos, and access to the artist for interviews, the DIY is more likely to find success with local media coverage. Morning and midday television news shows are great possibilities for exposure, but radio interviews are more difficult to land because of the value of morning drive time on commercial radio and the fact that many morning shows are now syndicated. College radio and other non-commercial stations (NPR) are the exception (see section on radio promotion in this chapter).

ADVERTISING

Advertisers determine where to place their advertising budget based on the likelihood that the advertisements will create enough of a sales increase to justify their expense. Advertisers must be familiar with their market and consumers' media consumption habits in order to reach their customers as effectively as possible.

Consumers are targeted through radio, television, billboards, direct mail, magazines, newspapers, and the Internet. *Consumer advertising* is directed toward potential buyers to create a "pull" marketing effect of buyer demand. *Trade advertising* targets decision makers within the industry, such as radio programmers, wholesalers, retailers, and other people who may be influenced by the advertisements and respond in a way that is favorable for the marketing goals. This creates a "push" marketing effect.

Advertising is crucial for marketing recorded music, just as it is for other products. The primary advertising vehicle in the recording industry for the major labels is local print sources, done in conjunction with retail stores to promote pricing of new titles, and is referred to as *co-op advertising*. But in addition to local print media, the record industry relies also on magazine, radio, television, outdoor, and Internet advertising. The impact of advertising is not easy to measure because much of its effect is cumulative or in conjunction with other promotional events such as live performances and radio airplay. *SoundScan* has improved the ability to judge the impact of advertising, but because marketing does not occur in a vacuum, the relative contribution of advertising to sales success remains somewhat of a mystery.

The most complex issue facing advertisers involves decisions of where to place advertising. The expansion of media has increased the options and complicated the decision. The chart in Table 3.1 represents a basic understanding of the advantages and disadvantages of the various media options.

Advertising is beginning to shift to the Internet for the second time. After initial efforts by advertisers to reach consumers on the Internet failed to produce the expected results, advertisers pulled out of the Internet, contributing to the Internet bubble burst in 2001. Now, advertisers are returning to the Internet as pay-per-click advertising has offered a new way for advertisers to pay only for those ads that lead to consumer action.

The developing DIY artist is not likely to benefit much from advertising of their recorded music in terms of cost per unit sold, but they can benefit greatly from advertising locally for live shows. The budget-conscious entrepreneur can parlay a small advertising budget into a marketing campaign. Suppose a club owner has a weekly budget of $350 to advertise their club and that week's performer. Instead of the standard club ad that normally runs each week, you can request that the ad contain additional information on your recorded products for sale—at the local store. Then, in exchange for promoting that store both in the advertisement and at the venue, the store owner may consider contributing to the advertising budget for the week and/or co-promoting the appearance. An "in-store" meet and greet can sweeten the deal and generate goodwill with the store manager, who may grant premium positioning of product in the store. Now, the original $350 budget has been parlayed into a marketing plan that is likely to draw more attention from the local media. (It doesn't hurt that you now have a more substantial budget to advertise in those media.) A word of caution though: in support of the local store selling your CDs, you should not undercut their retail price with your own sales at the gigs—keep your retail price consistent.

Table 3.1 Comparison of media

Media	Advantages	Disadvantages	DIY Artist
Television	Reaches a wide audience but can also target audiences through use of cable channels Benefit of sight and sound Captures viewers' attention Can create an emotional response High information content	Short life span (30–60) seconds High cost Clutter of too many other ads; consumers may avoid exposure May be too broad to be effective	Not likely to use television advertising, except as a participant in local festivals, fairs and other events
Magazines	High-quality ads (compared to newspapers) High information content Long life span Can target audience through specialty magazines	Long lead time Position in magazine not always certain No audio for product sampling (unless a CD is included at considerable expense)	DIY artists can get some coverage in niche music and local entertainment magazines
Newspapers	Good local coverage Can place quickly (short lead) Can group ads by product class (music in entertainment section) Cost effective Effective for dissemination of information, such as pricing	Poor quality presentation Short life span Poor attention getting No product sampling	Useful for advertising local shows

Media	Advantages	Disadvantages	DIY Artist
Radio	Is already music-oriented Can sample product Short lead, can place quickly High frequency (repetition) High-quality audio presentation Can segment geographically, demographically and by musical tastes	Audio only, no visuals Short attention span Avoidance of ads by listeners Consumer may not remember product details	Useful for advertising local shows
Billboards	High exposure frequency Low cost per exposure Can segment geographically	Message may be ignored Brevity of message Not targeted except geographically Environmental blight	Not likely to use except as a participant in local festivals, fairs and other regional events
Direct mail	Best targeting Large information content Not competing with other advertising	High cost per contact; must maintain accurate mailing lists Associated with junk mail	Was used until email provided cheaper channels
Internet	Best targeting—can target based on consumer's interests Potential for audio and video sampling; graphics and photos Can be considered point-of-purchase if product is available online	Slow modem speeds limit quality and speed Effectiveness of this new media still unknown Doesn't reach entire market Internet is vast and adequate coverage is elusive	Used frequently by indies and DIY artists
Mobile media	Good targeting, instant exposure to message, potential for viral effect, quick response	Limited content, limited call to action, for only specialty or "just-in-time" promotions	Not yet used much by DIY, but is changing

RADIO PROMOTION

The impact of radio airplay on record sales and artist popularity is still the most powerful singular force for breaking new artists. The reliance on radio to introduce new music to consumers causes record marketers to focus a lot of resources on obtaining airplay. This is done through the promotion department, where radio promotion people engage in personal selling to influence radio programming. Radio program directors make the key decisions on which music is played on the radio station and which is rejected. As a result, record labels and artists lobby radio program directors to encourage them to play their music. Decisions by radio programmers are the keys to the life of a record and have become the basis for savvy, smart, and creative record promotion people to carry out this lobbying effort.

Getting radio airplay is a challenge for all but the top music stars. Even the major labels spend a great deal of time and money to get airplay for their artists, sometimes without much success. Here are the steps based upon an article by Vivek Tiwary (2008) that an unsigned or indie label artist can take to generate some success:

1. *The list.* Start with a list of stations most likely to be responsive to your efforts to gain airplay. Commercial stations are the least likely to play music by unsigned artists. College radio stations are most likely to. Focus on stations in the area where the artist is performing. Other public radio stations (NPR) may be somewhat receptive to playing new music, especially if they have specialty shows for unsigned bands. Even commercial stations may have specialty music shows (such as "Sunday afternoon blues hour") that would be receptive to playing music more based upon its quality than popularity.

   Lists of radio stations can be found on the Internet, including
   http://radio-locator.com/
   http://www.npr.org/stations/
   http://www.starpolish.com/resources
2. *Servicing radio.* Great songs don't get airplay simply because they are great songs. The station must be courted (Farrish, 2009), which means making phone calls, providing them with CDs, swag, and concert tickets. The music director and/or the program director are the appropriate people to target. Most have posted "call times" during which they are available to answer phone calls from record promotion people. Be sure to observe those call times.

   Add, tracking and reporting. Getting "added" to a station's playlist is a good first step. The next goal is to get increased airplay, known as "rotations" (the number of times a song is played per day or week). Many college stations report their playlists to the College Music Journal (CMJ). Commercial stations playlists are reported via Broadcast Data Systems (BDS) from which the *Billboard* charts are compiled. Specialty shows on commercial and non-commercial stations may report to the various niche music charts, such as Americana or Latin.

   Tracking involves monitoring airplay and chart reporting. The charts serve to notify other stations of new songs on the rise, so chart position is very beneficial to artists at all levels. If your song is gaining popularity on some stations who report to charts, that information may be used by additional stations in determining which songs to add to the playlist for the upcoming week. By tracking the progress of your song on the charts, and documenting rotation activity at other stations, that information can be used in the sales pitch for the next round of calls to stations who have not yet added the song to their playlist.

3. *Interviews and performances.* As part of an artist's pitch to radio stations, they can offer to visit the station for an on-air interview and possibly a live acoustic performance in the station's studio. Another option is phone interviews, called "phoners," during which the artist calls the station for a live on-air interview. These appearances are more likely leading up to a local performance.
4. *Special promotions.* Working with local radio on special promotions can foster a relationship and increase the likelihood of getting airplay. Offer the station concert tickets, CDs, or other swag for on-air contest giveaways. Offer to perform for the station's remote promotional events. A "remote" is when the sales department of a radio station sets up a marketing campaign for a local advertiser that includes having the station broadcast remotely from the sponsor's place of business, and encouraging listeners to show up for the event. Offering your band's services is a great way to gain loyal support from the station at all levels.

New forms of radio broadcasting, including satellite radio and Internet radio, have opened up the possibility of getting airplay for lesser-known artists. The addition of several thousand new radio stations, many with specialty formats such as unsigned bands or regional music, has increased the opportunities for emerging artists. Many of these newer formats have less restrictive and longer playlists, so competing for airplay is not as difficult.

RETAIL PROMOTION

Radio is the most important marketing tool for influencing consumers to buy new music. So it would seem that music retailers would use radio as their primary advertising vehicle to promote their stores and product. But that is not the case. Print advertising is the music retailer's primary promotional activity.

> Consumers have been trained to look in Sunday circulars for sales and featured products on most any item. Major music retailers now use this advertising source as a way to announce new releases that are to go on sale on the following Tuesday, along with other featured titles and sale product.
>
> **Amy Macy, in Hutchison, Macy, & Allen, 2006, p. 213**

Promotional efforts inside record stores highlight different releases to motivate consumers to purchase the titles being promoted. It is often said that the last ten feet before the cash register is the most influential real estate for promotional activities. Whether it is listening stations near a coffee bar outlet or posters hanging in the front window, a brief encounter within the store's walls will quickly identify the music that the store probably sells (Hutchison et al., p. 214).

Featured titles within many retail environments are often dictated from the central buying office. As mentioned earlier, labels want and often do create marketing events that feature a specific title. This is coordinated via the retailer through an advertising vehicle called *cooperative advertising*. *Co-op advertising*, as it is known, is usually the exchange of money from the label to the retailer, so that a particular release will be featured. The following are examples of co-op (reprinted from Hutchison, et al., 2006, p. 214):

- *Pricing and positioning. P&P* is when a title is sale-priced and placed in a prominent area within the store.
- *End caps.* Usually themed, this area is designated at the end of a row and features titles of a similar genre or idea.
- *Listening stations.* Depending on the store, some releases are placed in an automatic digital feedback system where consumers can listen to almost any title within the store. Other listening stations may be less sophisticated and may be as simple as using a free-standing CD player in a designated area. But all playback devices are giving consumers a chance to "test drive" the music before they buy it.
- *POP, or point of purchase materials.* Although many stores will say that they can use POP, including posters, flats, stand-ups, etc., some retailers have advertising programs where labels can be guaranteed the use of such materials for a specific release.
- *Print advertising.* A primary advertising vehicle, a label can secure a "mini" spot in a retailer's ad (a small picture of the CD cover art), which usually comes with sale pricing and featured positioning (P&P) in store.
- *In-store event.* Event marketing is a powerful tool in selling records. Creating an event where a hot artist is in-store and signing autographs of his or her newest release guarantees sales while nurturing a strong relationship with the retailer.

As a larger percentage of record sales moves to the Internet—for physical sales as well as downloads—many well-known retailers have been forced to close their doors. Borders is one example. Other retailers have simply reduced the amount of floor space dedicated to recorded music and have begun to diversify if they had not already done so. In 2007, the National Association of Recording Merchandisers (NARM) awarded the large retailer of the year award to Amazon.com, the first for an online retailer. In 2008, iTunes surpassed Walmart as the #1 retailer in the U.S.

After years of cookie-cutter products in stores, many chain stores now allow the local store manager some discretion in determining product mix that appeals to local shoppers. The DIY artist should not hesitate to contact local managers of large chain stores in markets where there is potential for sustained sales. But remember, you should make sure the store's efforts are rewarded by supporting their cooperation with yours.

MUSIC VIDEOS

In 1981, MTV launched the first music video channel and gave record labels a reason to produce more of the new entertainment format. Other genre-specific channels soon followed, such as Black Entertainment Television (BET), VH-1 for adult contemporary music, and Country Music Television (CMT). It became evident early on that music video exposure was beneficial to developing artists' careers and promoting their music. Stars like Madonna and Michael Jackson owe a lot of their fame to video exposure. *Telegenics* became an important aspect in artists' careers, forcing record labels to concentrate on signing artists with visual appeal.

That being said, the costs of producing a music video are astronomical, and record labels must weigh the benefits of creating one against the additional costs. Additionally, the plethora of music videos crowded the airwaves by the early 1990s, so that producing a music video did not automatically guarantee airplay on the major video outlets.

Videos remain an important promotional tool, if not for consumer promotion, then to showcase the artist to booking agents for TV shows and concerts. The rise of YouTube and other online video outlets has spurred a new kind of video format that is more edgy and an alternative to the slick Hollywood-style videos on cable TV. It's called the viral video, but virility is but one component. To begin with, many of the successful viral videos were not actually made by hobbyists but are produced to give that appearance—like the movie the *Blair Witch Project*. Then, because the videos seem to come from "the street" but have interesting qualities, they are passed around as consumers get the impression that they "found it first" and want to share this nugget with their friends. If they had the impression it was a corporate-sponsored mass media product, they would probably not bother with grassroots "sharing" because they would assume everyone would be exposed to it soon enough. Some type of video presence is now recommended for all artists regardless of the level of their career or marketing budget. The success of YouTube and the shift in consumption of video entertainment programming from broadcast and cable TV to web-based on-demand viewing has led to a proliferation in video-on-demand services and opportunities for artists to place videos. High end sites like Vevo and Hulu offer programming from the major media companies. Independent artists may find more success with services like Vimeo and YouTube.

GRASSROOTS MARKETING

Grassroots marketing, sometimes called guerrilla marketing or street marketing, consists of using nontraditional marketing tools in a bottom-up approach to develop a groundswell of interest at the consumer level that spreads through word of mouth (WOM). *Diffusion of innovations theory* discusses the role of "opinion leaders" or trendsetters who are instrumental in the diffusion of any new product, trend, or idea.

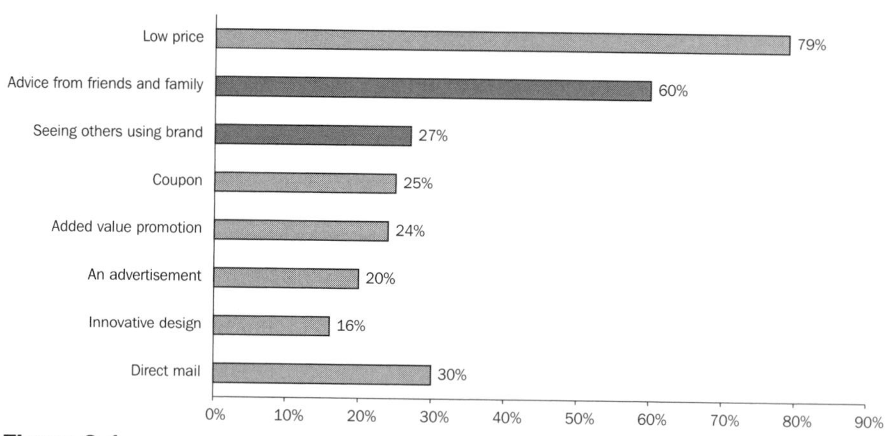

Figure 3.1
Factors influencing music purchases (source: Hutchison, et al., 2006)

In grassroots marketing, these opinion leaders are targeted, sometimes by hiring them to work for the artist or label, and their influence on their peers is exploited to promote new products and trends. Groups of trendsetters that become a part of the marketing establishment are called street teams. Sometimes it is enough to have the street team members adopt the new styles or consume the products visibly in the marketplace (such as drinking Sprite or wearing a new outfit). This peer-to-peer marketing is very influential, especially among younger consumers. Record labels commonly have grassroots marketing departments who manage street teams in various geographic locations around the country.

On the Internet, the concept of street team has been expanded to include using these same peer-to-peer marketing principles to reach out to consumers in chat rooms, on message boards, and in social networks (see Chapter 12).

INTERNET MARKETING

Internet marketing is the focus of much of this book. This section concentrates on the use of Internet-related sites in other aspects of the marketing plan. It is common these days for every television show, radio station, and print publication to host its own web site. Often these sites can be used for interactive marketing campaigns, such as contests. Contestants who appear on "American Idol" get their weekly affirmation from viewers who log on to the Internet or use their cell phone to cast their votes. Marketing for artists and record releases often incorporates a web component for each media event, whether it's to drive traffic to the media vehicle's web site or to encourage fans to sign up for information or prizes. A debut of a new video on BET can be coordinated with a promotion on www.BET.com where fans enter contests, get free music, purchase priority tickets, or participate in other promotions. In-store autograph signings by an artist can be publicized on the retailer's web site. Tour schedules can include links to each of the venues so the concert attendee can learn more about the venue and its location (where to sit, where to park, etc.).

So each aspect of a marketing plan, whether it's coverage in a particular media vehicle, tour support, or supporting an album at retail, should contain an online counterpart to make the experience interactive for consumers and to drive traffic to the media or retailer's site.

TOUR SUPPORT

Tour support is money or services that a record company provides to help promote an artist and ultimately sell more records. This is one aspect of marketing that is crucial for independent artists and those on indie labels. Generally, the marketing budget does not allow for retail product placement, radio airplay, or advertisements. So the indie artist must rely on touring to build a fan base and generate record sales. Tour support consists of contacting local media and retail in each market where the artist is to appear and providing any promotional materials and support necessary, including appearances. Aspects of tour support may be handled by publicists and the sales department of a label. The publicist is responsible for getting local media coverage and arranging for the artist to appear for interviews and impromptu performances on TV and radio. The sales department is responsible for setting up retail promotions and in-store appearances by the artist for autograph signings.

> Touring is one of the most important ways that an independent artist can spread the word about their music within both the industry and the general public. It can serve as an excellent catalyst of radio airplay; articles, stories and reviews; retail placement; on-air performances; in-store performances and other promotional opportunities. Touring is also expensive and exhausting for most independent artists.
>
> **INDIEgo.com**

SPECIAL MARKETS AND PRODUCTS

With traditional record sales on the decline, labels and artists are always looking for new ways to make money from selling music, often referred to as special markets and special products. Getting songs included in a movie, on a TV show, or in a commercial are some examples of special markets. Compilations and samplers are special products. Retailers create "branded" CDs that are special editions available only through the sponsoring retailer. This concept is popular with all types of retailers, not just record retailers, because it draws traffic into the stores. Hallmark, William-Sonoma, and Starbucks are nontraditional retailers who use special products to draw customers to their retail establishments. In 2007, the Eagles decided to release their new album only in Wal-Mart stores. In exchange for the exclusive opportunity to offer the album, Wal-Mart spent a lot of money promoting the album that otherwise would have been spent by the record label—or, in some cases, not spent at all.

Figure 3.2
CD cover for compilation album (courtesy of Del Suggs)

Local artists have been known to band together to create a sampler or compilation album featuring up to a dozen artists (Suggs, 2011). The concept helps artists share the costs of manufacturing and exchange their fans with those of other artists on the project. Generally, each artist puts up their portion of the initial expenses for manufacturing and covers their own production costs. Each artist then gets an equal number of CDs to sell at their gigs or local stores.

CONCLUSION

The promotional aspects outlined in this chapter represent the traditional marketing methods that record labels and independent artists have employed to sell their products and promote their artists. As is evidenced by the rest of this book, the old-school marketing techniques are giving way to a new type of marketing that incorporates the latest in communication and entertainment technology, and that is keeping up with the rapid pace of innovation. But, old habits die hard.

Glossary

Bio Short for biography. The brief description of an artist's life or music history that appears in a press kit.

Clippings Stories cut from newspapers or magazines.

Consumer advertising Advertising directed toward the consumer as compared to trade advertising. Generally, this audience is reached through mass media.

Co-op advertising An effort by two or more companies sharing in the costs and responsibilities. A common example is where a record label and a record retailer work together to run ads in a local newspaper touting the availability of new releases at the retailer's locations.

Diffusion of innovations theory The process by which the use of an innovation is spread within a market group, over time and over various categories of adopters.

Discography A bibliography of music recordings.

End cap In retail merchandising, a display rack or shelf at the end of a store aisle; a prime location for stocking product.

Grassroots marketing A marketing approach using nontraditional methods to reach target consumers.

Guerilla marketing Using nontraditional marketing tools and ideas on a limited budget to reach a target market.

Music director (MD) The person responsible for a radio station's playlist of songs.

Point of purchase (POP) A marketing technique used to stimulate impulse sales in the store. POP materials are visually positioned to attract customer attention and may include displays, posters, bin cards, banners, window displays, and so forth.

Press kit An assemblage of information that provides background information on an artist.

Press release A formal printed announcement by a company about its activities that is written in the form of a news article and given to the media to generate or encourage publicity.

Pricing and positioning (P&P) When a title is sale priced and placed in a prominent area within the store.

Program director (PD) An employee of a radio station or group of stations who has authority over everything that goes out over the air.

Publicity Getting media exposure for an artist in the mass media that is not in the form of advertising.

SoundScan A company owned by Nielsen that is responsible for monitoring and reporting the sales numbers for recorded music. Retailers, labels, managers, agents, and other industry people subscribe to the service and retrieve the sales data online.

Street teams Local groups of people who use networking on behalf of the artist in order to reach the artist's target market.

Tear sheets A page of a publication featuring a particular advertisement or feature and sent to the advertiser or public relations firm for substantiation purposes.

Telegenic Presenting a pleasing appearance on television.

Tour support Money or services that a record company provides to offset the cost of touring and help promote the artist.

Trade advertising Advertising aimed at decision makers in the industry, including people in radio, retail, and booking agents.

Bibliography

Farrish, B. (2009). Radio Airplay 101—Why you have to promote to radio. http://www.radio-media.com/song-album/articles/airplay15.html

Hutchison, Tom, Macy, Amy, and Allen, Paul (2006). *Record Label Marketing*. Focal Press, Oxford, UK.

Suggs, Del (2011). Personal interview.

Tiwary, V. (2008). Press and Publicity. StarPolish. http://www.starpolish.com/advice/article.asp?id=33

Chapter 4

Domains and Hosts

Nuts and Bolts

WEB SITE BASICS: HOW WEB SITES ARE CONSTRUCTED

The World Wide Web uses a system of hypertext markup language (HTML) to organize text, graphics, and multimedia in an orderly fashion so that web browsers (Internet Explorer, Firefox, etc.) can make sense of the files and put together the pieces of the puzzle. On the server,[1] a web site looks like a collection of files and folders, as shown in Figure 4.1. Each web page is represented as an HTML document. The pictures and other assets for the page are located in the folder of the same name. The web browser or interface used to access and compile these pages reads HTML programming and puts the various components of the web page in the proper perspective, if all goes well.

When a web surfer types in the uniform resource locator (URL) of this location, the index page (by default) loads into the browser, picking up the page assets from the "home" folder (in this example). Links from the index page to subsequent pages connect the other HTML documents together to form the complete web site.

Many web sites now use *active server pages* instead of basic HTML pages. Webopedia defines active server pages as:

Name	Size	Type	Date Modified
home		File Folder	1/26/2007 1:34 PM
page2		File Folder	1/26/2007 1:34 PM
page3		File Folder	1/26/2007 1:34 PM
page4		File Folder	1/26/2007 1:34 PM
index	1 KB	HTML Document	1/26/2007 1:34 PM
page2	10 KB	HTML Document	1/26/2007 1:34 PM
page3	14 KB	HTML Document	1/26/2007 1:34 PM
page4	18 KB	HTML Document	1/26/2007 1:34 PM
SPACER	1 KB	GIF Image	4/20/1999 8:16 AM

Figure 4.1
The arrangement of a web site's files and folder on the server

> abbreviated as ASP, [this] is a specification for a dynamically created Web page that utilizes ActiveX scripting—usually VB Script or Jscript code. When a browser requests an ASP, the Web server generates a page with HTML code and sends it back to the browser. So ASPs are similar to CGI scripts, but they enable Visual Basic programmers to work with familiar tools.
>
> **Webopedia definition, 2011**

Domain Name

The first aspect of building a web site involves registering a domain name. The domain name is the "web address" that your visitors will become familiar with and use to access your site. The URL is the means of identifying the web site location on the Internet. An Internet address (for example, http://www.yourname.com) usually consists of the access protocol (http), the domain name (www.yourname), and the top level domain (TLD), such as .com, .org, or .net. The URL will also contain the directory path and file name, such as /myband/index.html. The general default-loading file is named *index.html* or *index.htm*. This file name should be the name of your home page, because this is the page that the web browser looks for when your customers enter in the address. The file index.html is served by default if a URL is requested that corresponds to a directory on the server where your web site resides.

Master of Your Domain

What's in a name? It is important that the URL, or Internet address, for your artist be simple and easy to remember. A long URL will confuse customers and prevent them from finding the web site. There is also more possibility of error when the customer has to enter in a long name such as http://www.cheapdomainprovider.com/~yourcustomername/personalweb/bandname/index.html.

The domain name is the single most valuable piece of real estate on the Internet. Bands that have been around since before the Internet but who were not quick to register their band name when the Internet took off are sometimes forced to use a less-than-perfect URL address. For example, the band Van Halen does not own the web address www.vanhalen.com. A fan of the band owns that site. The official band web site is www.van-halen.com, and some fans might not know or remember to include the dash. In another example, the official U.S. White House web site is www.whitehouse.gov. A whimsical company that markets political swag (slang for stuff we all get) for a while set up shop at www.whitehouse.org.

Many services available on the Internet will register the artist's domain name. A quick online check with these providers can determine if the artist's or band's name is available. Go for the dot-com extension if it is available, and use a domain name that is memorable and simple. This will become your online "brand."

Branding is defined as creating a distinct personality for the product (in this case, the artist, not the label) and telling the world about it. As companies rely more and more on the Internet for promotion, the dot-com name becomes a brand name. So the URL should reflect the brand that the artist wishes to promote. What if the name you want is already taken? If the artist is not yet well established, it may even be wise to rename the band or change the spelling of the artist's name to accommodate the branding that goes along with having an Internet presence. As a result of the growth of the Internet, many companies have recreated their brand to reflect the style found on URL addresses, combining words together and using capital letters to distinguish between words. Examples include names such as SunTrust bank instead of Sun Trust. This trend had been extended to include abbreviations commonly used in text messaging and similar to the shorthand found on automobile license plates, where the words *in, be, to* and *for* are replaced by the letters and numbers N, B, 2, and 4 (e.g., NSYNC, Boyz2Men). Consider all possibilities when acquiring domain names.

BuyDomain.com suggests "When it comes to online branding, shorter is better. If you want your visitors to simply remember your domain name and your brand, keeping the domain name short is ideal" (2011). However, they point out one advantage of longer domain names that include valuable keywords: search engine rankings. Long domain names that include keywords such as "band" or "music" may actually help the web user find the site, if they are using a search engine to find the site rather than typing in the URL.

Consider the following tips when selecting a domain name:

1. *Make it easy to remember.* Hyphens are hard to remember and hard to communicate verbally when describing the site address to someone. If you have any peculiar symbols or spellings, be sure to emphasize that in your marketing materials.
2. *Keep it short and simple.* Also, avoid strange spellings, unless you intend to brand your artist with that same spelling (such as "z" instead of "s"). Remember that words may be spelled differently outside the United States or in non-English-speaking countries. Some companies and artists like to use acronyms for the domain name (e.g., www.hsx.com for Hollywood Stock Exchange). The problem is that most of the short acronyms are already taken. Dennis Forbes (2006) reported that all of the three-letter acronyms are taken and nearly 80% of four-letter *acronyms* are already registered. Less than 5% of five-letter acronyms are taken.
3. *Make it descriptive of the site.* It is easier to memorize and identify a site name that relates to the content or subject matter of the site. You can add *band, sings, songs,* or some other word to the end of your artist's name, especially if the simpler domain name is taken.
4. *Use the dot-com domain, if available.* Most web surfers are accustomed to using the dot-com extension and will default to that unless you make it clear that your TLD is dot-net or dot-org. You may even want to register all variations of your domain name with the different TLDs.
5. *Use a keyword in the name, if possible.* The name of the artist is good, but adding music or band helps search engines categorize the site based on these keywords. Consider registering the name both ways.
6. *Be consistent.* Even if you register several variations of your domain name to cover all bases, stick with just one for your marketing materials to avoid

confusing your customers. It is common to register all variations (www.arstistname.com, www.artist-name.com, www.artistnameband.com) and then forward all of these to the same domain.

7. *Your domain name should be your site name.* This is important because when people think of your web site, they will think of it by name. So have the domain name match the name of the site that appears in the masthead.

The bottom line is that the domain name is important. In the article "The Importance of a Quality Domain Name," John Stone (2006) stated, "Your domain name is your Internet phone number. Many of the same rules apply. Get a number that is easy to remember or spells your business name."

As of 2009, it was reported that for the dot-com extension, all three-letter possibilities were gone and all four-letter *words*. The most popular registered domain length is 11 characters. All of the top 10,000 family names are taken. Domain speculators have snatched up many of the one or two word domain names that are potentially useful and have practical application. This has caused startup web companies to get creative with naming and branding—hence the funny sounding company names, like Wiggio. Below are suggestions based upon articles from *Smashing Magazine* (2011) of some creative ways to name and "brand" your company or artist with an available domain name.

1. Use a discoverable name—real words and phrases so that they show up in search engine results. Many common phrases are already registered. The catch is to find one that is still available.
2. Use two or more words stuck together. The earlier example of SunTrust, or even YouTube would be an example of a compound name. Another example, JobSpice is a site for creating attractive résumés.
3. Use an acronym that is meaningful or easier to remember. Acronyms can be expanded. "Wiggio" stands for "working in groups" with the "gio" added on at the end of the acronym (see item 6).
4. Blend two words together. "Microsoft" blended the "micro" part of microcomputers with the "soft" part of the word software.
5. Tweak a word. Flickr is a well-known photo sharing site based on a modification of the word flicker, a photographic term.
6. Affix a word with a prefix or suffix. Friendster is one example. MiMedia is another, as is iTunes.
7. Domain hacking. This does not refer to ripping off someone's domain name, but rather to use a commonly available domain extension as part of your domain name. For example, a band called the "Eight of Us" adopted the domain name "8of.us." The extension ".fm" was originally designated for the Federated States of Micronesia but is commonly used by FM stations, such as Last.FM. YouTube now uses the "http://youtu.be" designation for specific videos to shorten the URL from its previous length.
8. An intentional misspelling of a common word that describes the site. "Knotebooks" is a collaborative notebook site for students to create wikis.
9. Make up a new word, like Odeo, Onovo, and Zukmo.

It is not necessary to actually possess a physical site with the address www.bandname.com. This name can be used as the URL, and then visitors can be redirected or *forwarded* to the actual web site, which may have a longer address. In other words, the actual physical site for an artist may be http://www.recordlabelname.com/artists/the_artist/index.html but fans only need to type in www.bandname.com. It is recommended that you register derivations of the artist's name to cover visitors who may not know the correct spelling.

Registering Your Name

Once you have selected a domain name or determined the domains that you would like to register and use, there are several *domain name registrar* services that will register any available domain name for $12 or less. The cost of registration has gone down in recent years, making it affordable to register more than one name or variation. Getting your domain name involves registering with Internet Corporation for Assigned Names and Numbers (ICANN).

Registration is for between one and nine years. Some domain registration services such as GoDaddy.com will automatically renew your registration annually and charge the cost to your credit card or PayPal account. Some web hosting services will include the cost of domain registration in the fee for hosting services. Some of the more popular domain registration services include Yahoo!, GoDaddy.com, Dotster.com, Network Solutions, Active-Domain and Register.com. You can also search for available domain names at these sites.

> ICANN is responsible for the global coordination of the Internet's system of unique identifiers. These include domain names (like .org, .museum and country codes like .uk), as well as the addresses used in a variety of Internet protocols. Computers use these identifiers to reach each other over the Internet.
>
> **www.icann.org**

Purchasing a Previously Registered Domain Name

If you must have a certain domain name and it is not available, it may be possible to purchase the name from the current owner. The first step is to type the domain name into the address box of your browser and see what page comes up. Many times, domain speculators will post a "this domain for sale" link to contact them. Services such as GoDaddy provide domain backorder services that will monitor a domain name and notify you if and when it comes up for renewal or resale. They will also serve as your buyer's representative if you want to purchase a name through a domain auction process. WHOIS is an internet function used to search for domain registration information. It identifies who owns the domain name and how to get in touch with them. Sometimes you can deal directly with the owner.

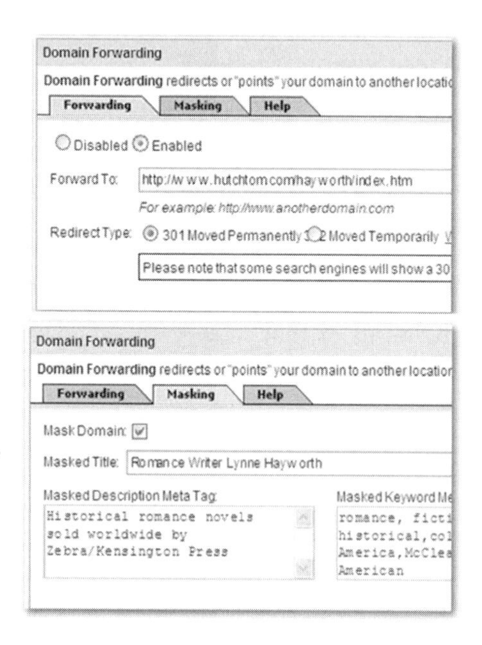

Figure 4.2
Example of domain forwarding

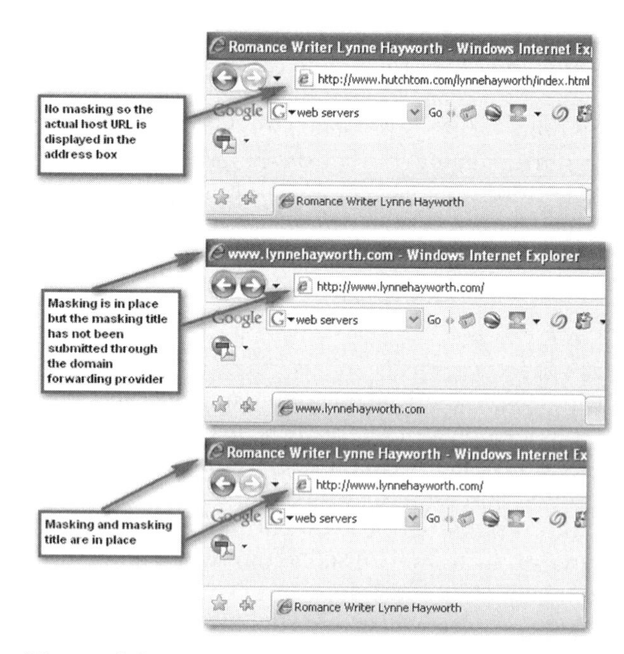

Figure 4.3
Example of domain masking and title masking

FORWARDING AND MASKING

Domain forwarding or *URL redirection* redirects all Web traffic for a domain name that you have registered to the specific URL where the site is located. When someone types your domain name into a browser, the system will automatically forward or redirect that person to whatever host you specify that contains the folder with the web site in it. For example, writer Lynne Hayworth's web site is located at www.hutchtom.com/hayworth/index.html. Web surfers looking for that site can type in www.lynnehayworth.com and automatically be redirected to the proper site.

Usually the service provider with which you register your domain name will offer domain forwarding and *masking*. When setting up domain forwarding, it might be beneficial to employ *domain masking*, a process that keeps your registered domain name in the URL box instead of displaying the actual URL address for the server and folder where the site is located. This could eliminate some confusion for the web visitor and perhaps protect the true identity or location of the web site. With domain masking, you are requesting that the service in charge of the forwarding "mask" the actual address by keeping the registered domain name in the address bar. In addition to masking, it will be necessary to submit a masked title, keywords, and a description.

QUICK RESPONSE CODES, MICROSOFT TAGS AND SHAZAM TAGS

New to mobile communications is the concept of using tags in advertisements to direct consumers to a web site using their mobile or tablet device. The idea is that, for print media, a tag similar to a bar code is printed in the advertisement. The consumer simply uses an application and their device's camera to scan the tag, which instantly takes them to the URL encoded in the tag.

The idea for radio and television is similar, with companies like Shazam providing the bridge between an inaudible audio signal coded to the web page and the user's mobile device. The user activates the microphone in the device via an application that scans and identifies the web address encoded in the signal during the show or commercial. Embedded in the commercials for the 2011 movie *Transformers 3* were Shazam tags giving users access to download a free Linkin Park song and a chance at movie tickets.

Will quick response (QR) codes, Microsoft tags and Shazam tagging make snappy domain names irrelevant? The ability of consumers to access web sites quickly upon viewing an advertisement or promotion in either print or television media certainly makes easy-to-remember domain names less important. There is speculation that the new codes will decrease the value of sought-after domain names. Shane Cultra of DomainShaneDotCom states

> The purpose of domains was to create a memorable alternative to IP addresses that were merely a series of numbers with little distinction from one to another. QR codes provide this same use. By taking a photo or scan this lets the user get to a site or information without having to type or recall a web address.
>
> **Cultra, 2011**

But keep in mind that U.S. smartphone adoption in 2011 was only at 40%, and these codes are only useful for mobile devices scanning from a traditional media source. Everyone else uses hot links with a simple link name, such as "click here."

For the small business owner or do-it-yourself musician, acquisition of a QR or Microsoft tag is simple and free. There is a plethora of QR code generators on the Web—easily found by just Googling QR code generators. Facebook has a code generator at http://www.facebook.com/QRcode.generator?sk=app_4949752878. Some of the premium generators also sell a subscription that will provide analytics on tag users. For Microsoft tags, you only need to sign up for a Windows Live account and visit http://tag.microsoft.com/home.aspx. With both types, you type in the URL that you want to have the tag direct users to, and an image is generated. You can copy the image for use on CD booklets, posters, business cards and other print materials. In addition to creating tags for web sites, you can create tags for phone numbers, emails and other uses.

Figure 4.4
QR code for the Facebook page for generating QR codes

Figure 4.5
Servers

WEB HOSTING

Web sites are stored on a physical server, or host. *Web hosts* are companies that provide space on a server they own for use by their clients. They also provide Internet connectivity to access that server. Most services provide basic functions for storing web pages, storing files, and offering email. If a record label has its own server, it can use URL redirection to forward the artist's URL to the label's server. If the label does not have a host provider, many commercial companies offer "space" to host a web site. There are companies that offer free web site hosting. These services should be avoided for a professional site such as that for an artist or record label because the free services tack on banner advertisements and popup ads. These advertisements annoy customers and discourage them from visiting the site. It is worth the money to pay for a cleaner site, without all the ads. There are two considerations when selecting a web host: *storage space* and *bandwidth.*[2]

How Much Space?

It is important to ensure the hosting site has enough data storage space for all the functions a music-based site needs. Personal web sites, with photos, blogs, and text, use less than 10 megabytes (MB) of server space. One GB of space will probably be enough to hold graphics, photos, videos, and sound files, although one MP3 file can take up to 4 MB. A thorough site should include a bio, photos, band news, a tour itinerary, a place to sign up for a mailing list, press and media coverage (reprints), a discography, lyrics, audio files, contests and giveaways, merchandise, links to other favorite sites, video, and contact information. It is vital to include some way for fans to purchase the artist's recordings—either directly from the web site or by directing them to an online retailer. The site will require a virtual shopping cart if the artists want to sell directly from their site (discussed later in the book) and some scripting.

Unlike just a few years ago, the trend now is to house some, or most of the assets (site content) on third-party sites, such as YouTube, and use embedded players to stream that content through the artist's web site. In other words, the web visitor never leaves the artist's page but still can watch a video streaming from YouTube. Generally this is accomplished by using a widget or copying and pasting a piece of code. With all these options, it is now possible to get by with less hosting space just as hosting providers are offering more space for less money.

The cost of hosting services continues to drop, or in some cases, hosting companies maintain their pricing structure but increase the amount of hosting space. Hosting service GoDaddy.com used to offer 5 gigabytes (GB) of space and 250 GB of bandwidth (monthly) in its economy plan for less than $5 per month. That was in 2007. By 2011, GoDaddy was offering the same $4.99 per month plan, but doubled capacity to 10 gigabytes of space and unlimited bandwidth (GoDaddy web site).

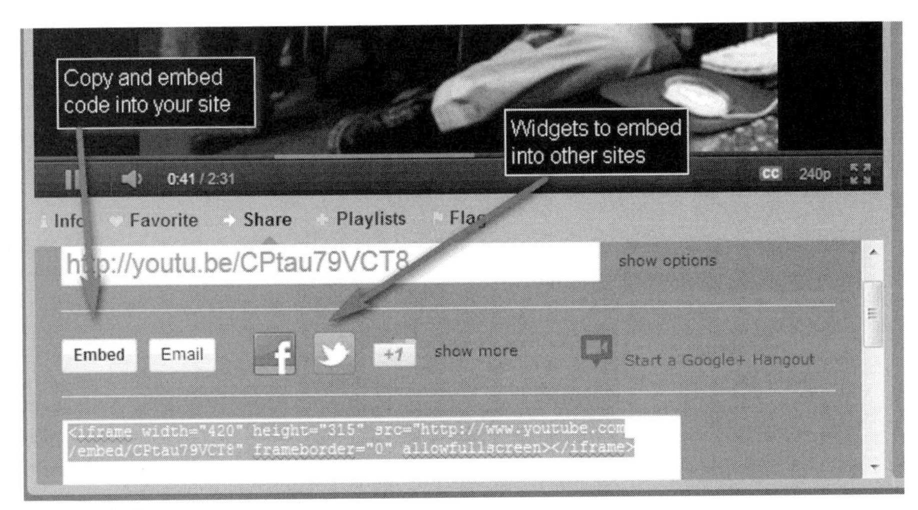

Figure 4.6
Example of using embedded player code from YouTube

Bandwidth and Uptime

Sufficient *bandwidth* needs to be allocated to accommodate numerous visitors to the site. Bandwidth is the amount of data that can be transmitted in a fixed amount of time—it is the size of the "highway." If a lot of users access a server simultaneously, sufficient bandwidth is required to transmit the information from the server to the users in a timely fashion and to avoid Internet congestion. It is wise to purchase more storage space and bandwidth than you think you will need. Some host companies are like cell phone companies: If you exceed your limit, the additional charges can add up in a hurry. The free services sometimes impose a bandwidth limit and will temporarily shut down sites that exceed this limit. That would be disastrous for a business that depends on Internet traffic for income.

Another factor is *uptime*, the percentage of time that the host is accessible to potential customers attempting to visit your web site. A hosting provider should have close to 100% uptime so that your customers can access at any time without receiving error messages.

File Transfer Protocol

To communicate with the hosting service FTP can be used to move files back and forth from the host server to your personal computer. Most high-end web design software has an FTP function. Files can then be updated on the home computer and then uploaded to the server. Dreamweaver and Microsoft Expression Web both have FTP features but also allow for server-side editing of the files on the host server. Good hosting services will also have an FTP function as one of their dashboard features (the control panel).

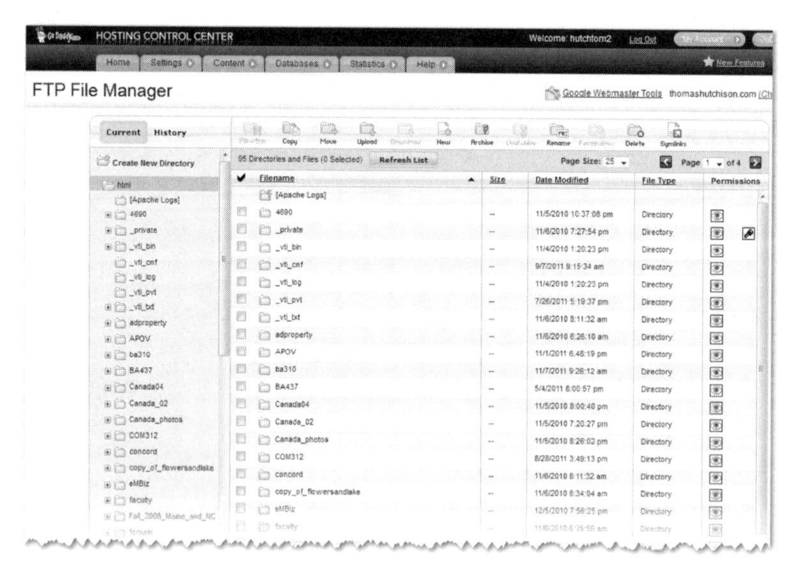

Figure 4.7
GoDaddy FTP file manager

Choosing a Host

Often a web hosting package will include other features such as email accounts with the address of info@yourbandname.com, web traffic statistics, subdomain access, and database and forms management. Before deciding on a host and a hosting package, it is a good idea to determine what services you will need as the web site grows in popularity. Christopher Heng (2004), in his article "How to Choose a Web Host," listed the following criteria as important:

1. *Advertising.* Most free hosts impose advertising to offset their costs. It is not advisable to go with a site that is advertising competing or other products. The distractions are not attractive to the visitor.
2. *Amount of web space.* The popular services such as GoDaddy.com offer reasonably priced packages with 10 GB of space for less than $5 per month.
3. *FTP access, or file transfer protocol.* This is the ability to easily move files between the server and your hard drive. It is important that the software used to create the web site has the ability to "talk" to the server and engage in uploading and managing the site.
4. *Data transfer, including traffic and bandwidth.* The amount of bandwidth you are allowed to use to both upload materials and provide access for web site visitors is usually restricted by the web hosting package. Heng stated that a new site is likely to use less than 3 GB of bandwidth per month. Go Daddy's economy hosting plan allows for unlimited monthly traffic.
5. *Technical support.* This will allow your web site to provide reliable, consistent accessibility to visitors and prevents downtime caused by technical difficulties.
6. *SSL, MySQL, and the shopping cart.* If you plan to conduct transactions through your site, SSL (secure socket layer) guarantees encryption for credit card

numbers and other sensitive information; it will be discussed in the section on e-commerce. MySQL is an open source relational database management system that allows information from online forms to be processed through a database. According to Webopedia, "A shopping cart is a piece of software that acts as an online store's catalog and ordering process." Services like PayPal offer this feature, so it may not be necessary for an artist to provide it on the artist's web site.

7. *Control panel.* This allows the webmaster to easily manage aspects of the web account features.

CONCLUSION

The cost of setting up and maintaining a web presence is within reach for every musician, songwriter, and singer. The most important web possession is the domain name, and the domain name should be registered early in the process. There are several popular web hosting and domain registration services. Make sure the hosting plan is sufficient to serve the artist as the career develops.

Glossary

Active server pages (ASP) A web server technology from Microsoft that allows for the creation of dynamic, interactive sessions with the user.

Bandwidth The amount of information or data that can be sent over a network connection in a given period of time. Bandwidth is usually stated in bits per second (bps), kilobits per second (kbps), or megabits per second (mps).

Domain forwarding Redirecting requests on the Internet to a different Internet address. For example, domain forwarding allows multiple domain names to be registered, all of which point to the same web site.

Domain masking Also referred to as "masking" or "cloaking," works with web forwarding to keep your custom web address (e.g., www.yourdomain.com) in the browser address bar while visitors browse different pages in your site. This can be used to hide the real addresses of your web pages, either because those addresses are long and complicated (e.g., http://members.bigcompany.com/~username/page.html) or because it gives your web site a more professional appearance. Your visitors will see a cleaner, more memorable address in the browser address bar.

Domain name A name that identifies one or more Internet protocol (IP) addresses.

Control panel Included in web hosting packages is an online web-based application that allows you to easily manage different aspects of your account. Most control panels will let you upload files, add email accounts, change contact information, set up shopping carts or databases, view usage statistics, and so on.

Embedded object An object created with one application and embedded into a document created by another application. Embedding the object, rather than simply inserting or pasting it, ensures that the object retains its original format. In fact, you can modify the embedded object with the original program.

FTP File transfer protocol. This is the language used for file transfer from computer to computer across the Web.

Internet Corporation for Assigned Names and Numbers (ICANN) ICANN is responsible for the global coordination of the Internet's system of unique identifiers.

Masthead An alternate name for the nameplate of a magazine or newsletter.

MySQL Pronounced "my S-Q-L" or my-sequel, MySQL is an open source relational database management system (RDBMS) that uses Structured Query Language (SQL), the most popular language for adding, accessing, and processing data in a database. MySQL is the most popular database system for web sites.

Quick response tags A type of matrix bar code developed for the automobile industry but now popular with mobile devices for communicating information.

Shazam The name of a popular music-discovery application for the Apple iPhone device. Shazam is a free application that helps users figure out the name of a catchy song being played on the radio, television or other place. Users simply hold their iPhone device up to the speaker playing the music and the Shazam application will attempt to identify the tune's album, artist and song title. The Shazam application is available for the iPhone, Android, and BlackBerry smartphones.

Server side With server-side scripting, completing an activity involves sending information to another computer (server) across the Internet. The server then runs a program that processes the information and returns the results, typically a web page.

SSL Short for secure sockets layer, a protocol developed by Netscape for encrypting communications on the Internet.

Top level domain The last part of an Internet domain name—that is, the letters that follow the final dot of any domain name such as .com, .net or .edu.

URL Uniform resource locator, the global address of documents and other resources on the Web. The first part of the address is called a protocol identifier, and it indicates what protocol to use; the second part is called a resource name, and it specifies the IP address or the domain name where the resource is located.

Web host A business that provides server space, web services, and file maintenance for individuals or companies that do not have their own servers.

Notes

1 A web server is the computer program (housed in a computer) that serves requested HTML pages or files.

2 This is only true if a straight, static HTML site is created, which most today are not. If not, there are also other considerations to make such as installed server-side software packages. These must correspond to how the web site is created and what coding tools are used. For example, if any database tools are used (for mailing list signup, etc.), the most popular database for web processing is MySQL. MySQL must be installed on the host in order for the web site to function properly. There is also the LAMP (Linux-Apache-MySQL-PHP) versus .NET (Microsoft's standard) differences to consider. A server with the proper capabilities for each must be chosen depending upon programmer preference.

Bibliography

BuyDomain.com. Suggestions for domain names. http://domaininfo.buydomains.com/ec/domain-name-basics/longest-domain-names/

Cultra, S (2011). Are QR codes going to have an effect on the value of domains? DomainShaneDotCom. http://domainshane.com/are-qr-codes-going-to-have-an-effect-on-the-value-of-domains/

Forbes, Dennis (2006). Interesting facts about domain names, www.yafla.com/dennisforbes/Interesting-Facts-About-Domain-Names/Interesting-Facts-About-Domain-Names.html.

Heng, Christopher (2004). How to choose a web host, www.thesitewizard.com.

Medeiros, João (2011). How Shazam pivoted its business to help U.S. television broadcasters. *Wired Magazine*. http://www.wired.co.uk/magazine/archive/2011/10/start/discovery-channeller

Smashing Magazine (2009). The effective strategy for choosing right domain names. May 2. http://www.smashingmagazine.com/2009/05/02/the-effective-strategy-for-choosing-right-domain-names/

Stone, John (2006). The importance of a quality domain name, http://tools.devshed.com/c/a/Domain-Name-Tips/The-Importance-of-a-Quality-Domain-Name.

www.webopedia.com

Chapter 5
Creating the Web Site

WEB SITE GOALS

Before one can begin designing a web site, it is necessary to determine the purpose of the site and outline the goals for the web site. For a musician or singer, these goals should include the following:

- Creating and reinforcing brand awareness
- Advertising and promoting products (recordings and concert tickets)
- Generating direct sales (mail order and e-commerce)
- Creating a sense of community among an artist's fans
- Creating repeat traffic

The design will need to incorporate aspects that support and enhance these goals. It should be attractive and suit the artist's image. It must also be designed to appeal to the target market. For a small business, such as a recording studio, venue, booking agency, public relations firm, or management company, the goals will be somewhat different, since those businesses are more concentrated on providing a service. Booking agencies will want to feature the artists they represent, and the success stories that go with them. The Agency Group, Ltd. features a list of artists, the contacts for booking them, and their availability. Recording studios will also want to feature well-known artists who have recorded there, the facilities and equipment, and rental fees. Record labels may or may not provide an online e-commerce section of their web site, however smaller indie labels are more inclined to sell directly to consumers.

Clubs and bars should have a web page that represents the image of the club. The same logo as on the signage: a color scheme that is repeated on all signage, printed materials and the web site. Before posting a photograph of customers, the web site developer should get release forms from all recognizable persons in the photo. Most important for a club is to provide a schedule of upcoming events with as much detail as possible. This would include artists or entertainers with photos, short bios, and links to samples. The site should also include admission prices (cover charge), restrictions, and directions to the venue. Contact information should be included.

Now, let's examine each goal for an artist's web site.

Branding

Branding is described as creating a distinct personality for the product (in this case, the artist, not the label) and telling the world about it. Artists who create a well-known brand can parlay that into endorsement deals, an acting career, or the position as a spokesperson for a worthy cause. Artists Madonna and Missy Elliott appeared in TV ads for the Gap, Queen Latifah has moved into films, and Bono of the rock group U2 has become the spokesperson for solving the problems of Third World debt and global trade. Therefore, artist branding should be an important part of any web site. Creating an easily identifiable logo and maintaining consistency in style and design can help support the brand. This extends to the style of the web site, which should convey the image that you want to project for the artist.

Promoting and Selling Products

The web site should create a demand for the artist's products. The primary products are recorded music and tickets to live shows. Secondary products include T-shirts, posters, bumper stickers, and other souvenirs. Enticing the web visitor to purchase the recordings, or even just creating a desire to purchase, should be an important aspect of the site. This goal can be achieved through offering music samples on the site and background information about the music and the recordings. Links to online retailers will encourage the visitor to follow through on purchase intentions, creating the impulse buy. Tour schedules, concert photos, samples of live recordings, and links to sites that sell tickets can create a demand for concert tickets. Maps to the venues also encourage concert attendance. The web site is a great place for the merchandising of T-shirts and other mementos. Photos of these items are important (see the section on e-commerce).

SENSE OF COMMUNITY

Artist web sites are a good place for like-minded people to connect with each other. These fans probably have much in common—especially their interest in the artist. A sense of community can be created through message boards, online chats with or without the artist, blog comment sections, and by including photos of fans at concerts. These devoted fans can become opinion leaders (the virtual street team[1]) by influencing their friends to become fans of the artist and encouraging their friends to visit the web site. Facebook and MySpace have provided an excellent opportunity for fans to engage in community social behavior centered on a common interest in an artist.

The web site should also provide materials for professional journalists looking for information on the artist, as well as contact information for both journalists and booking agents. Some sites provide print quality publicity photographs and in-depth biographical information suitable for reprint in mainstream publications.

Creating Repeat Visitation

It is important to provide elements on the web site that will keep visitors coming back as new developments occur in the artist's career. Certain attributes can be included in a web site that encourage repeat visits; these are outlined in Chapter 7. These features include frequently updated blogs, future tour dates, contests, news, new music samples, message boards and activity from social network sites. Many services offer plug-in widgets that can add value to your web site and are updated regularly from an outside source.

Web Site Design

Web site design is defined as the creation and arrangements of web pages that make up a web site. It's part art and part commerce. The first page is the home page, although some web sites will have a splash page with a welcome message or graphic image that sets the tone for the web site (see the section on splash pages). Aspects of the design include content, usability, appearance, and visibility (the ability of users to find your site on the Web). A good web site is one that is attractive, uncluttered, quick to load, and easy to navigate. The site must offer something of value to the consumer—information, products, and freebies. It also helps to make the web site fun, keep content fresh, and give people a reason to return. Effective web sites avoid large flash programs, flashing text, animation, and large graphic files. Many Internet users do not yet have cable modems, and a web site that is slow to load is sure to fail.[2]

The home page identifies your company or brand and extols the benefits of the site and its products. The site description, news, and a logo or image should be featured on the home page. This page should be updated frequently to reflect changes and notify the visitor of new and interesting developments. The home page often posts announcements, although a "What's New" page can also contain news briefs and updates. The home page should also contain links and navigational tools for the rest of the site. A good web page should balance text with graphics or images and sport a layout that is inviting and interesting.

Basic Design Rules

Once you have established the goals and determined the audience for your web site, it is time to decide on layout and content for the web site in general and then for each page. This is called storyboarding; it is when "the organized content is used to develop a diagram or map" (www.amacord.com/services/storybrd.html). The storyboarding process consists of developing a sketch of the site's structure, a detailed structural outline (which is the URL of each page listed in an organized format), and a detailed sketch of each page.

It is important to first get a sense for the overall site, how many pages will be needed, and what materials to include on each page. Content should be categorized according to user needs and organized in a way that takes into consideration the audience characteristics, their information preferences, specifications of the majority of computers, and the audience's web experience.

It might be helpful to use blank index cards, one for each web page, and focusing only on content, list the information that should be included on each page. Then for layout, use blank letter-sized sheets to create a diagram for each page. By determining the scope of the site first, the designer can then get an idea of the navigation needs and how to set up menus and style sheets (see the WYSIWYG section).

1. *Design web sites, not pages.* Determine the overall look of the site and the scope of information before attempting to design any one particular page. This will help the designer determine how many pages are needed, what should be on each page, and the overall consistent look of the site.
2. *Keep page size to a reasonable length.* The general rule is to design the page so that it is no longer than twice the length of the computer screen, especially for the home page. Readers do not like to scroll down and tend to become lost or discouraged by long web pages. The exception to this may be blogging, which tends to be more linear than most other web elements. When long pages are used, be sure to have plenty of bookmarks that can direct the user quickly back to the top of the page. Never use horizontal scrolling.
3. *Use appropriate graphics.* Don't load a lot of photos on one page—it will slow the loading time. Too many photos overload both the page and the eye. Make sure all photos are appropriate for that particular web page. If you want to include an album of photos, relegate it to a special photos page. Use thumbnails to present examples of photographs that can be enlarged at the click of a mouse, or alternatively, embed a slide show of images from SlideShare or Flickr. Web visitors are now accustomed to thumbnails but should be reminded that they can click for an enlarged version of the photo. Sometimes this is done with a "mouse-over" command, where a text message "click to enlarge" appears as the visitor runs the mouse over the thumbnail.
4. *Clearly specify the purpose of the web site and why the user should visit regularly.* The home page should have a clear, simple headline description of the site. If that is not suitable, then the masthead and dominant graphic or photo should clearly illustrate the site's purpose or products. Stick to the subject in the web site, keeping your goals in mind. If you want to include something that is not germane to the topic, consider linking to a separate web site. Keeping the site current and changing out content will encourage repeat traffic.
5. *Keep it simple and easy to read.* Nothing screams amateur more than a site that is cluttered and hard to read. Avoid backgrounds that conflict with the text and confuse the eye. Set up a color scheme that is appealing, consistent, and does not induce eye fatigue. Yellow or red text on a white or busy background never works.
6. *Balance the design elements.* Don't go too heavy on either text or graphics. Just like for a magazine layout, white space can be very appropriate when used properly. Basic principles of desktop publishing also apply to web design.
7. *Content is king.* Content should drive the design. Make your design appropriate for the theme and brand of the artist or product. Use content to give visitors a reason to explore the site and to return.

8. *Make navigation easy.* It is important to link pages in a consistent, well-planned manner that is intuitive to users. Navigation bars are usually found across the top, under the masthead, or on the left side of the page. Web visitors are used to this design, which was originally created because web pages load from the top and left side, and navigation information was sure to appear on the page regardless of the end user's browser or computer screen. Keep your navigation bar in the same location on every page, with a clearly identified link back to the home page.
9. *Keep your site up to date.* Visitors will not return to a site that looks abandoned, with old news, outdated information, out-of-style design elements, and cobwebs. If you plan to update on a regular basis, state that intention on the site and then be sure to follow through! If you state you will provide a weekly podcast, then be sure to have a new one available at the same time each week.
10. *Be consistent.* Keep the basic look and theme the same throughout the site. Use the same color scheme, background, and fonts on each page. Do not use many different fonts. For a more consistent look, use the same masthead on every page—with slight variations if necessary to identify individual pages. Stick with standard layouts. The reason the three-column layout is so common is that it works (Kyrnin).

Approach your web design from your customer's point of view. Too many web sites are designed to appeal to the desires of company insiders instead of focusing on what the consumer wants, needs, and expects. You need clarity: The purpose of

Table 5.1 **Red flags of an unprofessional web appearance**

Red flags of an unprofessional web appearance
Poor browser compatibility
Animated bullets
Too many graphic and line dividers
Multiple banners and buttons
Menu bar placed inappropriately
Poor use of tables
Too much advertising
Large welcome banners
No meta tags
Under construction signs
Scrolling text in the status bar
Large scrolling text across the page
Poor use of mouse-over effects

Source: www.phoebemoon.com

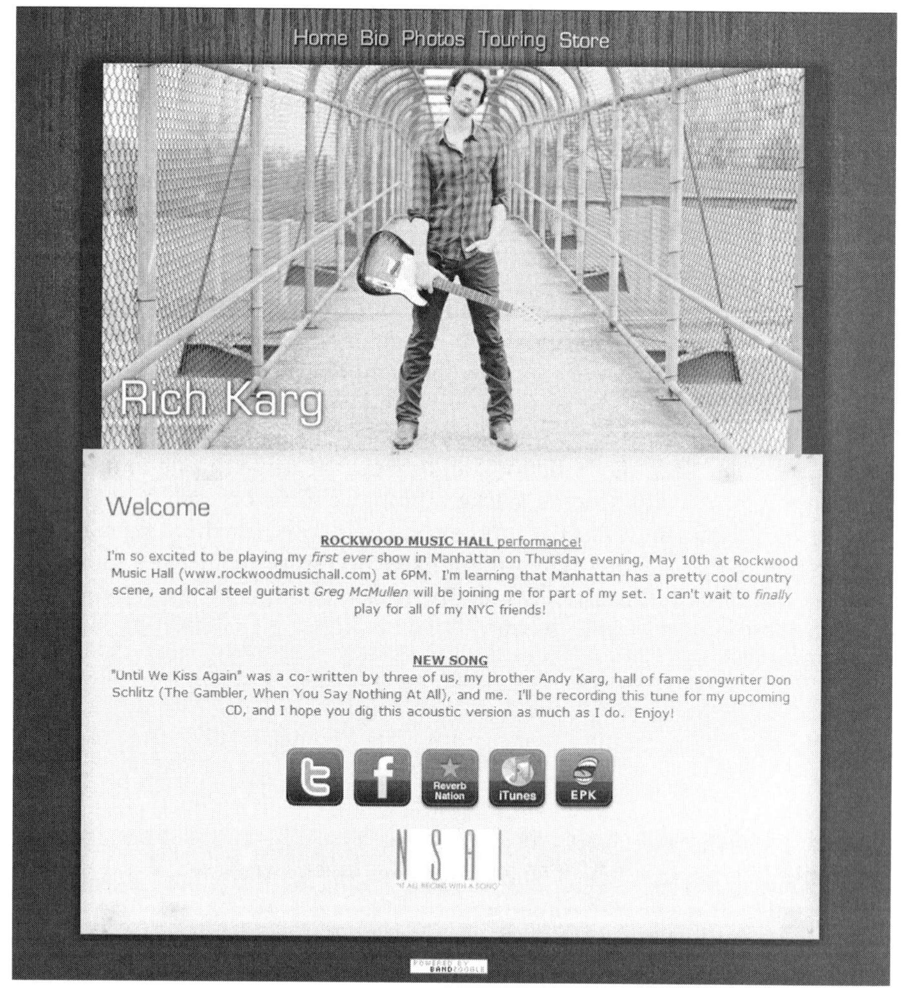

Figure 5.1
Example of an artist web site from www.richkarg.com

the web site needs to be upfront and clear. What are you selling? Why should the customer stay on your site instead of visiting the competition? You need appeal: the web site should look fresh and contemporary, not cluttered and dated. In his book, *Don't Make Me Think*, author Steve Krug recommends minimizing visual noise, taking advantage of what web visitors expect (such as where you place the menu buttons), breaking up pages into clearly defined areas, making it obvious what is clickable, and creating a visual hierarchy on each page. His advice is to use a common sense approach; make everything self-evident so the user doesn't have to stop and think about where to go next or how to get there. You need flexibility: Most web visitors are just browsing until something catches their attention. Give control over to your visitors. Layering information (click for more detail) allows the user to make the choice to delve deeper or continue browsing.

SPLASH PAGE

A splash page can be used to enhance the image of the web site. This is the first page a web surfer is directed to when accessing the site. Usually a splash page will have the logo and elaborate graphical layout, but not much information. The visitor is encouraged to enter the site by clicking on the page. Automatic splash pages direct the visitor to the home page after a few moments.

Do not overload the visitor with large flash programs that take a long time to load. An effective splash page has only a graphic image and the brand logo. Web Design Services India states that splash pages can be useful for entertainment web sites such as those for movies, video games, children, music, photographers, and travel sites.

If you use a splash page, Client Help Desk (www.clienthelpdesk.com, 2004) advises one to

> consider dropping a cookie into each site visitor's computer that automatically skips the splash screen on subsequent visits. Even people with the patience to deal with a splash screen once will be tested upon seeing it repeated each time they return to the site.

Always offer the visitor the option to "skip the intro" and move on quickly to the home page.[3]

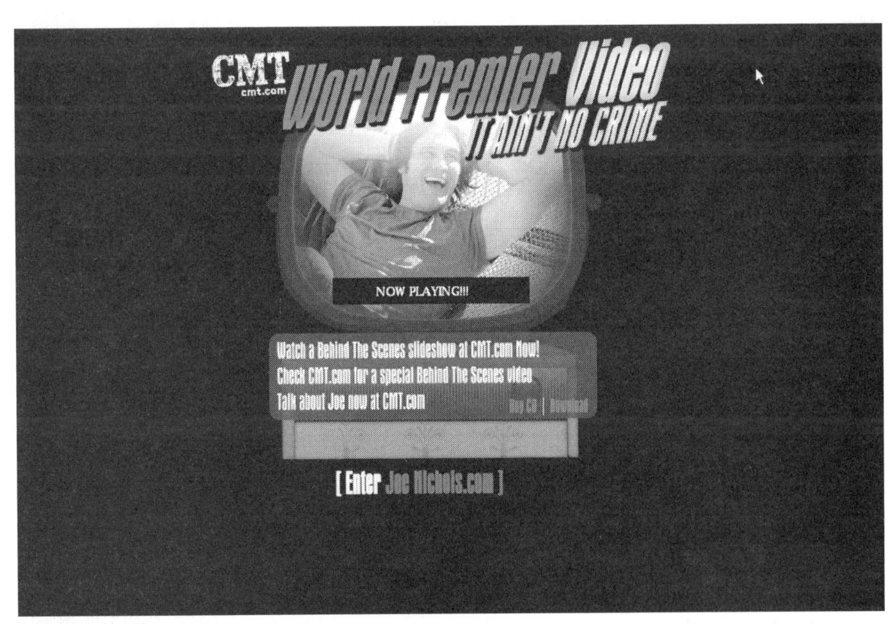

Figure 5.2
Splash page from Joe Nichols web site (courtesy of Joe Nichols and CountryWired.com)

WYSIWYG or WYSINOT?

> You can't expect your designs to look absolutely identical on every computer. Web design is just not like that. There are situations where they can break completely—elements shift to a different place on the page or disappear completely.
>
> **Joe Gillespie, www.wpdfd.com**

Many web design software programs feature a process called WYSIWYG, which means "what you see (on the screen) is what you get." One of the first idiosyncrasies web designers encounter is the fact that a web site's look will vary from one computer to another. Not all web visitors use the same browser software and hardware, so what may look right on one computer may look completely different on another, or not work at all. Some of the variables that affect consistency are type of software browser, monitor size, monitor resolution, and the end user's font settings. Web designers must not make the mistake of assuming that a layout will look as good on other people's computer as it does on theirs. The standards for HTML agreed upon by the World Wide Web Consortium (W3C) are helping to create some consistency among various browsers, but not all users will have the latest browser version.

The screen captures shown in Figures 5.3, 5.4, and 5.5 illustrate the variations in a web site's look based on browser and screen resolution. They all show the exact same web page.

Figure 5.3 was created on a liquid layout, where the size of the web page adjusts itself to fit the screen. On larger screens with higher resolution, the fixed elements (photos) remain the same size, whereas the liquid aspects (text arrangement) vary from computer to computer and are set up to conform to a certain percentage of the computer screen. The same web page shown on a screen with a higher resolution (Figure 5.4) has the menu bar on the right side at a distance from the main photograph. To minimize this problem, use fixed-width tables then set the elements inside specified-width boxes in the table. For years, the use of tables was popular until the introduction of cascading style sheets and universal standards for web pages.

On the high-resolution screen shot in Figure 5.4, note how the menu bar is far to the right of the photograph, rather than blended in, as was the case with the lower resolution Microsoft Internet Explorer screen shot. On the Netscape screen grab (Figure 5.5), the menu is unintentionally beneath the photo rather than on the right side of the screen.

It is best to check out a new web design on different systems before launching the site. There are several online services (such as www.browsercam.com) that will provide screen shots of your web site loaded with different browsers, operating systems, and monitor resolutions.

Almost all contemporary web designers have adopted the fixed width layout, rather than the liquid layout illustrated in the previous figures. Jennifer Kyrnin describes the fixed layout as

Figure 5.3
Example of web page with intended layout (courtesy of www.kargboys.com)

Figure 5.4
Example of web page on wide screen (courtesy of www.kargboys.com)

Figure 5.5
Example of web page in an older Netscape browser (courtesy of www.kargboys.com)

> ... layouts that start with a specific size, determined by the Web designer. They remain that width, regardless of the size of the browser window viewing the page. Fixed width layouts allow a designer more direct control over how the page will look in most situations.
>
> **(Kyrnin, 2012c)**

The advantage of the liquid layout is that the entire screen is used, and allows for flexibility with font sizes. The disadvantage is that it will not look the same on all computers. The advantage of the fixed width layout is consistency across all browsers and platforms, but can waste space on larger monitors by having the formatted web page appear only in the center, with lots of background on either side.

Try this experiment on your favorite web sites: open the web page in your browser as it is maximized on your monitor (taking up the whole monitor). Then, at the top right of the page, click the square between the minus sign and plus sign. On a Mac, it will be the yellow (center) circle in the upper left. Then grab one edge of the reduced-size window and move it in and out. Watch how the shape of the web page changes—or doesn't change as you vary the page width.

Early web designs incorporated frames: generally a wide banner frame at the top of the page, a long horizontal frame on the left for the menu, and the larger content frame beneath the banner and to the right of the menu. As the web visitor scrolled down the window with the content, the menu frame remained unchanged. When web browsers evolved, frames gave way to tables and nestled tables, which allowed for more content organization. Tables allow for greater control over page layout and create more visually interesting pages. Tables are used to define and separate elements in the document, such as navigation bars, masthead, side bars, text, and photos. The example in Figure 5.6 has four columns and five rows showing. Note how some of the cells have been merged to create a masthead or banner that spans more than one column or text that spans more than one row. The use of tables allows for a fixed width layout by setting the table width in pixels instead of percentage of the page.

In Figure 5.7, the borders of the cells have been made visible to demonstrate a "tables" document. By reducing the border width to zero, the cell borders become invisible while still constricting the cell content to the specified area. Cells can be created to float (the liquid layout) and expand (to fill the screen), or set to a specific width and height. Cells can also have different backgrounds to set them apart from the others.

Cascading Style Sheets

Recently, cascading style sheets have become popular. "Style sheets were developed as a means for creating a consistent approach to providing style information for web documents" (Wikipedia). Cascading style sheets (CSS)

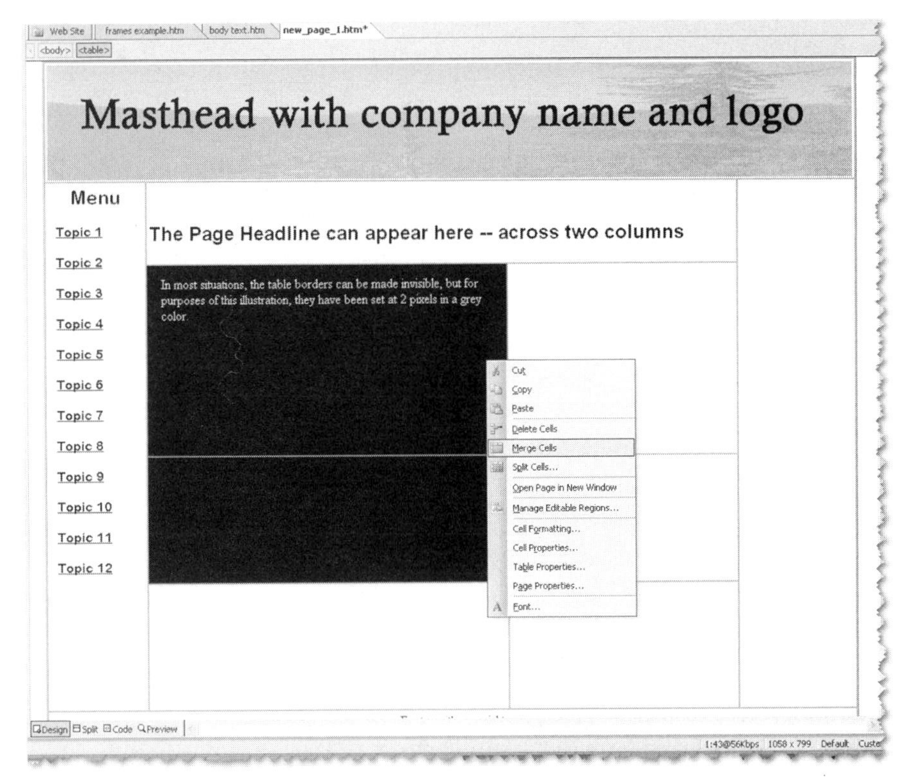

Figure 5.6
Example of tables layout

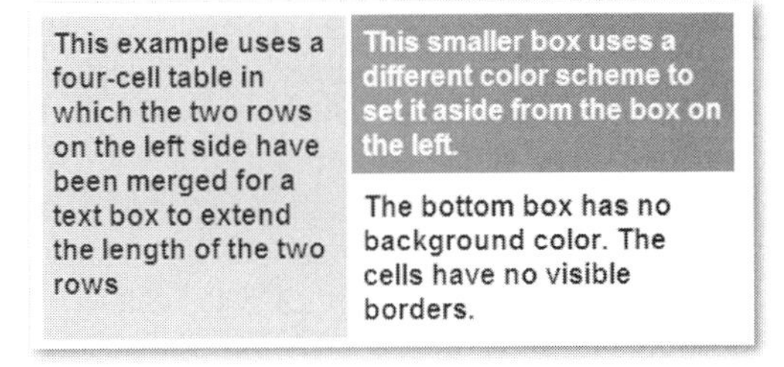

Figure 5.7
Example of tables with no cell borders

allow web developers to control the style and layout of multiple web pages at once, just by editing one master template (the CSS document). Style sheets are the technical specifications for a document—the formatting. CSS is used to define colors, fonts, and layout, and is designed to separate document content (written in HTML) from document presentation (written in CSS). The coding of the style sheet is separated from the content, unlike HTML, where formatting code is embedded within the content. Style sheets are saved in external .css files. The appearance and layout of multiple pages can be changed all at once by

Figure 5.8
Example of web page using cascading style sheets (courtesy JoDeeMessina and CountryWired.com)

editing the style sheet. In addition to the superior look and feel of web pages, CSS is also used to define how a page should look on other platforms and formatting for the print version of the page.

Here comes the cascading part. CSS has various levels, each building on the last by adding new features. The main style sheet will affect the overall look of the web site, keeping it consistent from page to page, and influencing each page. Subsequent levels of style sheets (cascades) will apply to sub-sections of the site and add additional customization for specified subgroups of pages.

Jennifer Kyrnin states "there are two parts to cascading style sheets: the styles and the stylesheet." The stylesheet is a separate document from the content-filled web pages. Externally linked stylesheets can be applied to all pages of the site through linking (<link rel="stylesheet" type="text/css" href="styles.css">). By removing the formatting language from the content pages, you can put the important content first on a page, thus improving search engine results (see Chapter 7 on search engine optimization). The "style" part contains the descriptive tags that determine formatting and layout for the pages referenced in the link command. More in-depth information on CSS tags is presented in Chapter 6.

Graphics

Graphics are an important part of any web site. There are some general rules to follow when using graphics on a web page. The three most popular formats for using graphics on a web page are JPEG, GIF, and PNG. JPEG is short for Joint Photographic Experts Group. It is good for photographs and supports 16.7 million colors. The compression actually throws out data to create a smaller file. Sharp edges may appear blurred, so JPEG is not recommended for graphics that contain sharp lines or drastic color changes. It also does not support transparency, so if you want the background of your image to be transparent, JPEG is not the format to use. A progressive JPEG file presents a low-quality image at the first moment of download, and then over several passes it improves the quality.

The GIF (graphics interchange format) format is excellent for graphics that have large areas of the same color and abrupt color changes. The format supports a maximum of 256 colors and thus is not the best to use for photographs. Gradual changes in color may show up as progressive rings of color changes, as evident in Figure 5.10. The GIF format supports transparency, allowing the graphic designer to remove the background of the image. The newest format, PNG (Portable Network Graphics), has images that always look great and offer good compression ratios. PNG was designed to offer the best features of the GIF format, but with millions of colors. PNG also allows for transparency, but it is not supported by older browsers and AOL.

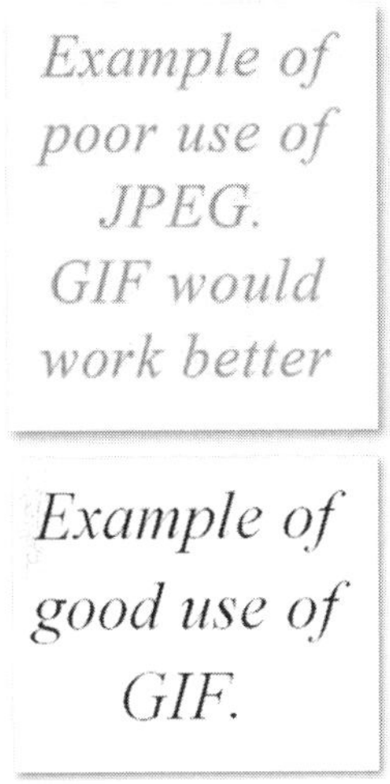

Example of poor use of JPEG. GIF would work better

Example of good use of GIF.

Figure 5.9
Example of a poor use of JPEG and cleaner lines with GIF

Editing Graphics

Adobe Photoshop is the industry standard for image manipulation (editing graphics and photographs). Even the novice user is easily able to crop pictures, create layers, overlay text, rotate images, modify colors and contrast, convert formats, and brush out unwanted elements of the photograph. For the amateur photograph editor, there are some shareware and free programs available for basic photo editing. In 2008, Adobe Systems launched a free online version of Photoshop called Photoshop Express. This web-based version works with any type of computer and operating system. Adobe also offers a boxed consumer version of Photoshop Express (Fedh, 2008).

Figure 5.10
Example of a poor use of GIF. This should be a gradual change from dark to light

The online site Pixlr Express offers users the opportunity to upload and manipulate images online. Pixlr Editor has the options to build an image from scratch, with some neat effects. www.pixlr.com.

FastStone offers a freeware version of its image viewer, which contains quite a few image editing and manipulation functions, such as those basic functions listed earlier, drop shadow, sepia tone, red-eye reduction, special effects, sharpen or blur, and resample compressed photos. The downloadable software is available at www.faststone.org. A commercial version is also available for a reasonable fee, and FastStone offers several other useful products.

In an article titled "Top 8 Free Photo Editors for Windows," Sue Chastain mentioned the following free photo editors. Her list has been updated to include a couple other programs that are either shareware or offer a free trial period, including several for the OS platform.

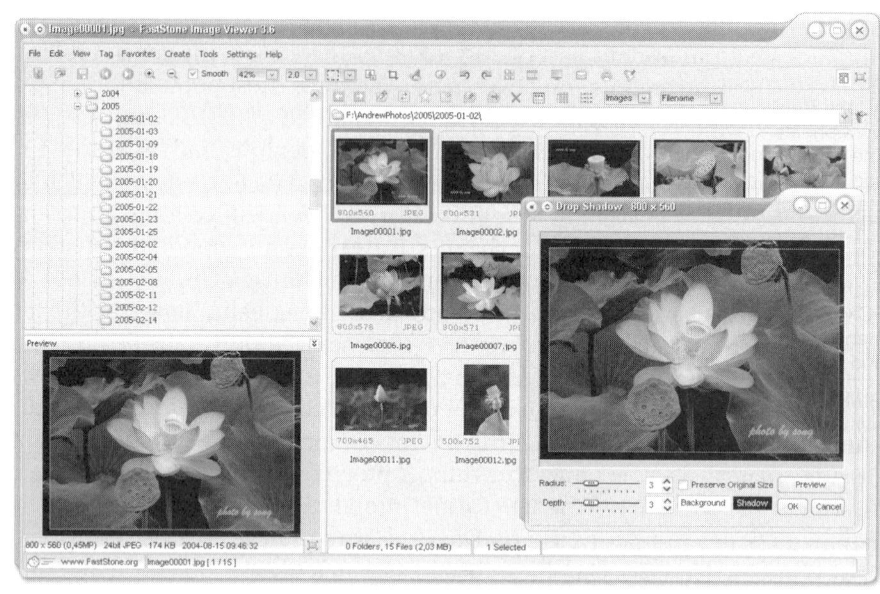

Figure 5.11
FastStone image editor (courtesy of FastStone, www.faststone.org)

Creating Thumbnails

Thumbnails of photographs are popular on web pages. A thumbnail is a small version of a picture that opens a larger graphic file of that picture when the user clicks on it. Most web design programs will automatically create thumbnails with a simple command, but the concept is simple enough to create with a bit of HTML text. First, open the photograph in any photo editing program. Reduce the size of the photograph to the ideal thumbnail size (maybe 150 pixels wide by 200 pixels high). Be sure to resample the resized photograph (thumbnail) for best results. Save the new photograph under a different name so that it won't overwrite the original. A good rule of thumb is to use the same name but add "_small" to the name or "_thm" so you can distinguish it from the original. Then when creating the web page layout, include the smaller thumbnail-sized photo on the page. Create a hyperlink that will lead to and open the larger graphic file. The HTML may look something like that shown in the box below.

HTML Language for Thumbnails

```
   <body>
   <p><a href=http://www.kargboys.com/images/KBCover.jpg
target="_blank"><img src="http://www.kargboys.com/images/
KBCover_small.jpg" alt="Karg Boys cover" width="150"
height="160"></a></p>
   </body>
```

Table 5.2 Photo editing software

Adobe Photoshop Express online
www.photoshop.com/tools/overview
The free version of Adobe Photoshop Express is available online, but not for download, and has hobbyist features but lacks the high-end professional features of the full program. It's not platform dependent.

FastStone Image Viewer and Editor for Windows
www.faststone.org
FastStone supports all major graphic formats including BMP, JPEG, JPEG 2000, GIF, PNG, PCX, TIFF, WMF, ICO, and TGA. It has a nice array of features such as image viewing, management, comparison, red-eye removal, emailing, resizing, cropping, color adjustments, and musical slideshow.

GIMP for Windows; GIMP for Mac OS X
www.gimp.org
GIMP is an image manipulation program, a freely distributed piece of software for photo retouching, image composition, and image authoring. It works on many operating systems.

Serif PhotoPlus
www.freeserifsoftware.com/software/PhotoPlus/default.asp
PhotoPlus is a photo editing software that enables users to fix and enhance digital photos, create stunning bitmap graphics, and even produce web animations.

Paint.net
www.getpaint.net
Paint.NET is free image and photo editing software for computers that run Windows. It features an intuitive and innovative user interface with support for layers, unlimited undo, special effects, and a wide variety of useful and powerful tools.

Pixlr
http://www.pixlr.com/
Pixlr has web-based photo editing tools for existing photos that you upload to the site. The editor also allows the creation of images from scratch.

Pixen for Mac
http://pixen.en.softonic.com/mac
Pixen is a free (donation-ware) graphics editor for Mac OS X. It has been specially designed for pixel artists, but is also suitable for other types of pixel-based illustration and animation.

Photoscape
www.photoscape.org/ps/main/index.php
Photoscape is a fun and easy photo editing software that enables you to fix and enhance photos.

XnView
www.download.com/XnView/3000-2192_4-10067391.html?tag=lst-0-5
XnView supports red eye correction, crops and transforms JPEG images without loss of quality, generates HTML pages and contact sheets, and provides batch conversion and batch renaming.

Saint Paint Studio
www.download.com/Saint-Paint-Studio/3000-2192_4-10066321.html?tag=fd_sptlt
The Saint Paint Studio paint package is designed to be the essential base tool for editing photos, web graphics, icons, images, and animations.

For more photo editor software, try
www.focalpress.com/cw/Hutchison
www.photo-freeware.net
www.download.com
www.softpedia.com

The alt="Karg Boys cover" denotes text that will be used to fill in if the image does not load into a user's browser. The target command will open the photograph in a new window. If you want it to open in the same browser window, the target command would be target="_self". The alt tag is also used by screen readers, making it critical that it contains an accurate description.

There are also several shareware programs that will create thumbnails quickly if you have many photos to process, including Easy Thumbnails. Easy Thumbnails is a popular free utility for creating accurate thumbnail images and scaled-down/up copies from a wide range of popular picture formats (www.fookes.com/ezthumbs).

Web Design Software

Most software programs provide WYSIWYG editing, meaning you can manipulate components on the web page while seeing exactly what the finished product will look like (theoretically). This has made web designing much easier, especially for the novice web developer. In addition to WYSIWYG, most programs also provide a code-based view for HTML tweaking. For most popular web development programs, it is recommended that the novice user start with ready-made templates (see the section on templates), many of which are available for sale or for free by third-party developers. The most commonly used programs to create web sites are Adobe's Dreamweaver and, less so, Microsoft's Expression Web.

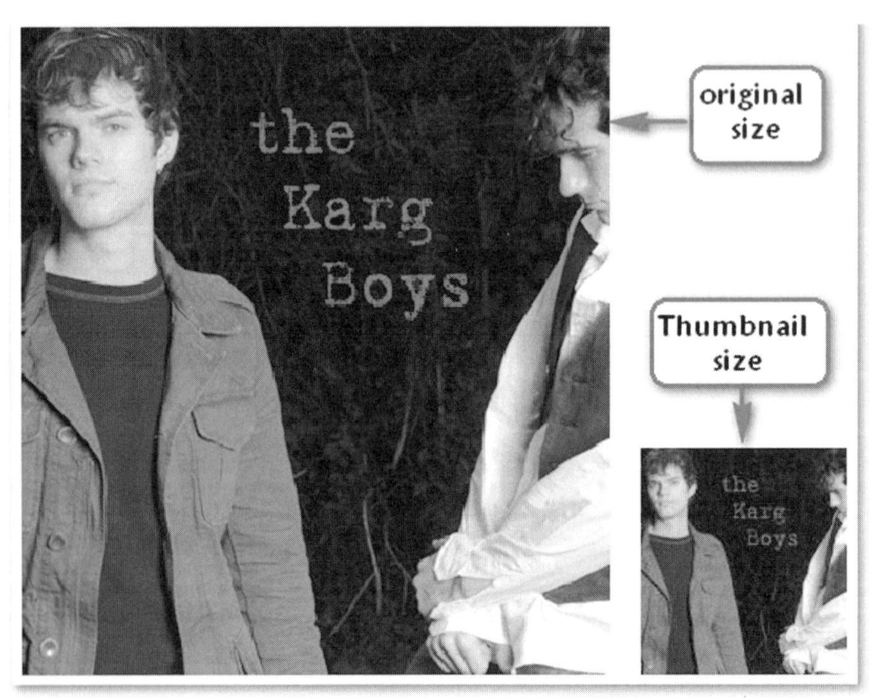

Figure 5.12
Example of photo and thumbnail

ConsumerSearch.com states that

> Dreamweaver is the best (albeit expensive) choice of web authoring tools for Windows and Mac users. Adobe Dreamweaver CS5.5 is the industry standard, providing both visual and code-level capabilities for creating standards-based web sites and designs for the desktop, smartphones, tablets, and other devices. It includes strong support for Cascading Style Sheets (CSS) and all major scripting languages.
> **www.consumersearch.com/www/software/web-design-software**

Microsoft's current web design software is Expression Web, replacing the popular but dated Microsoft FrontPage. Expression Web 4.0 is the newest version, with features like SuperPreview that allows the designer to view a web design as it would appear in a variety of browsers. It supports a broad range of scripting (see Chapter 6) and server-side technologies. Bloggers and reviewers compare it more closely to Dreamweaver than to the earlier FrontPage program and say it was built from the ground up rather than being an upgrade of FrontPage.

CoffeeCup is an inexpensive, easy-to-use program for beginners, with detailed tutorials. Their Visual Design Editor touts itself as the easiest to use allowing for WYSIWYG editing in the "visual editor" mode or HTML editing in the "code editor" mode. In the visual editing mode, you can drag and drop images, text, and tables, and edit on the fly. (The full version is $49 in 2012.) CoffeeCup offers an à-la-carte selection of web site creation programs such as a site designer, shopping cart creator, form builder, and more.

Coffee Cup Free HTML Editor 9.6 is a trimmed-down, free version of the CoffeeCup HTML Editor, and does not contain the Visual Editor, CSS Menu Designer, Image Mapper, and Website Color Schemer. CoffeeCup also offers a host of free supplemental software for transferring files to the server (FTP), creating fonts, viewing images, and more. http://www.coffeecup.com/freestuff/

Web Studio 5.0 is the priciest of the midrange software at $149. It is easy to use for the newbie and includes add-ons like a library of templates, photos, buttons, and special effects. Its features include a WYSIWYG design with drag-and-drop components, FTP uploading, and a built-in graphics generating tool, but it lacks an HTML code view for editing, and adding external HTML code can be a challenge.

Serif's Web Plus, according to ConsumerSearch, has features comparable to WebStudio. Jennifer Kyrnin (2012d) states "Serif WebPlus is a web site editor that is perfect for the small business owner who doesn't have time to learn HTML or web design." It offers a range of templates, smart objects, and automatic navigation bars and has a free version, Web Plus Starter Edition. The company offers templates, wizards, and tutorials to help newcomers create attractive web sites. Web Plus X5 is the commercial version.

Web Easy Professional 8.0 is a nice, easy Windows-based program that would work well for the novice web designer. According to ConsumerSearch.com, it

> features WYSIWYG with drag-and-drop, FTP uploading, CSS support, real simple syndication (RSS) feed support, and multimedia support including

Flash animation and scripting, and it has a nice widget tool for building forms or creating scripts. There is a library of templates and thousands of clip-art images, as well as a built-in graphics generating tool. However, it lacks a spell checker and has no HTML editing capability—Web Easy Professional is strictly WYSIWYG.

Web Easy is highly rated by Top Ten Reviews.com. Web Easy Express 6.0 is a freeware version.

WYSIWYG Web Builder 7 is a low-cost, easy-to-use, but limited design program. It features drag-and-drop components including flash and forms, but offers no way to access and edit the code. Web Builder features online tutorials, and a fully functional 30-day trial version is available.

PC Magazine describes SJ Namo WebEditor 6 Suite as "an affordable entry-level program that can grow along with a user's skill level and development needs." The magazine gives it a high rating and touts its easy-to-use tools. David Nevue, in his book *How to Promote Your Music Successfully on the Internet,* stated that it is his personal favorite of the WYSIWYG editors and he personally uses it. The full version sells for less than $100, and the company offers hosting plans starting at $4.99 per month. The negatives listed by both *PC Magazine* and CNET review are that it lacks sufficient tutorials and the features are buried inside dialog boxes.

SiteSpinner is a drag-and-drop software program mentioned in Nevue's book as recommended by one of his readers. Top Ten Reviews.com rates it as average, with nice text and image editors, all for less than $50. SiteSpinner offers tutorials, user guides, and online support, as well as a 15 non-consecutive-day free trial period and free hosting for the first year.

Sandvox Standard 2 is an easy-to-use Mac-based web design program, with a wide variety of templates, widgets and a WYSIWYG editor. Their site boasts a simple step-by-step process of: "pick a look, mix and match, customize and expand, and publish."

RapidWeaver is another popular entry-level Mac-based web design program for under $100 from RealMac software. The software has 45 themes to choose from, and easy-to-use customization features. A MacWorld article by Adam Berenstain (2011) states "you can build media-rich, template-based sites with the added flexibility of inserting HTML code to customize your pages. However, interface quirks may make some design features difficult for users who prefer a more freehand approach."

Nvu is an open-source free program with the basic tools needed to create a web site and an interface that is reminiscent of earlier FrontPage programs. It claims to support many of the same easy-to-use features that make Dreamweaver and FrontPage so popular. Nvu includes some more advanced bells and whistles like JavaScript coding and CSS support. It supports forms and has a file management system (FTP). It provides a WYSIWYG edit view, a code view, and preview mode, similar to FrontPage, and it is compatible with both Windows and Macintosh platforms.

Software as a Service (SaaS) Web Design Software

One of the newer trends in computer applications is called software as a service (SaaS). Instead of downloading and installing a program on your computer, the program resides on the server (server-side), and you access it remotely to use it. Even Microsoft Office and Adobe Acrobat offer SaaS versions on their web sites. SaaS services are growing by 20% per year. One of the advantages of SaaS use is the ability to access it from anywhere, so you are not restricted to one computer to use the software to make changes to the web site. One disadvantage is that generally, a monthly or annual subscription fee is required, so it is not a one-time purchase. SaaS web design offerings are still in the early stages of development.

There are several SaaS options for designing and creating web sites. Most of them require a monthly service fee that includes web hosting. They are generally easier to learn and ideal for the novice web creator because they are modular, often contain flexible templates, and are seamlessly integrated into the hosting process.

Intuit is now offering a SaaS web site service touting that you can build a web site in three easy steps: pick a design, customize it, show the world. Features include hundreds of templates, hosting, analytics, shopping carts, blogs, email addresses, and hosting. Plans start at $7.99 per month.

Jigsy (formerly Viviti) is a web site builder that allows you to easily build your own dynamic web site, either by selecting a premade theme or making your own theme simply by using HTML and CSS. Everything is hosted on their server and all management and creation is done in your browser. Pricing starts at about $10 per month.

BaseKit includes all the tools, widgets and templates that you need to build a beautiful web site. BaseKit enables you to build web sites as soon as you have signed up so you can have your web site live in a matter of minutes. It is also priced at $10 per month.

Weebly is another free host-based web site service. The service offers drag-and-drop web site building, widgets for audio and video files, a blog feature, and an easy form builder. www.weebly.com.

Wix is a web-based site builder that offers a free version and a premium version. However, the free version is advertiser-supported, and not recommended for professional application. The premium, ad-free version starts at less than $5 per month. One advantage of such a tiered system is that you can actually build your site for free and decide whether to keep the site and pay for the upgrade, or to move on to a different platform.

Edicy is a super simple tool for building web sites. Just sign up and your site will be online in minutes. Edicy boasts that it is multilingual, providing members with up to 20 languages, and support for mobile browsers. Edicy offers fewer template options, but has a free version or pricing plans that start at less than $10 per month.

Google Sites. Google offers a free, basic web site tool with a limited number of adjustable templates. They also offer a premium service with 10 gigabytes of storage and seamless integration of Gmail, Google Calendar, Google Docs and Google Talk.

Web Development Software

The major software packages are available at many online and physical retail locations.

- Microsoft Expression Web, www.microsoft.com/expression/products/overview.aspx?key=web
- Adobe Dreamweaver, www.adobe.com/products/dreamweaver
- CoffeeCup, www.coffeecup.com/software
- Google Sites, www.google.com/sites/help/intl/en/overview.html
- Nvu, http://nvudev.com/index.php
- Namo WebEditor, www.namo.com
- RapidWeaver for Mac, www.realmacsoftware.com
- Sandvox, http://www.karelia.com/sandvox/
- WebStudio, www.webstudio.com
- SiteSpinner, www.virtualmechanics.com
- Web Easy, http://www.usa.webeasy.avanquest.com/
- Web Builder, www.wysiwygwebbuilder.com
- WebPlus, www.freeserifsoftware.com/software/WebPlus/default1.asp

SaaS Web Software

- BaseKit, http://www.basekit.com/tour
- Edicy, www.edicy.com
- Intuit, http://www.intuit.com/websites/
- Jigsy, http://jigsy.com/
- Wix, www.wix.com.

Templates, Anyone?

Templates are predesigned web pages that feature basic elements (placeholders) such as text boxes, backgrounds, buttons, a header, and images. The idea is that you take the template and replace the generic elements with your own. For the novice web designer, using a ready-made template may be a good place to start. Even if you end up throwing out much of the original template and replacing it with your own features, it creates a baseline to work from. The downside of using a template is that your web site might not look unique if others also use the template with little modification. But generally, the template can serve as a guideline because designers with more experience created them. It also provides some of the basic elements such as buttons, backgrounds, dividers, and text boxes. In her article "Why and How to Use Templates Effectively," Jacci Howard Bear suggested that templates can save time, provide consistency from page to page, and be a less expensive alternative to hiring a designer. She suggested the following steps to modify a template and make it your own:

1. Select the right template, one that closely suits the subject matter so that fewer modifications need to be made.
2. Change the graphics from the generic stock photos and graphics to your own.

3. Change fonts to your own, but keep them suited for the style and image of the site.
4. Change text formatting, perhaps adding bullet points or subheadings.
5. Change the template color scheme. The background, text colors, and other elements can be changed to reflect the image of the site.
6. Change the template layout. Although the original purpose of having a template is to provide a ready-made web page design, alterations such as flipping to a mirror image or swapping out sections can help to personalize the template.

There are templates available that include CSS code instead of using tables. In her article "Free Web Templates" Jennifer Kyrnin provides links to free templates and categorizes them based upon design structure, such as: 3-column centered flexible, 3-column layout with footer, and so forth. Related articles guide the user through the process of setting up a CSS site using the free templates offered in the article (http://webdesign.about.com/od/websitetemplates/a/bl_layouts.htm).

Lorem ipsum dolor, what? Many templates come with dummy or placeholder text of nonsense designed as filler until the web designer can replace it with actual text for the web site. In her article "Lorem ipsum dolor," Jacci Howard Bear suggested working with the placeholder text to experiment with font types and colors and background colors to get an idea what the final product will look like and to give you an idea how much text to write for the web page. You can cut and paste the placeholder text over and over until it approximates the length of the replacement text. It comes in handy if you need to design a web page or a newsletter before the actual article copy is ready for insertion. Just don't forget to replace the dummy text before publication. You can generate your own dummy text at http://www.lipsum.com/.

Many of the web site packages listed in this chapter contain some basic templates. For those with no prior web design experience, it is good practice to experiment with those templates before deciding on a look and layout for your web site. Of course, these packaged templates are the least unique because everyone using that software has access to them. Many third-party companies offer templates. Some are software specific and allow for a more seamless interaction with the program. Others are general HTML coded templates that are designed to work reasonably well with many web design software programs. Some templates are available for free, whereas others charge a fee for either one

Placeholder text

Lorem ipsum dolor sit amet, consectetuer adipiscing elit, sed diam nonummy nibh euismod tincidunt ut laoreet dolore magna aliquam erat volutpat. Ut wisi enim ad minim veniam, quis nostrud exerci tation ullamcorper suscipit lobortis nisl ut aliquip ex ea commodo consequat. Duis autem vel eum iriure dolor in hendrerit in vulputate velit esse molestie consequat, vel illum dolore eu feugiat nulla facilisis at vero eros et accumsan et iusto odio dignissim qui blandit praesent luptatum zzril delenit augue duis dolore te feugait nulla facilisi.

template or a series. A good place to start looking is TemplatesReview (www.templatesreview.com).

Template Monster is a site that is often cited in blogs and reviews (www.templatemonster.com). The service gets high marks from reviewers and users. You can use the search function to specify what types of templates you are looking for (music) and what feel you want to them. For less than $50, the search returns several possibilities. The site also offers a host of free sample templates and free clip art. 4Templates is another site that offers low-cost templates, starting at less than $22 (www.4templates.com). FreeWebsiteTemplates offers a wide selection of templates at no cost, including Flash templates (www.freewebsitetemplates.com).

Fonts

Browsers will only support a limited variety of fonts, so when selecting a unique and rare font for a masthead, logo, or wordmark, it is wise to create the item in the original editing software, and then save it as a GIF image for use on the web site. The down side to that is that the text in the image will not be read by search engines, but the wordmark will look consistent across platforms and on all computers. Failure to convert a wordmark can result in font substitutions that often do not turn out well. Logos and wordmarks can be copyrighted to protect your brand image. So the design of a unique logo can be an important aspect of any web design. Some of the most recognizable wordmarks and logos (that can't be displayed here for copyright reasons) include eBay, Disney, Coca-Cola, FedEx, and Canon (http://breezycreativedesign.com/2010/02/26/10-best-wordmark-logos-of-all-time/).

All text-editor programs, such as those found in Microsoft Office and OpenOffice.org, have a variety of fonts. Even many of these fonts are not supported by all browsers, so when using them, consider the conversion to an image. Microsoft PowerPoint in Office 2007 or above has many advanced features that can be used to create unique logos, mastheads and wordmarks. Start with a blank page. Set the background to the same color as the background on your web site. Insert a text box, select your font, and type in the words. Then double click on the box to open the "drawing tools." Under WordArt styles, you can choose from shadow, reflection, bevel, 3-D rotation and transform. By highlighting the text and clicking the right mouse button, you can bring up a menu that adds "3-D format." You can select the bevel style, contour, material type (such as plastic, metal, matte) and lighting style. It is wise to set these before working with 3-D rotation. But keep in mind that some of the best wordmarks are the simplest.

You can also use PowerPoint to create reflections of images. First, insert an image. Double click on that image to bring up "picture tools" that are similar to the "drawing tools" function. The reflection function is in the picture effects menu.

There are also cool fonts you can download to add to Microsoft Office. Urban fonts is a site that provides a plethora of contemporary and creative true type fonts. The fonts can be downloaded and will self-install into Microsoft Office. Once installed, they will appear in the pull-down font menu the next time you open the program (www.urbanfonts.com). Be sure to convert your wordmark to a graphic before incorporating it into the web site.

For a more flexible, dynamic solution, FontsLive offers web-friendly fonts and provides the code for embedding them into your site. Pricing starts at $40 per year. (www.fontslive.com)

TypeFront is a font distribution platform that gives designers and font sellers the tools they need to take advantage of the new wave of downloadable font support in web browsers. (www.typefront.com)

More options for finding fonts can be found at http://download.cnet.com/windows/fonts/?tag=nav.

28 DAYS LATER
Acid Dreamer
Graffiti
Frutopia

Figure 5.13
Examples of urban fonts

Elements and Content for an Artist Web Site

An artist's web site should contain elements that help achieve the goals set out at the beginning of this chapter: branding, promoting products, creating a sense of community and generating repeat traffic. As a part of the branding process, the overall look of the site should reflect the taste of the artist and the expectations of the target market.

An artist's web site should contain the following basic elements:

1. *A description and biography of the artist.* The home page should have some information or description, but a separate biography page should be created for in-depth information about the artist.
2. *Photos.* Promotional photos, concert photos, and other pictures of interest. This can include shots of the artist that capture everyday life, photos of the fans at concerts, and other photos that reflect the artist's hobbies or interests.
3. *News of the artist.* Press releases, news of upcoming tour dates, record releases, and milestones such as awards. This page should be updated often, and outdated materials should be moved to an archive section.
4. *Information about recordings.* Discography and liner notes from albums to increase interest in the recorded music of the artist. Comments from the artist on the recordings. Information on the recording process.
5. *Song information.* Lyrics and perhaps chord charts (again to increase interest in the recorded music).
6. *Audio files.* These may be located on the purchase page to encourage impulse purchases. They may be 30- to 45-second samples or streaming audio rather than downloadable files, to protect against piracy. Third-party music sampling and shopping widgets will be discussed in Chapter 8.
7. *Videos.* Music videos and/or videos of the artist performing have become an important marketing tool for artists. After all … this is the entertainment business. Club owners want to see what an artist's onstage performance is like before contracting them. Fans are more engaged in the music when the visuals are accompanying it. (See Chapter 8 for more information on videos.)
8. *Membership or fan club signup page.* Allows visitors to sign up for your newsletter or to access more exclusive areas of the site. This will help you

build an email list and allow for more control over content posted on message boards in restricted areas of the site.

9. *Tour information.* Tour dates, set lists, driving directions to venues, photographs/video from live performances, touring equipment list.
10. *E-store.* Merchandise page for selling records, T-shirts, and other swag.[4]
11. *Contests or giveaways.* To increase repeat traffic and motivate fans to visit the site. These can be announced at concerts.
12. *Links.* To other favorite sites, including links to purchase products or concert tickets, venue information, the artist's personal favorites, e-zines (online magazines), etc. Ensure that all your offsite links open in a new window so the visitor can easily return to your site.
13. *Contact information.* For booking agencies, club managers, and the media. This could also include print-quality images for the press.
14. *Message board or chat rooms.* This allows the fans to communicate with one another to create a sense of community. This can be an area restricted to members only.
15. *Printable brochures or press kits.* Electronic versions of any printed materials that the artist uses in press kits and to send out should be made available on the web site. (see Glossary). On the link to retrieve these items, be sure to mention if they are in the PDF format—some older browsers and slower modems lock up when attempting to open a PDF file in a browser window. Sonicbids is an online site that offers users the ability to create professional-looking electronic press kits for artists (www.sonicbids.com).
16. *Links to social networking pages.* Be sure to have direct links to your Facebook fan page, a link to sign up to your Twitter feed, and direct links to any other social network sites you actively maintain.
17. *Widgets* for assisting web visitors in posting your information to their social network pages.

By including these elements you will cover the goals of the web site, generate traffic and repeat visits, and provide an around-the-clock source of information and entertainment for fans. Coupled with an aggressive web promotion campaign, a well-designed web site can increase the visibility and popularity of an artist at any stage of the artist's development and career. Once you have determined what content should be included on a web site, decisions must be made regarding the design and layout—what colors, fonts, images, and so on should be used.

The best advice is to scour the Internet to find web sites and web components that are appealing and serve as examples and inspiration for building the perfect site. If a design expert is to be employed, these examples will illustrate to the designer what is expected of the new site. For the do-it-yourselfers, templates are available for most web design programs. Some are offered as part of the software package, some offered as free downloads, and some offered for sale (often for a small fee) through commercial software web sites. It may be worth spending a few dollars on a template with a professional and contemporary look. All templates can be modified or customized—the idea is to keep the elements that work and replace those that do not. A good web site can also be the product of evolution. Each time the site is updated, it is tweaked with minor improvements until it finally has the intended look of professionalism and success.

In conclusion, the web site is considered the home base from which promotions are launched. All marketing and promotional materials and campaigns can then direct fans to the web site for more of what they like about the artist. But the web site is just the beginning of the Internet presence for an artist. Chapter 7 addresses how to maximize the web site to increase traffic and sales. Chapters 10 through 12 outline how to promote the web site and the artist on the Internet.

Creating Content for the Web Site

An artist web site is one part press kit, one part electronic store front, and one part social networking site. One of the most challenging aspects of creating a new web site is building up the assets that will be used on the site. The earlier section on basic design rules discussed how to take the content and design a site using storyboarding. The previous section included a list of potential items that may be incorporated into the web site, but how are these assets developed? The publicist is the best person to create these assets. He or she knows how to craft stories and select photographs that depict the artist in the image that best suits the artist's career. Do not leave it to the webmaster to create and select these assets—this is not the webmaster's area of expertise.

If the artist has a press kit, that is a good place to start—using electronic versions of text files and graphics. These items should be developed in advance of designing the web site and should be developed for both online and offline use. An artist biography is usually written by a trained journalist or publicist, who interviews the artist, reads up on the artist's background, understands the market, and then creates a compelling story via the bio. A professional photographer should be hired to take publicity photos. Publicity shots are not the same as a publicity photo. A publicity shot is one taken backstage with other celebrities or at events. The publicity photo is the official photographic representation of the artist. Publicity photos should be periodically updated to keep current with styles and image. But once a photo is released to the public, it is fair game for making a reappearance at any time in the artist's career, even for artists who have moved on and revamped their image.

The press release is another asset that should be incorporated into the web site. Press releases are a standard tool in public relations, one that works better than letters or phone calls (Spellman, 2000). The press release is used to publicize news and events and is a pared-down news story. Here are some examples of when a press release should be used:

1. To announce the release of an album
2. To announce a concert or tour
3. To publicize an event involving the artist or label
4. To announce the nomination or winning of an award or contest
5. To publicize other newsworthy items that would be appealing to the media

The press release should be written with the important information at the beginning. Today's busy journalists don't have time to dig through a press release to determine what it is about. They want to scan the document quickly to determine whether the information it contains is something that will appeal to their target audience.

The press release needs to have a slug line (headline) that is short, attention grabbing, and precise. The purpose or topic should be presented in the slug line. The release should be dated with contact information including phone and fax numbers, address, and email. The body of text should be double spaced. The lead paragraph should answer the five Ws and the H (who, what, where, when, why, and how). Begin with the most important information; no unnecessary information should be included in the lead paragraph (Knab, 2003). In the body, information should be written in the inverse pyramid form—in descending order of importance.

CONCLUSION

The beginning of this chapter stated how a web site is important to other entities in the music business: record labels, booking agents, venues, and studios. For each of these business types, the first consideration should be the audience. What do your customers like and expect? What are they looking for? What are your goals?

A web site should be created after determining the goals of the site, the image of the brand, and the expectations of the visitor. The elements for inclusion should be decided in advance, and the site should be planned before work begins on development. Assets or content for the site should be developed independently with consideration given for elements that will be used both on the web site and in other promotional materials. Publicists should work with web designers and marketers to ensure that the web site meets the goals. In addition to assets, web designers[5] and developers need tools, including a web design program and an image editing program.

Glossary

Banner A typically rectangular advertisement placed on a web site, either above, below or on the sides of the web site's main content and linked to the advertiser's own web site.

Blog Short for web log, a blog is a web page that serves as a publicly accessible personal journal for an individual. Typically updated daily, blogs often reflect the personality of the author.

Branding Creating a distinct personality for a product (in this case the artist, not the label) and telling the world about it.

Cookie A message that a web server gives to a web browser. The browser stores the message in a text file. The main purpose of cookies is to identify users and possibly prepare customized web pages for them.

CSS Short for cascading style sheets, a new feature being added to HTML that gives both web site developers and users more control over how pages are displayed. With CSS, designers and users can create style sheets that define how different elements, such as headers and links, appear. These style sheets can then be applied to any web page.

FTP File transfer protocol, the protocol for exchanging files over the Internet.

THE KARG BOYS K-B-

News & Contact Info

Boys in the news...

Billboard
August 13, 2005

Industry News
Richard and Andy Karg sign with Universal Music Publishing Group/ Nashville for a world-wide publishing deal/ artist development deal...

The TENNESSEAN
Saturday, April 21, 2007

Brother Duo... On Garrison Keillor's Show *A Prairie Home Companion*
By BONNA de la CRUZ, Staff writer

A brother duo from Nashville trying to make it on Music Row wants to be named the most talented "People in their Twenties" by national public radio show A Prairie Home Companion...hosted by Garrison Keillor.
"You have to look at it as a huge audience—around 4 million people listen in," brother Richard said. "So we've already won."
They both attended Middle Tennessee State University and majored in the recording industry Program. MTSU professor Tom Hutchison Submitted a sample of their songs to the radio show's contest. They were a top pick out of 740 entries.

Contact the Boys...

Myspace:
WWW.MYSPACE.COM/KARGBOYS

Web Site:
WWW.KARGBOYS.COM

e-mail:
KARGBOYS@YAHOO.COM

The Boys perform as ...
- An acoustic duo
- An acoustic duo with percussion
- A five-piece band

To book the Boys for...
private parties, conferences, weddings, special events, conventions, holidays, coffee shops, fairs, festivals, clubs, anytime, anywhere...
please call:

615 - 497- 3315

REFERENCES AVAILABLE UPON REQUEST

"Let me just say this about the night...you know you're good when you get crowd participation without asking or prompting for it at a sports bar with a major game on a big screen television."

THE KARG BOYS K-B-

Press & Booking Info

the Karg Boys

"These are two boys that epitomize brother-based harmonies."
- Phil Everly

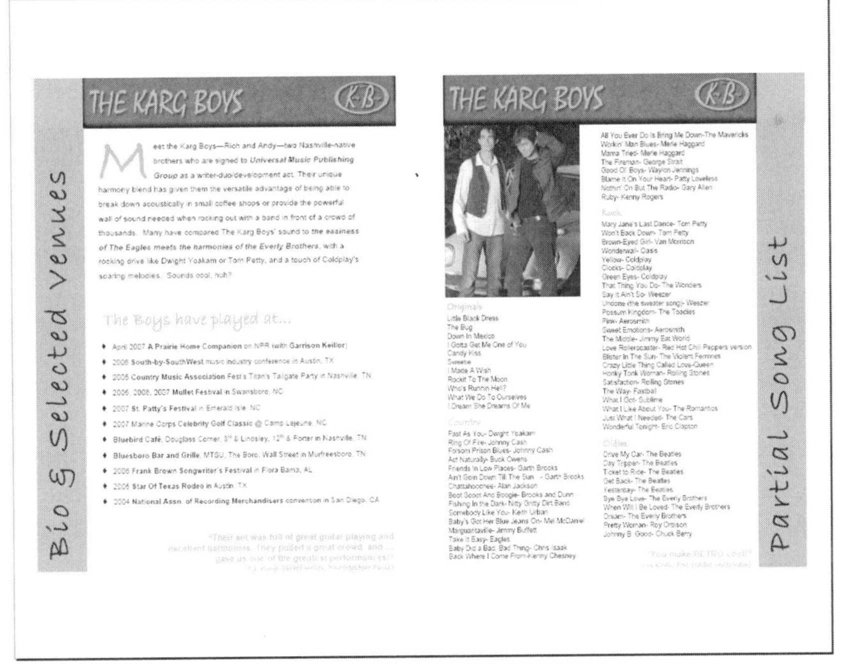

THE KARG BOYS K-B-

Bio & Selected Venues

Meet the Karg Boys—Rich and Andy—two Nashville-native brothers who are signed to ***Universal Music Publishing Group*** as a writer-duo/development act. Their unique harmony blend has given them the versatile advantage of being able to break down acoustically in small coffee shops or provide the powerful wall of sound needed when rocking out with a band in front of a crowd of thousands. Many have compared The Karg Boys' sound to ***the easiness of The Eagles meets the harmonies of the Everly Brothers***, with a rocking drive like Dwight Yoakam or Tom Petty, and a touch of Coldplay's soaring melodies. Sounds cool, huh?

The Boys have played at...

- April 2007 **A Prairie Home Companion** on NPR (**with Garrison Keillor**)
- 2006 **South-by-SouthWest** music industry conference in Austin, TX
- 2005 **Country Music Association** Fest's Titan's Tailgate Party in Nashville, TN
- 2006, 2006, 2007 **Mullet Festival** in Swansboro, NC
- 2007 **St. Patty's Festival** in Emerald Isle, NC
- 2007 **Marine Corps Celebrity Golf Classic** @ Camp Lejeune, NC
- **Bluebird Café**, Douglass Corner, 3rd & Lindsley, 12th & Porter in Nashville, TN
- **Bluesboro Bar and Grille**, MTSU, The Boro, Wall Street in Murfreesboro, TN
- 2006 **Frank Brown Songwriter's Festival** in Flora Bama, AL
- 2006 **Star Of Texas Rodeo in Austin**, TX
- 2004 **National Assn. of Recording Merchandisers** convention in San Diego, CA

"Their set was full of great guitar playing and excellent harmonies. They pulled a great crowd and ... gave us one of the greatest performances!"

THE KARG BOYS K-B-

Partial Song List

Originals
Little Black Dress
The Bug
Down In Mexico
I Gotta Get Me One of You
Candy Kiss
Sweetie
I Made A Wish
Rockit To The Moon
Who's Runnin Hell?
What We Do To Ourselves
I Dream She Dreams Of Me

Country
Fast As You- Dwight Yoakam
Ring Of Fire- Johnny Cash
Folsom Prison Blues- Johnny Cash
Act Naturally- Buck Owens
Friends In Low Places- Garth Brooks
Ain't Goin Down Till The Sun - Garth Brooks
Chattahoochee- Alan Jackson
Boot Scoot And Boogie- Brooks and Dunn
Fishing In the Dark- Nitty Gritty Dirt Band
Somebody Like You- Keith Urban
Baby's Got Her Blue Jeans On- Mel McDaniel
Marguaritaville- Jimmy Buffett
Take It Easy- Eagles
Baby Did a Bad, Bad Thing- Chris Isaak
Back Where I Come From-Kenny Chesney
All You Ever Do Is Bring Me Down-The Mavericks
Workin' Man Blues- Merle Haggard
Mama Tried- Merle Haggard
The Fireman- George Strait
Good Ol' Boys- Waylon Jennings
Blame It On Your Heart- Patty Loveless
Nothin' On But The Radio- Gary Allen
Ruby- Kenny Rogers

Rock
Mary Jane's Last Dance- Tom Petty
Won't Back Down- Tom Petty
Brown-Eyed Girl- Van Morrison
Wonderwall- Oasis
Yellow- Coldplay
Clocks- Coldplay
Green Eyes- Coldplay
That Thing You Do- The Wonders
Say It Ain't So- Weezer
Undone (the sweater song)- Weezer
Possum Kingdom- The Toadies
Pink- Aerosmith
Sweet Emotions- Aerosmith
The Middle- Jimmy Eat World
Love Rollercoaster- Red Hot Chili Peppers version
Blister In The Sun- The Violent Femmes
Crazy Little Thing Called Love-Queen
Honky Tonk Woman- Rolling Stones
Satisfaction- Rolling Stones
The Way- Fastball
What I Got- Sublime
What I Like About You- The Romantics
Just What I Needed- The Cars
Wonderful Tonight- Eric Clapton

Oldies
Drive My Car- The Beatles
Day Tripper- The Beatles
Ticket to Ride- The Beatles
Get Back- The Beatles
Yesterday- The Beatles
Bye Bye Love- The Everly Brothers
When Will I Be Loved- The Everly Brothers
Dream- The Everly Brothers
Pretty Woman- Roy Orbison
Johnny B. Good- Chuck Berry

"You make RETRO cool!"

Figure 5.14
Printable press kit available from the artist's web site as PDF file (courtesy The Karg Boys)

GIF Pronounced giff or jiff, GIF stands for graphics interchange format. GIF is limited to 256 colors and is more effective for scanned images such as illustrations than it is for color photos.

Home page The main page of a web site. Typically, the home page serves as an index or table of contents to other documents stored at the site.

JPEG Short for Joint Photographic Experts Group, and pronounced JAY-peg. JPEG is a compression technique for color images, used for most photographs on the Web.

Liquid layout Layouts that are based on percentages of the current browser window's size. They flex with the size of the window.

Navigation bar A set of buttons or graphic images typically in a row or column used as a central point that links you to major topic sections on a web site.

PDF Portable document format (PDF) is a file format that has captured all the elements of a printed document as an electronic image that you can view, navigate, print, or forward to someone else. PDF files are created using Adobe Acrobat, Acrobat Capture, or similar products (whatis.com).

PNG Short for Portable Network Graphics, and pronounced ping, a new bit-mapped graphics format similar to GIF.

Splash page The page on a web site that the user sees first before being given the option to continue to the main content of the site. Splash pages are used to promote a company, service, or product or to inform the user of what kind of software or browser is necessary to view the rest of the site's pages.

Storyboarding A storyboard is an expression of everything that will be contained in a program or web site—what menu screens will look like, what pictures (still and moving[6]) will be seen including when and for how long, and what audio and text will accompany the images, either synchronously or hyperlinked.

Thumbnail A miniature display of a page or photo that can be enlarged when clicked. This enables a web page design that can hold many small images. The thumbnail is simply a smaller version of the picture connected by a hyperlink to the larger photo.

Virtual street team A group of Internet marketers, generally selected from the target market, who actively promote the artist on the Web, generally from the perspective of the artist's avid fans.

Webmaster A person or group of people responsible for the design, implementation, management, and maintenance of a web site.

Wordmark A standardized text logo or graphic representation of the name of a company, institution, or product name used for purposes of identification and branding.

WYSIWYG Pronounced WIZ-zee-wig and short for "what you see is what you get." A WYSIWYG application is one that enables you to see on the display screen exactly what will appear when the document is printed or published to the Web.

Notes

1 A street team in the music business is a local group of people, generally members of the target market, who use networking on behalf of the artist in order to reach their target market. A virtual street team serves this same purpose on the Internet and is not necessarily local.

2 Some web sites offer the visitor "high-speed" and "low-speed" versions or "Flash" and "Non-Flash." The visitor can make the decision at the splash screen and proceed accordingly. Of course, this adds to the complexity and the cost of the web site.

3 Some music sites with splash screens also play music while the splash screen loads and after. While the use of music is discouraged on the splash screen or home page, feature, you should also include a mute button in addition to the "skip intro" feature if you insist on including music with the splash screen (or any screen).

4 The e-store can be either integrated into the web site or feature a link to the artist's products on a third party selling site.

5 There is a difference between web designers and web developers. A designer would not need a web design program, they work only with the graphic design and image editing tools, mostly Photoshop. The developer would be using the site design tools such as Dreamweaver, although

most pro developers use other things. Occasionally these two people are one and the same, but in most pro web shops, they're at two different desks.

6 Technically, these are called static and dynamic. In web page terms, "static" is a page with fixed text and standard images, "dynamic" is a page with variable information, moving or rotating pictures (or animation) and possibly some visitor interaction.

Bibliography

Bear, Jacci Howard (2012). Lorem ipsum dolor. http://desktoppub.about.com/cs/pagelayout/a/lorem.htm

Bear, Jacci Howard (2012). Why and how to use templates effectively. http://desktoppub.about.com/cs/pagelayout/a/use_templates.htm

Berenstain, Adam (2011). RapidWeaver product review. http://www.macworld.com/article/158172/2011/05/rapidweaver51.html

Chastain, Sue (2011). Top 8 free photo editors for Windows. http://graphicssoft.about.com/od/pixelbasedwin/tp/freephotoedw.htm

CNET (2006) Web Editor Reviews. Namo WebEditor Suite. http://reviews.cnet.com/web-graphics/namo-webeditor-suite-2006/4505-3637_7-31752255.html

ConsumerSearch.com. Web design software reviews. www.consumersearch.com/www/software/web-design-software

Fedh, Amanda. (2008, March 27). Adobe launches free version of Photoshop. *The Tennessean*, business section, p. 1.

Gillespie, Joe. www.wpdfd.com

Knab, Christopher (2003, November). How to write a music-related press release. http://www.musicbizacademy.com/knab/articles/pressrelease.htm

Krug, Steve (2006). Don't make me think: A common sense approach to web usability. Berkeley, CA: Pearson.

Kyrnin, Jennifer (2012a). Basics of web design. http://webdesign.about.com/od/webdesigntutorials/a/aa070504.htm

Kyrnin, Jennifer (2012b). Fixed width layouts versus liquid layouts. http://webdesign.about.com/od/layout/i/aa060506.htm

Kyrnin, Jennifer (2012c). CSS step by step: How to add CSS stylesheets to your web pages. http://webdesign.about.com/od/css/a/aa010702a.htm

Kyrnin, Jennifer (2010d). About Serif WebPlus. http://webdesign.about.com/od/windowshtmleditors/fr/serif-webplus-profile.htm

Kyrnin, Jennifer (2010e). Free web templates: These free web templates will help you build an XHTML+CSS website. http://webdesign.about.com/od/websitetemplates/a/bl_layouts.htm

Moon, Phoebe. www.phoebemoon.com.

Nevue, David (2005). *How to Promote Your Music Successfully on the Internet*. www.promoteyourmusic.com

PC Magazine Software Reviews. SJ Namo WebEditor 6 Suite. www.pcmag.com/article2/0,1759,1607976,00.asp

Spellman, P. (2000, March 24). Media power: Creating a music publicity plan that works: Part 1, www.harmony-central.com/Bands/Articles/Self-Promoting_Musician/chapter-14-1.html

Top Ten Reviews. SiteSpinner. http://website-creation-software-review.toptenreviews.com/sitespinner-review.html

Top Ten Reviews. Web Easy. http://website-creation-software-review.toptenreviews.com/web-easy-professional-review.html

Web Design Services India. www.webdesignservicesindia.com
Web Graphics. www.sph.sc.edu/comd/rorden/graphics.html
Web Style Guide. http://webstyleguide.com/page/frames.html
www.pcmag.com/encyclopedia_term
www.amacord.com/services/storybrd.html
www.webopedia.com
www.wikipedia.com
www.theclienthelpdesk.com

Chapter 6

HTML and Scripts

HTML: THE BASICS AND WHY YOU NEED THEM

Hypertext markup language (HTML) is the predominant authoring language for the creation of web pages. HTML defines the structure and layout of a web document by using a variety of tags and attributes to denote formatting of certain text as headings, paragraphs, and lists. HTML is written in the form of tags bracketed by the greater than and less than symbols, such as < tag >. Most tags come in pairs, the opening tag is listed as <tag>, and the closing tag is </tag>, which denotes the end of the previous command. For example, if you wanted to italicize a word in a sentence, you would precede the word with the tag for italics <i> and follow the word with the end tag </i>. Failure to include the end tag would result in everything from that point forward being presented in italics. A web visitor's browser examines the HTML for instructions on how to display the graphics, text, and other multimedia components. Tutorials can be found at www.w3schools.com.

When someone types in a URL or clicks on a web page link, the browser requests a document from a web server via the hypertext transport protocol, or HTTP. The server then sends the document back to the user, which is displayed on the browser. The things that are contained in the document (text, photos, audio and video files, etc.) were all put there using HTML structure.

Most web design programs will allow the novice to build a web site without any knowledge of HTML language, but a basic understanding of HTML code can come in handy for tweaking web pages and widgets. For example, YouTube now provides options for designing how embedded videos will look on your site—you can select the border color, adjust the size of the thumbnail, and adjust the size of the video window, within limits (see Chapter 8). With a bit of understanding, you could then modify the code even more to fit your specific needs.

The Backstory

As browser markup language evolved, new needs arose. HTML-laden documents became very code-heavy. Not all browsers respond the same way to the language, and thus web pages may look very different from one browser to another. The evolution brought about XML, HTML 4, and then XHTML. HTML 4 is a markup language designed for displaying text and documents across different platforms and machines. TechTarget describes HTML 4 as:

> The final version of the Hypertext Markup Language (HTML) before the Extensible Markup Language (XHTML) and remains the set of markup on which most large web sites today are based. Among new features introduced in HTML 4.0 were: 1) The cascading style sheet, the ability to control web page content at multiple levels, 2) the ability to create richer forms, 3) support for frames (which is already supported by the major browsers), 4) enhancements for tables that make it possible to use captions to provide table content for Braille or speech users, and 5) the capability to manage pages so that they can be distributed in different languages.
>
> **TechTarget definition**

XML (Extensible Markup Language) is a markup language that defines a set of rules for encoding documents in a format that is readable by both humans and machines. XML is not a replacement for HTML, but is a complement to HTML. XML was created to structure, store, and transport information while HTML is designed to display data, with an emphasis on how the data looks in a browser. XHTML is a stricter and cleaner version of HTML. It stands for Extensible HyperText Markup Language. It has more rigorous standards than HTML (tags must have an end tag, and be case-sensitive). But done correctly, it allows continuity among browsers and platforms, including mobile devices.

All of these standards are set by the World Wide Web Consortium, led by Web pioneer Tim Berners-Lee. According to their web site "W3C's primary activity is to develop protocols and guidelines that ensure long-term growth for the Web. W3C's standards define key parts of what makes the World Wide Web work."

HTML5 is the fifth revision of HTML and is designed to support a variety of platforms, both computer-based and mobile-based, and support audio and video without the need for third-party software. HTML5 was still under development in early 2012, with complete adoption predicted by 2014.

Getting Started with HTML

All HTML documents start with the command <html>. This lets the browser know to read and interpret the commands as HTML. The last tag on the web page should be the end tag </html>. This tag tells your browser that this is the end of the HTML document. Beneath the <html> tag at the top of the page, you will find header information between the <head> and </head> tags. This information is not displayed on the page. The head element contains information about the document, some of which helps search engines catalog and describe the site. Head tags include those shown in Table 6.1.

Here is an example of the source code for a heading for a web page head that was shown in Chapter 5 (Figure 5.1) (www.sauceboss.com):

```
<!doctype html public "-//w3c//dtd html 4.0 transitional//en">
<html>
<meta http-equiv="Content-Type" content="text/html;
charset=iso-8859-1">
```

Table 6.1 Source code for head tags

Tag	Description
<!DOCTYPE>	Defines the document type; goes before the <html> start tag
<head>	Defines information about the document
<title>	Defines the document title
<base>	Defines a base URL for all the links on a page
<link>	Defines a resource reference
<meta>	Defines meta information (as in meta tags—see the text)

```
<meta name="DESCRIPTION" content="Bill Wharton (Sauce Boss) has
cooked gumbo for over 140000 people for free during his high
energy concerts with his band–The Ingredients">
<meta name="GENERATOR" content="Mozilla/4.73C-CCK-MCD {C-UDP;
EBM-APPLE} (Macintosh; U; PPC) [Netscape]">
<meta name="keywords" content="blues music hot sauce gumbo
slide guitar Bill Wharton datil pepper habanero Liquid Summer
recipe contest Sauce Boss Jimmy Buffett Parrothead Planet Gumbo
Podcast Florida Blues">
<link rel="stylesheet" href="main.css" type="text/css"
media=screen><title>Sauce Boss</title>
</head>
```

META TAGS

Meta tags are author-generated HTML commands that are placed in the head section of an HTML document. These tags help identify the content of the page and specify which search terms should be used to list the site on search engines. Popular meta tags can affect search engine rankings and are generally listed in the HTML code sections titled "Meta Keywords" and "Meta Description." Search engines often use the description meta tag and display it in the results. A meta tag can be generated automatically by the site www.submitcorner.com. The example tag shown here will let a search engine know to categorize the above web site under blues music, slide guitar, and by the artist's other endeavor, Louisiana-style cooking.

The web page title is also significant and should reflect the nature of the site. The <TITLE> tag is the caption that appears on the title bar of your browser and is the name on the clickable link listed in the search engine results. (Example: <title>Sauce Boss</title>.)

Figure 6.1
Google search results for Sauce Boss, including tags (courtesy of Bill Wharton)

The table shown in Figure 6.2 is from www.makemyownwebpage.com and describes the various types of meta tags. The author describes meta tags as "where the search engines go to see what your priorities are."

BODY TAGS

The body of an HTML document uses tags to format text and graphics. There are numerous tags to denote font size, color, style, and other features, and there are others that denote paragraph formatting. Figure 6.3 indicates some of the body text formatting tags commonly used, including break, alignment, and justified.

Other tags may denote font style and color. It is recommended that you use standard, commonly used fonts. If the web visitor's browser or computer is not set up for the fonts you have chosen for your body text, the substitution fonts may disrupt the formatting of the page and cause elements to shift out of place.

FONTS AND COLORS

HTML tags also indicate characteristics of the text appearance such as font size, color, and formatting.

In Figure 6.4, Arial and Times Roman fonts are used, along with underline, italics, bold, and font colors. The <i> tags indicate italics, <b> is for bold type, <u> is for underline, and the font tag can include, among other things, face, color, and size.

META Tags

The <meta> tags are where the search engines go to see what your priorities are. A couple of other resources are also looking at meta tags for other content. You may have several Meta tags for a single web page.

Syntax

```
<META NAME="item-name" CONTENT="items">
<META HTTP-EQUIV="items" CONTENT="item-coding">
```

Modifiers

NAME="DESCRIPTION" CONTENT="*description*"

This description is what most search engines tell the world about your web page.

NAME="KEYWORDS" CONTENT="*key, word, keyword*"

Keywords help define your topic for what the search engines are looking for. Separate each keyword or phrase with a comma.

NAME="ROBOTS" CONTENT="*index,follow*"

Tells the Search Engine Robots how to look at your site. The possible choices here are first *INDEX* or *NOINDEX*. This tells the search engines whether or not you want them to scan this page and include it in their indexes. Second term is *FOLLOW* or *NOFOLLOW*, which tells the search engines to follow the links on the page and index those or not. Not all search engines recognize this term.

NAME="RATING" CONTENT="*general*"

This helps define the level of objectionable materials within this web page. Possible choices here are *GENERAL* (general audience... anybody), *MATURE* (Adult materials... may be objectionable), *RESTRICTED* (18 years old and up) and *14 YEARS* (14 years old and up).

NAME="AUTHOR" CONTENT="*mtg@makemyownwebpage.com*"

Lists the author's name and/or email address.

NAME="COPYRIGHT" CONTENT="*Copyright 2002-12*"

Displays the copyright, trademark or any intellectual property information about the web page.

NAME="GENERATOR" CONTENT="*notepad*"

Displays the publishing tool used to make my own web page.

HTTP-EQUIV="refresh" CONTENT="*0*; URL=*index.htm*"

This allows you to forward the visitor to a new page as listed by the URL term. The timing (just before URL) is set to how many seconds to pause before forwarding.

Figure 6.2
Example meta tags (with permission of www.makemyownwebpage.com)

Table 6.2 Commonly used HTML formatting commands

Basic text commands		
 	Break	Same as carriage return.
<p>	Paragraph	Carriage return plus adds a blank line.
<b> </b>	Bold	Text between the two tags is displayed in bold.
<i> </i>	Italics	Text between the two tags is displayed in italics.
<u> </u>	Underline	Text between the two tags is displayed in underline.
<font size=?></font>	Font size	Sets size of font from 1 to 7.
<font color=?> </font>	Font color	Sets font color, using name or hex value.
<font face=?></font>	Font typeface	Sets the typeface such as Times Roman, Arial, Tahoma, etc.
<font size="3" color="#FF000" face="arial">The actual text affected by the commands</font>	Example of font size, color and face.	
Images		
<image src=url>	Insert an image	url would be the address where the image is found and generally ends with .jpeg or .gif.
Image subcommands	Width = 120	The picture width will be displayed as 120 pixels wide.
	Height = 60	The picture width will be displayed as 60 pixels high.
	Align = left	The picture will align to the left of text.
	ALT = "text"	Tells the browser what text to insert on the screen if the image is not available.
	Border = 4	The border around the image will be 4 pixels.
	Hspace = 4	The horizontal space between the image and surrounding text will be 4 pixels.

	Vspace = 4	The vertical space between the image and surrounding text.
<img border="0" src="imagename.jpg" width="75" height="92" align="left" vspace="2">	Example of image with size, alignment, and text space.	
Links		
<a href="mailto:EMAIL"> </a>	Email link	Creates a mailto link that automatically opens up the user's default email program and inserts the email address.
<a href="newpage.html"> Link to New Page</a>	Hyperlink to another site or location	The "Link to New Page" portion of the command is what is displayed on the screen as the hot link.
Color Commands		
Hex HTML colors are indicated by a six-digit series of letters and numbers but can also be indicated by a color name.		
bgcolor="#000000"	Background color	Used to tell the browser what color of background to use.
text="#ffffff"	Text color	The standard color to use on general text that follows.
link="#004000"	Link color	The standard color to use on unvisited links that follow.
vlink="#44aaff"	Color for visited links	
alink="#ff00ff"	Color for mouse-over	As the cursor passes over the link, this is the momentary color the link changes to while the user is hovering over the link with the cursor.
background="file.gif"	Background image	If you prefer to use an image as the background, this command specifies the image.

```
<body>
This is line one followed by a break.<br>
And here is line two with a break. <br>
And a third and final line.<p> </p>

<p>
This is line one followed by a break.<br>
And here is line two followed by a paragraph tag.</p>
<p>And a third final line.</p>

<p align="left">Left Justified Text</p>
<p>Here is a small block of text enough text to cause
the sentence to wrap around and illustrate the effect of
the "left" tag on word wrap.</p>

<p align="right">Right Justified Text</p>
<p align="right">Here is a small block of text, enough
text to cause the sentence to wrap and illustrate the
effect of the "right" tag on word wrap.</p>

<p align="center">Center Justified Text</p>
<p align="center">Here is a small block of text where
the sentence wraps and illustrates the effect of "center"
on word wrap.</p>

<p align="justify">Justify Justified</p>
<p align="justify">Here is a small block of
fully-justified text, enough to wrap and illustrate the
effect of "justify" on word wrap.</p>
```

This is line one followed by a break.
And here is line two with a break.
And a third and final line.

This is line one followed by a break.
And here is line two followed by a paragraph tag.

And a third final line.

Left Justified Text

Here is a small block of text, enough text to cause the sentence to wrap around and illustrate the effect of the "left" tag on word wrap.

Right Justified Text

Here is a small block of text, enough text to cause the sentence to wrap and illustrate the effect of the "right" tag on word wrap.

Center Justified Text

Here is a small block of text where the sentence wraps and illustrates the effect of "center" on word wrap.

Justify Justified

Here is a small block of fully-justified text, enough to wrap and illustrate the effect of "justify" on word wrap.

Figure 6.3
Example of page body formatting

```
<font face="Arial">Arial 12 point font</font><p><b><i>
<font face="Times New Roman" size="4">Times Roman 14 point type
in bold and italics</font></i></b></p>
<p><font face="Times New Roman" size="4">
This sentence contains <u>underlines</u>,
<i>italics,</i> <bold>bold,</b> and
<font color="0000FF"> a color change to blue</font>
```

Arial 12 point font

Times Roman 14 point type in bold and italics

This sentence contains underlines, *italics,* **bold,** and a color change to blue

Figure 6.4
Example of font formatting

CASCADING STYLE SHEETS

The concept of cascading style sheets (CSS) was introduced in Chapter 5. Now, we will address some of the specific code and discuss how to incorporate them into your web site. Once the style sheet is formed, it has to be linked to the HTML documents and/or code, so that is a good place to start. There are several methods for "linking up" the style sheet with the HTML language: linking to an *external style sheet*, embedding the CSS information in the document, inlining, importing, and a few others. To link the CSS sheet to the HTML document, the <LINK> tag is placed in the document HEAD, between the <head> and </head> tags.

```
<link rel="stylesheet" type="text/css" href="mystylesheet.css"
media="screen" />
```

The REL attribute defines the relationship between the linked file and the HTML document. The TYPE attribute is used to specify a media type—text/css for a cascading style sheet—allowing browsers to ignore style sheet types that they do not support (WDG, no date). The SCREEN command relates to your computer screen. If the CSS file is in the same folder as your HTML file then no path is required (like the example above) but if it is saved in a folder, then it must be specific, such as

```
href="foldername/mystylesheet.css".
```

The alternative is to *embed* the style sheet *within* the HTML document. You must add the code to the head of the document (between the <head> and </head> tags and within the <style> and </style> tags. For example:

```
<style media="screen" type="text/css">
Insert all your style sheet commands here
</style>
</head>
```

Creating the style sheet is next. There are three parts: the selector, the property, and the value. The *selector* which tells the browser which part of the document is affected by the command, such as (in the example: "body"). The *property* specifies what aspect of the layout is being specified (in the example, "color" and "background-color" are both properties). The *value* gives the value for the style ("blue" and "#ffff00").

While most web design programs, especially with the use of templates, can make understanding and writing CSS code optional, it is useful to have an understanding of how it works—just in case you want to make modifications to the design code. One example of CSS code is shown in Figure 6.5. For the novice developing a web site from scratch using CSS, one approach would be to use CSS code generators. In his article "Ten Really Awesome CSS Generators," author AJ provides description and links to write specific CSS code for functions such as buttons, borders, grids, fonts and others (http://www.aoclarkejr.com/10-really-awesome-css-generators.html).

<html>
<head>
<title>Example Style Page</title>
<style type="text/css">
body {
color: blue;
background-color: ##ffff00 }
</style>
</head>
selector
property
value

Figure 6.5
Example of CSS style sheet code

CSS CODE GENERATORS

http://www.cssportal.com/
http://css3generator.com/
http://www.quackit.com/css/codes/

CGI SCRIPTS

Common Gateway Interface (CGI) scripts are defined as script files executed on a web server in response to a user request. They are commonly used to process data sent when a form filled in by a user is sent back to the web server. A CGI program is executable, and is basically the equivalent of letting visitors run a program on your system. The most common form found on artist web sites is the email registration form that gathers email addresses from visitors so that they may be added to the mailing list. The "Tell a friend" form is another example of a GCI script.

Forms

Forms are a common and popular feature on web sites. The email signup and tell-a-friend scripts mentioned later are examples of simple forms. But forms are used for all kinds of purposes, from gathering information to processing e-commerce orders. Most high-end web development programs offer some type of features for creating forms. One of the important aspects of creating a form is determining where the information will go—either stored as a database on a server or sent as an email to the appropriate person, or both. The first step involves creating the form and specifying the action to be taken upon submission (where the data go). Then, the particular form fields can be created to elicit information from the web visitor. There are also companies online who provide form or survey features. Some start with a basic, free plan with many of the more advanced features reserved for the paid plans. These services have user-friendly web sites that allow even the novice to create a form, save the data, and analyze the results.

The most popular of these is **Survey Monkey**. The company hosts the forms on its site. Survey Monkey offers a free version, with a limit of ten questions and only displays the first 100 responses. The creative survey designer can gather a lot of information in ten questions by using the matrix multiple choice question

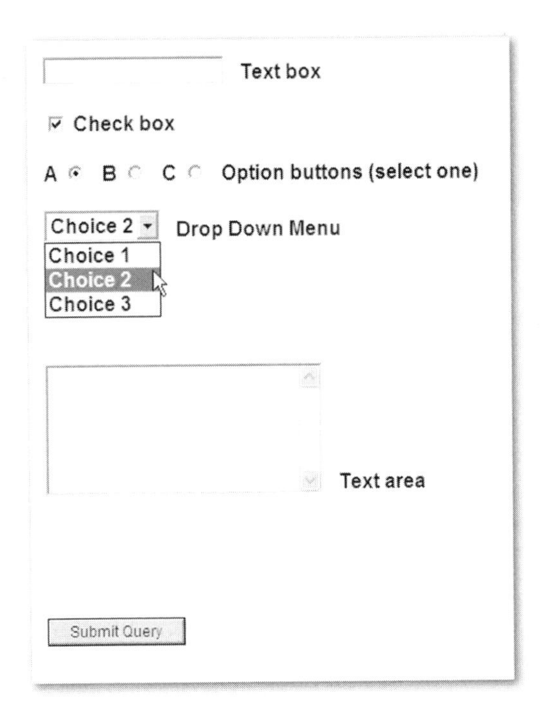

Figure 6.6
Example of forms

type. Monthly plans start at $17. New to Survey Monkey is the opportunity to embed your surveys into your own web site. This has several advantages: it maintains the look and feel of your site, it keeps visitors on the site longer, and visitors are more likely to answer the survey if they can view it before committing to taking it. Survey Monkey stores the data—the filled-out forms—on its web site with convenient features for analyzing or downloading the data.

Another such service, Freedback, does not host the form but helps users create the form and then capture the HTML code to place on your web site. With Freedback, you can instantly receive notifications of new form submissions via email or SMS. Freedback sends the data to the email address specified on the account. The free version is advertiser-supported. Paid, ad-free packages start at $9 per month for one form.

Google Forms is the latest popular entry into the forms arena, first appearing in 2008. Google Forms is part of Google Docs and can be accessed through the user's Google account. According to blogger Andy Seidl (2008), "Google Docs provides a flexible mechanism for email notification, sharing access with other users, online backup, and access to data in multiple formats." The data are stored as a "Google Doc," in spreadsheet form, and can be downloaded and manipulated like any other spreadsheet.

GUESTBOOK SCRIPT

Here is a simple CGI script for creating a form for signup for a mailing list:

```
<FORM ACTION"http://www.artistwebsite.com/mailinglist/
subscribe.pl" METHOD="post"><INPUT TYPE="hidden" NAME="account"
VALUE="hutchtom"><INPUT TYPE="hidden" NAME="body" VALUE="<BODY
BGCOLOR=white TEXT=black LINK=blue VLINK=darkblue>"><INPUT
TYPE="hidden" NAME="action" VALUE="subscribe"><TABLE
BORDER="2" CELLPADDING=0 CELLSPACING=0><TR><TD><TABLE
BORDER="0" CELLPADDING=2 CELLSPACING=0 BGCOLOR="#eeeeee"><TR
BGCOLOR="#cccccc"><TD><B>Your email:</B></TD><TD><INPUT
TYPE="text" NAME="email"></TD></TR><TR><TD COLSPAN="2"><INPUT
TYPE="submit" VALUE="Join mailinglist!"></TD></TR>
</TABLE></TD></TR>
</TABLE>
</FORM>
```

Most commercial web entities want more information than just the person's email address, so more elaborate scripts ask for additional information, including demographics. This script involves storing the information in a database to be accessed by the webmaster.

TELL-A-FRIEND SCRIPT

The tell-a-friend script is illustrated in Figure 6.7. As a result of automated spamming programs that are abused by unscrupulous web marketers, many such forms now include a component to verify that a human is filling out the form, instead of an automated software program.

JAVA AND JAVASCRIPT

Whereas CGI is for server-side programming, often referred to as back-end programming, JavaScript is used for client-side programming, often referred to as front-end programming (although it can also be used for server side programming). Server side programming runs on the host's server, whereas client side programming runs in the user's browser. Wikipedia has the following paragraph to explain the difference:

> Server-side scripting is a web server technology in which a user's request is fulfilled by running a script directly on the web server to generate dynamic HTML pages. It is usually used to provide interactive web sites that interface to databases or other data stores. This is different from client-side scripting where scripts are run by the viewing web browser, usually in JavaScript. The primary advantage to server-side scripting is the ability to highly customize the response based on the user's requirements, access rights, or queries into data stores.

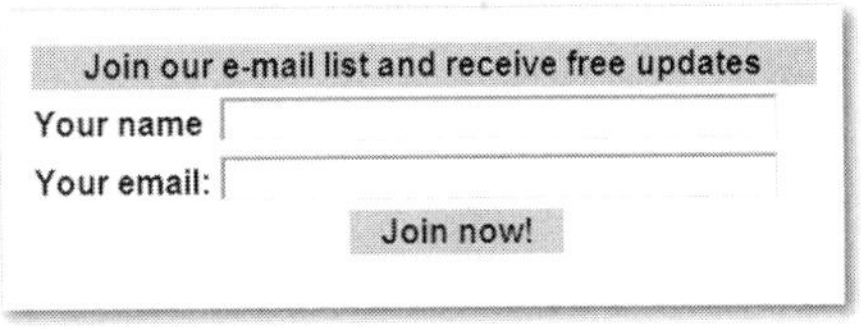

Join our e-mail list and receive free updates
Your name
Your email:
Join now!

Figure 6.7
Join Now! form

```
<FORM ACTION="http://www.javascript.nu/ogrifree/tellafriend.asp" METHOd="post">
<INPUT TYPE="hidden" NAME="yoururl" VALUE="http://www.hutchtom.com">

<TABLE BORDER="2" CELLPADDING=0 CELLSPACING=0>
<TR><TD><TABLE BORDER="0" CELLPADDING=2 CELLSPACING=0 BGCOLOR="#cccccc">
<TR BGCOLOR="#cccccc"><TD COLSPAN="2"><B>Tell a friend about this site</B></TD></TR>
<TR><TD>Your name:</TD><TD><INPUT TYPE="text" NAME="name"</TD></TR>
<TR><TD>Your email:</TD><TD><INPUT TYPE="text" NAME="from"</TD></TR>
<TR><TD>Friend's email:</TD><TD><INPUT TYPE="text" NAME="to"</TD></TR>
<TR><TD COLSPAN="2">INPUT TYPE="submit" VALUE="Tell a Friend"></TD</TR>
</TABLE></TD></TR>
</TABLE></FORM>
```

The code above will result in this form:

Tell a friend about this site
Your name:
Your e-mail:
Friend's e-mail:
Tell a Friend!

Figure 6.8
Tell-a-friend form (courtesy of CGI4Free.com)

According to Ibama Tmunotein in his article *Client-side and Server-side JavaScript*, "Server-side JavaScript is ideal for creating web applications that can be run on any platform, on any browser, and in any (programming) language."

Java is a programming language developed by Sun. NetScape responded by creating JavaScript. Microsoft then added its own version of JavaScript to Internet Explorer, called JScript (D. Smith, 1998). The difference between Java and JavaScript is that Java can stand on its own whereas JavaScript must be placed inside an HTML document to function. JavaScript is text that is fed into a browser that can read it and then is enacted by the browser. It can be modified on the fly. Java, on the other hand, creates a "standalone" application—the Java "applet" (a small application), which is a fully contained program. Java needs to be recompiled if it is modified, and then reinserted into the web page.

Resources for Scripts

- Online HTML Code Generator, http://htmlcode.discoveryvip.com
- Click & Go drop down list generator, www.webdevtips.co.uk/webdevtips/codegen/clickgo.shtml
- Meta tag and SEO generators, www.submitcorner.com
- Guestbook and mailing list script generators, www.javascript.nu/cgi4free
- HTML and CSS scripts, www.hypergurl.com/htmlscripts.html
- Updated information available at www.focalpress.com/cw/Hutchison.

WEB WIDGETS

Web widgets are defined as a portable piece of code that an end user can install and execute within any separate HTML page. They often use DHTML, Adobe Flash, or JavaScript programming language and wrap it up in a nice user interface. Often these widgets are incorporated into social networking pages, blogs, and personal web sites, installed by the user. Not all widgets are compatible with all systems. In 2007, known as the year of the widget, the top social networking sites began to open up their platforms to widgets such as iLike developed by third parties.

Marketers are creating new ways to use widgets to advertise and sell products. Here is how widgets such as ReverbNation's Music Player work. The widget company hosts a web site where the end user can construct the personalized parameters of the widget. The user-friendly interface allows the user to customize and personalize the widget, such as create a playlist, upload favorite photographs, and so forth. Then with a mouse click and a password, users upload the widget to their social networking page, which seamlessly and transparently adds the special feature to their social networking page or web page. This allows for content to be dynamic as the widget updates information based on the user's choices and activity—such as monitoring iTunes listening activity to create a playlist to be posted on a social networking site, or whenever the user adds photos to the web site that provides the service and the widget. Alternatively, widget creators will provide coding language that you can paste in to your own site.

For example, the web site www.flickr.com allows users to upload pictures and create slide shows that they can share with other web users. The slide.com site creates widgets that work with MySpace, Facebook, Pinterest, Tumblr, Twitter, Blogger, and WordPress. Flickr has joined a host of other Web 2.0 services who no longer require the user to cut and paste a widget, but the widget is automatically posted to the outside account by signing in to that account with your username.

CAPTCHA

In today's age of spamming, web site managers have adopted techniques to prevent automated programs from performing functions that are supposed to be performed by human visitors to the site, such as posting messages on a message

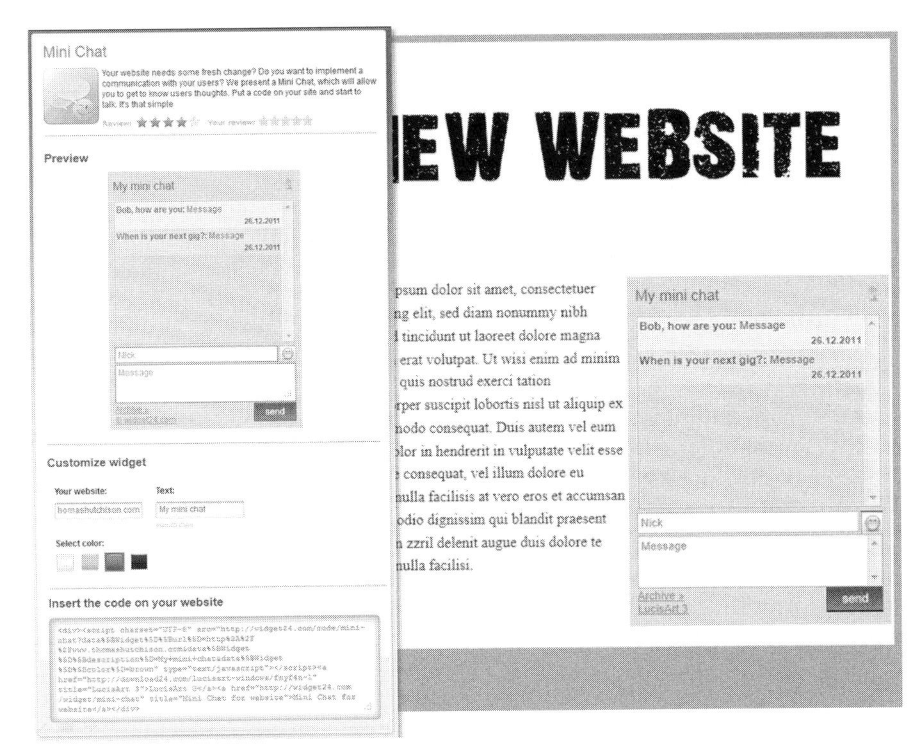

Figure 6.9
Example of widget code and embedded widget

Figure 6.10
Share features common with many Web 2.0 services

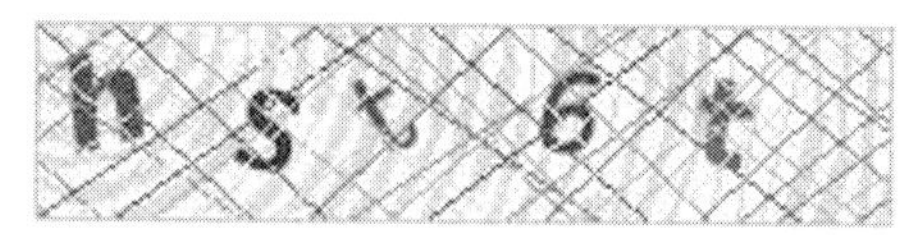

Figure 6.11
A typical graphic verification image

board or sending emails. Typically, this is achieved through a CAPTCHA. CAPTCHA is a program that can tell whether its user is a human or a computer. It is a loosely contrived acronym meaning "completely automated public Turing tests to tell computers and humans apart." CAPTCHAs are graphics presented with distorted text found at the bottom of web registration forms. Many web sites use CAPTCHAs to prevent abuse from "bots," or automated spamming programs. No computer program can read distorted text as well as humans can, so bots cannot enter sites protected by CAPTCHAs. Thus, email accounts and message boards can be protected through the use of CAPTCHA programs. It can also protect forms and online polls. The process involves installing a program that can generate and grade tests—in this case, an easy test of repeating the letters and numbers that appear in the distorted graphic. The concept was developed at Carnegie Mellon's CyLab, and the code is now offered for free at http://www.captcha.net/

FLASH

Flash, a popular authoring software developed by Macromedia and now owned by Adobe, is used to create graphic/animation programs with navigation interfaces, graphic illustrations, and simple interactivity in a resizable file format that is small enough to stream across a normal modem connection (although most Flash content over a "normal modem connection" is disturbingly slow to load). It is platform independent and gives web designers the ability to add bells and whistles to animate any web page. Wikipedia states,

> Flash technology has become a popular method for adding animation and interactivity to web pages; flash is commonly used to create animation, advertisements, various web page components, to integrate video into web pages, and more recently, to develop rich Internet applications.
>
> **wikipedia.com**

The end user must have Adobe's Flash player, a free software download, installed to run the Flash programs. Adobe sells the software program to develop Flash programs to add to web sites.

In the article "What is Flash, when and why to use it," author Stefan Mischook weighed the pros and cons of using Flash, emphasizing that it's not the best option for all applications, stating that Flash should be used "selectively to enhance an HTML-based site." One major advantage is that there are no compatibility issues with the various browsers and platforms on the market. Aycan Gulez, in his article "Top five reasons for limiting Flash use," stated that the Adobe Flash plug-in is installed in over 95% of web users' browsers, so just about all users can display Flash content. But, on the flip side, he emphasized the limitations of Flash including limited navigation ability (you may need to wait until it's finished loading to move on), slow load times, poor rendering of text, and the inability to print out or search through text. The general rule on Flash-heavy splash pages is to offer the web visitor the option to skip the Flash introduction.

There are other, less-expensive programs that can create the same effect, including mix-fx (www.mix-fx.com). CoffeeCup.com offers software that will create particular Flash effects. Glogster.com offers limited online templates from which you can develop a multimedia Flash-type presentation at no charge (www.glogster.com).

Steve Jobs was not a fan of Flash and refused to incorporate it into the iPhone or iPad system because it is vulnerable to attack from malware and was a power hog on mobile devices. In 2010, Jobs wrote an essay in which he stated "Symantec recently highlighted Flash for having one of the worst security records in 2009." He also listed the battery life issues and that Flash did not support touch-screen technology and relies heavily on mouse commands.

HTML5, the open standard backed by so many companies (including Adobe) has made Flash redundant and expendable. In November 2011, Adobe announced it was abandoning plans to develop mobile Flash applications and will instead focus on HTML5 (Arthur, 2011). In preparation for the changeover from Flash to HTML5, Google launched the beta version of a service in July 2011 called Google Swiffy that will convert Flash SWF files to HTML5 versions. The software is free and available at http://www.google.com/doubleclick/studio/swiffy/.

RSS FEEDS

Wikipedia describes *really simple syndication* (RSS) as

> a family of Web feed formats used to publish frequently updated content such as blog entries, news headlines or podcasts. An RSS document, which is called a "feed," "web feed," or "channel," contains either a summary of content from an associated web site or the full text. RSS makes it possible for people to keep up with their favorite web sites in an automated manner that's easier than checking them manually.

RSS was first invented by Netscape, which wanted a way to get news stories and information from other sites and have them automatically added to its site. The RSS feed starts with the XML command (Extensible Markup Language), then the RSS command, some commands for formatting, and the location (link) for the feed.

> We found early on the use of RSS feed to help market and promote our clients in a unique way. Using the system to push news, tour dates, and blogs and journals to other sites and subscribers, we have developed

Figure 6.12
Illustration of how RSS feeds update fan sites

> those efforts, while the Internet as a whole has adapted its use. With the inventions of widgets and sidebars for Windows, Macs and even more so Vista, we are able to take the RSS feeds we have been using for several years and hit a whole new target easier (the desktop of the consumer). We can now send snippets of code that will show the latest news and tour information right on their desktop or email box without requiring them to install something new.
>
> **Stephanie Orr-Buttrey, www.CountryWired.com**

The music business uses RSS feeds to keep fans informed of upcoming events or to provide updated content for fan-based web pages. This actually allows the "feeder" (the artist's web manager) to control content that appears on other web sites quickly and easily with one feed.

TESTING ACROSS BROWSERS AND PLATFORMS

As illustrated in Chapter 5 on web design, a web layout that looks correct and normal on one computer may look completely different on another computer. The layout of a web site varies with the following characteristics:

- Type of browser (Firefox, Internet Explorer, Safari, etc.)
- Version of browser
- Browser settings
- Computer operating system
- Monitor resolution

In his article "Browser Compatibility Tutorial," Tom Dahm explained why a web site looks different depending on the preceding factors. He described the browser as "a translation device … it takes a document written in the HTML language and translates it into a formatted Web page." The standards for HTML and browser compatibility are setup by the World Wide Web Consortium (W3C) that publishes these standards. But as HTML evolves, the older browsers, and even some of the newer ones, fail to keep up and support all the newest bells and whistles. Generally, the newer browser versions are more standardized than the older versions. That presents problems for web designers who want to use all the latest features, some of which are not supported for the portion of web visitors who are still using the older versions.

Font availability and size can be problematic. A web site may be created in a font style that is not available to many of the site's visitors. Fonts reside on the user's computer and are put into place based on the HTML instructions. If that font is not available on that computer, the browser substitutes another font, sometimes with grave consequences. Font size can cause problems also. Many browsers allow users to customize their default font size. Some users prefer to increase font size to reduce eyestrain. This may lead to a web page that is out of proportion, with text dominating the page. It is wise to use standardized fonts when creating a web page. If unique fonts are to be used, they should be converted to graphic files (preferably GIF files) so that consistency will be maintained across platforms.

Once the web site has been created and uploaded, it is wise to check out compatibility by either visiting the site on a variety of different computer setups or using one of the online services that simulate or capture different browser experiences. Other tips include the following:

- Don't build your web site entirely with Flash, and use a dedicated web design program instead of Microsoft Word or Microsoft Publisher.
- Check your web page HTML at http://validator.w3.org or www.anybrowser.com/validateit.html.
- Check your page's/site's CSS at http://jigsaw.w3.org/css-validator.
- Check your links at http://validator.w3.org/checklink or www.anybrowser.com/linkchecker.html.
- Check your images for accessibility issues at http://juicystudio.com/services/image.php.
- Check your content for readability at http://juicystudio.com/services/readability.php.
- Check to see if your CSS's text and background colors have sufficient contrast at www.accesskeys.org/tools/color-contrast.html.
- Check how your site is viewed on different systems at www.anybrowser.com/siteviewer.html, www.anybrowser.com/ScreenSizeTest.html, or http://www.browsercam.com/
- Check your page's performance and web page speed at www.websiteoptimization.com/services/analyze.
- Check several things at once with www.netmechanic.com/products/HTML_Toolbox_FreeSample.shtml.

These performance checks should be supplemented with human browser tests: ask friends with different hardware and software systems to test your site and report back to you before launching your site.

CONCLUSION

There is a variety of scripting codes that have been developed and evolved as web browsers and content have become more sophisticated. Now, efforts are underway for an inclusive scripting language (HTML5) that has the power and flexibility to be all things to all web users, whether on a Mac, a PC, or a mobile platform.

Glossary

Applet A small Java program that is cross-platform compatible and can be embedded in the HTML of a web page. Web browsers, which are usually equipped with Java virtual machines, can run the applets to perform interactive graphics, games, and so on.

Browser A software application used to locate and display web pages. Contemporary browsers are graphical browsers, meaning they can display graphics as well as text and can present multimedia information, including sound and video, though they require plug-ins for some formats.

CAPTCHA A program that can tell whether its user is a human or a computer. The process involves installing a program that can generate and grade tests—in this case an easy test of repeating the letters and numbers that appear in the distorted graphic that humans can read and software programs cannot.

CGI scripts Common Gateway Interface (CGI) scripts are defined as script files executed on a web server in response to a user request. Used for user-generated forms.

Client-side JavaScript (CSJS) JavaScript that enables web pages and client browsers to be enhanced and manipulated.

Client-side programming Occurs on the end-user side of a client-server system—these programs are executed by your browser (the client).

CSS Short for cascading style sheets, a feature being added to HTML that gives both web site developers and users more control over how pages are displayed. With CSS, designers and users can create style sheets that define how different elements, such as headers and links, appear. These style sheets can then be applied to any web page.

Flash A bandwidth-friendly and browser-independent animation technology. As long as different browsers are equipped with the necessary plug-ins, Flash animations will look the same. With Flash, users can draw their own animations or import other vector-based images (Webopedia).

Hypertext markup language (HTML) The predominant authoring language for the creation of web pages. HTML defines the structure and layout of a web document by using a variety of tags and attributes.

Java A client-side programming language with a number of features that work well on the Web. Small Java applications are called Java applets and are downloaded from a web server and run on the user's computer by a Java-compatible web browser.

JavaScript A server-side scripting language embedded in the HTML language of a web page that adds interactive functions to HTML pages. JavaScript is easier to use than Java, but it is not as powerful and deals mainly with the elements on the web page.

Java Virtual Machine (JVM) An abstract computing machine, or virtual machine, is a platform-independent execution environment that converts Java byte code into machine language and executes it.

Meta tags Author-generated HTML commands that are placed in the head section of an HTML document. These tags help identify the content of the page and specify which search terms should be used to list the site on search engines.

Plug-in A hardware or software module that adds a specific feature or service to a larger system.

Server-side JavaScript (SSJS) JavaScript that enables back-end access to databases, file systems, and servers.

Server-side scripting Scripting that runs on the server side of a client-server system. CGI scripts are server-side applications because they run on the web server compared to programs that run in the user's browser.

Web widgets A small application that can be ported to and run on different web pages by a simple modification of the web page's HTML.

XML The Extensible Markup Language is a powerful tool for creating documents using structured information. XML is a software- and hardware-independent tool for carrying information.

Bibliography

AJ (2010). Ten really awesome CSS generators. http://www.aoclarkejr.com/10-really-awesome-css-generators.html

Arthur, Charles (2011). Adobe kills mobile Flash, giving Steve Jobs the last laugh. http://www.guardian.co.uk/technology/2011/nov/09/adobe-flash-mobile-dead

Burns, Joe (2005, January 4). Java vs. JavaScript: HTML goodies. www.htmlgoodies.com/beyond/javascript/article.php/3470971

Conger, Cristin (2011). What is HTML5? http://news.discovery.com/tech/what-is-html5-111007.html

Dahm, Tom (2010). Browser compatibility tutorial. www.netmechanic.com/products/Browser-Tutorial.shtml

Gulez, Aycan (2001). Top five reasons for limiting Flash use., www.wowwebdesigns.com/power_guides/limiting_flash_use.php

Jobs, Steve (2010). Thought on Flash. http://www.apple.com/hotnews/thoughts-on-flash/

Kyrnin, Jennifer (2012). What is RSS? http://webdesign.about.com/od/rss/a/what_is_rss.htm

Mischook, Stefan (1996). What is Flash, when and why to use it. www.killersites.com/articles/articles_FlashUse.htm

Orr-Buttrey, Stephanie (2012). Interviews with author, February 2012

Rouse, Margaret (2005) Flash. CIO-Midmarket. http://searchcio-midmarket.techtarget.com/sDefinition/0,,sid183_gci214563,00.html

Seidl, Andy (2008). Google Docs replaces Survey Monkey. http://faseidl.com/public/item/212067.

Smith, Dori (1998). What is JavaScript? *MacTech*, v14, 5. www.mactech.com:16080/articles/mactech/Vol.14/14.05/WhatisJavaScript/?%2Findex.html

Smith, Mike (2008). A guide to HTML and CGI scripts. www.it.bton.ac.uk/~mas/mas/courses/html/html.html.

Taylor, Matthew James (2009). Four methods of adding CSS to HTML. http://matthewjamestaylor.com/blog/adding-css-to-html-with-link-embed-inline-and-import

TechTarget. (2005) Definition of HTML 4. http://searchsoa.techtarget.com/definition/HTML-40

Tmunotein, Ibama Supreme (2004, October 20). Client-side and server-side JavaScript. www.devarticles.com/c/a/JavaScript/Client-side-and-Server-side-JavaScript

WDG (Web Design Group). (n.d.) Linking style sheets to HTML. http://htmlhelp.com/reference/css/style-html.html

Wikipedia. (n.d.) CAPTCHA http://en.wikipedia.org/wiki/Captcha

www.freedback.com.

www.humanverify.com.

www.iLike.com.

www.slide.com.

www.webopedia.com.

www.w3c.org.

www.w3schools.com/xml/xml_whatis.asp

Chapter 7

Optimizing and Monitoring Your Web Site to Increase Traffic

Building a good web site is imperative for being successful on the Internet, but there are also certain modifications to a web site that can generate more traffic. Part of that involves adding elements to the web site that improve search engine rankings, called *search engine optimization* (SEO). Increasing traffic to the web site includes not only SEO, but also creating elements of the web site that will (1) bring in more visitors, (2) retain visitors longer, and (3) create more repeat traffic. Web surfers visit particular web sites because they are looking for something of interest. For music fans, information about music and the music itself play an important part, but it is also about the community and fun. Visitors need a reason to stay and a reason to return. This chapter discusses ways in which a web site can be modified to improve traffic to the site.

SEARCH ENGINE OPTIMIZATION

Search engine optimization is the practice of guiding the development or revamping of a web site so that it will naturally attract visitors by gaining top ranking on the major search engines for selected search terms and phrases. Chapter 12 discusses the process of submitting your web site to search engines to have them include your site in their directory. But you can improve site relevance, which helps determine how prominent your site is in the search engine results, by including a few extras in the site. The goal is to have your web site listed among the top results when visitors use search engines to find your site. Search engine optimization involves three key components and several minor, but effective, techniques—all of which add up to better rankings.

Keywords

Search engines look for *keywords* when sorting and ranking sites. This includes the keywords found in meta tags (Chapters 6 and 11), but also text that appears in the first couple paragraphs of the page to catalog "content-rich" sites. Because search engine placement is about beating the competition for the top slot in search results, the best place to start is to imagine what keywords your customers are likely to use when looking for sites such as yours, and by looking at the competition and what keywords they are using. Use one of the popular search engines and type in keywords that members of your target market would typically use to look for your artist or music. Don't be concerned about thinking of *all* relevant keywords at first—that will come later. As the results of the search are displayed, look at the first few listings to see if those web sites are indeed targeting the same market and are considered competitors. You can actually use these web sites to improve your own. Look at the keywords they use by using your browser's "view source" function to look at their meta tags. Look at the text appearing on their home page and the titles used. Consider any words or phrases that you find on these sites that you may not currently be using or have thought about. Internet consultant Bruce Clay stated, "Proper Search Engine Optimization requires that you beat your competition, so knowing the keywords and criterion used by your competition is the most important first step" (Clay, 2007).

Several web companies specialize in SEO, and for a small expense, you can have the experts handle this. There are also web sites that will assist you in evaluating which keywords are most successful in generating traffic and top search engine placement. Some of these sites provide a basic service for no charge. There are two types of tools for managing keywords: (1) keyword generators and (2) keyword verifiers. Digital Point Management provides a keyword tracker tool that determines keyword popularity. Google AdWords provides a keyword tool to help AdWord users determine the appropriate and most popular keyword phrases to use for their ads. By typing in a keyword statement or combination of keywords and clicking "get keyword ideas," the system displays results and indicates popularity for each suggested keyword phrase. Although these suggested phrases are for the benefit of AdWords users, the keywords generated by the program are also useful for inclusion in meta tags and in the text on the page. These and other programs help predict which keywords will be most successful for inclusion in your web site. Other software programs can test the effectiveness of your keywords already in use. By typing in your domain name and the keywords in question, the verification software can determine whether your web page shows up near the top of the search results for various search engines.

Keyword Verification Tools

- Google AdWords keyword tool, https://adwords.google.com/select/KeywordToolExternal
- Digital Point Management keyword tool, www.digitalpoint.com/tools/suggestion
- SEO Tools, www.seochat.com/seo-tools/keyword-suggestions-google
- Keyword verification tool, http://www.searchbliss.com/seo-tools/verify.asp

Updated information available at www.focalpress.com/cw/Hutchison.

Page Titles

The title tag of each page tells both users and search engines what the topic is for that particular page. It is wise to create a unique title for each page on your site, but with information that ties it all together. For example, for an artist site, the home page may be titled the artist's name and then band, music, etc. Subsequent pages should have a title that includes the specific information found on that

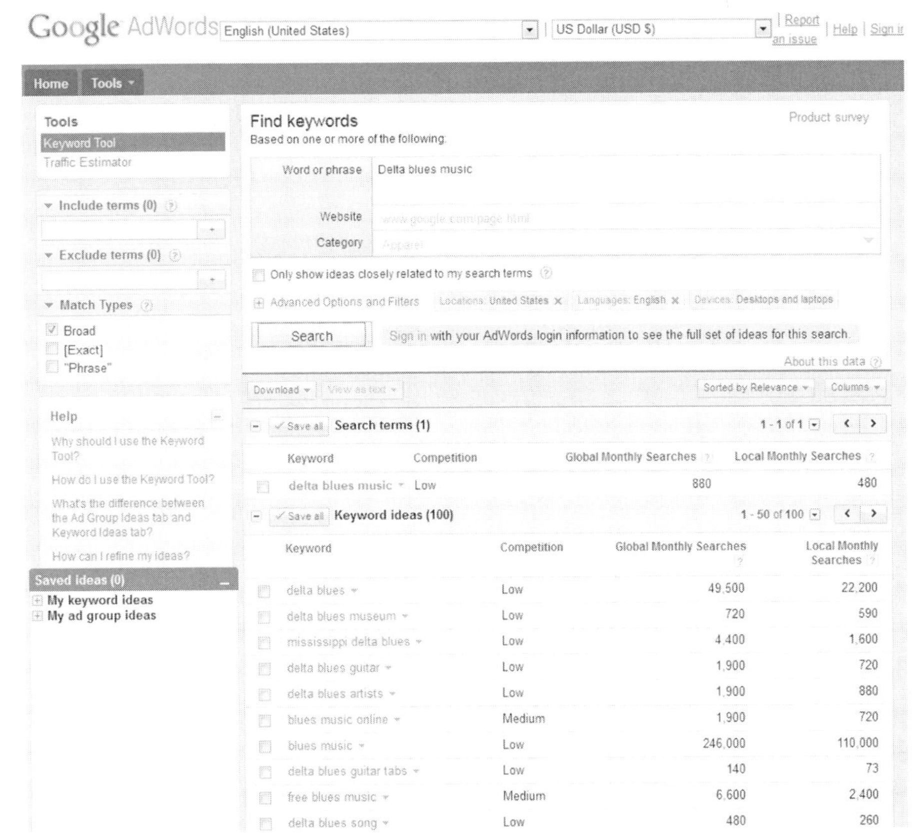

Figure 7.1
Google keyword tool (used with permission)

page, such as "Artist's Name—Bio" or "Artist's Name—Tour Dates." The reason for this is because search engine results generally list the page title in capital letters as the first link of the result.

Link Popularity

Another characteristic that search engines use to rank keyword results is *link popularity*: how many other sites think your site is important enough to link to it. The quantity (popularity) and the quality (relevance) of links to your site are used to determine ranking status. Link quality is defined as those from other sites with high page rankings for relevant search terms. Search engines use this information because they go by the assumption that the most important and relevant sites will have lots of other sites linking to it and also because it is hard for webmasters to fake or fool the search engine into giving a higher ranking than deserved. SearchEngineWatch.com states, "link analysis gives search engines a useful means of determining which pages are good for particular topics."

As outlined in Chapter 11, you can improve your link popularity by contacting webmasters at other relevant sites and asking them to place a link on their site

Sauce Boss
www.sauceboss.com/
Hot Sauce and Scorching Slide Guitar--Florida Bluesman Bill Wharton --The **Sauce Boss**--makes Liquid Summer Hot Sauce and has made gumbo for over ...
Schedule - Music - Gumbo - Press kit

Sauce Boss Tour Dates
www.sauceboss.com/concert.htm
Hot **Sauce** and Scorching Slide--Florida Bluesman Bill Wharton makes Liquid Summer Hot **Sauce** and has made gumbo for over 175000 people for free during ...

Figure 7.2
Title tag and description meta tags search results

that will lead visitors to your site. Use the search engines to find appropriate sites to request link placement. Type in the relevant keywords, and visit pages that appear at the top of the results. Ask those webmasters to add your link. Sometimes this is achieved through a link exchange. Chances are that a well-known music star is not going to agree to link to a site of an unknown artist, with the exception of major artists who are fans of the particular emerging artist. The best candidates for link exchange are similar artists and members of your fan base. Ask fans to link to your site, and perhaps offer a contest or incentive (a free download) for all those who comply.

When press releases or other written materials are disseminated, include several hot links to your artist's web site. As these articles are posted on web sites, e-zine sites, and blogs, the embedded links are spread virally and will appear in the electronically published version, thus increasing incoming link popularity.

Several free tools are available to measure link popularity. These tools search Google, AltaVista, Bing, Ask, AOL, and other search engines to determine how many pages are linking to your web page.

Keyword Popularity Tools

- Market leap link popularity tool, www.marketleap.com/publinkpop
- http://www.freewebsubmission.com/keyword-popularity.html

Google's Search Engine Optimization Starter Guide outlines several other web site alterations and specifications that will help with search engine ranking, and will help search engines correctly list and categorize your web site.

DESCRIPTION META TAG

Google uses portions of the description meta tag in search engine results to describe the site. Google states that they employ a number of strategies to select the portion of the description meta tag. You want to make sure they are not

Figure 7.3
Illustration of keyword bolding in Google search results

too short, too long or a duplication of other content (such as the title tag, page content, etc.). Google may instead use a relevant portion of the page's content if it matches up well with a user's query, although they prefer to use a portion of the descriptive meta tag. Meta descriptions that are simply a long list of keywords are less likely to be used by Google than text within the page that better describes the page's content (Krishnan, 2007). The Google Guide states "Adding a description meta tag to each of your pages is always a good practice in case Google cannot find a good selection of text to use in the snippet." Google also advises: to avoid a description that has no relation to the content on the page, avoid generic descriptions like "home page" and not to paste the entire contents of the page into the tag. The description meta tag provides both search engines and users with a summary of the page contents.

IMPROVE THE STRUCTURE OF YOUR URLS

Google recommends using file names that provide information and use folder names that are relevant to the content of the site. "If your URL contains relevant words, this provides users and search engines with more information about the page than an ID number or oddly named parameter would." Keep in mind that the URL is displayed as part of the search results. Each result is only three or four lines long, so any useless or redundant information reduces the amount of relevant information you can provide.

Organize your content so that it is structured logically, with folder names and file names that provide meaning. Google encourages the use of punctuation for clarity and advises to use hyphens instead of underscoring. When possible, shorten URLs by trimming unnecessary parameters.

MAKE THE SITE EASY TO NAVIGATE

Search engine bots have the same difficulty as humans in deciphering a web site with confusing organization and navigation. It helps the search engine

My Favorite Artist – Tour Dates
www.myfavoriteartist.com/concerts/tour-dates
Tour schedule for My Favorite Artist. One of the top blues guitar players in New Orleans, My Favorite Artist performs for thousands of blues fans from the Mississippi Delta to Hawaii.

correct

My Favorite Artist cat id=391
www.**myfavorite**artist.com/folder5/ ?cat_id=391
Tour schedule for My Favorite Artist. One of the top blues guitar players in New Orleans, My Favorite Artist performs for thousands of blues fans from the Mississippi Delta to Hawaii.

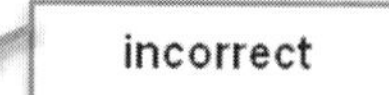

Figure 7.4
Illustration of importance of correct folder and file naming for search results

understand which content the webmaster thinks is most important. When planning, think about how the visitor will move from the home page through the site. Use *breadcrumbs*, when necessary, such as "my site > albums > the latest album > third song." That will allow the visitor to quickly navigate back to a previous section.

If the site is large enough, use a sitemap, a dedicated web page that categorizes all the other pages on the site by category. Google suggests creating one sitemap for users and another for search engines: an XML Sitemap file, which can be submitted to Google. Google's open source Sitemap Generator Script can assist in creating a Sitemap file. This beta software tool is available at http://code.google.com/p/googlesitemapgenerator/. According to the description, "Google Sitemap Generator is a tool installed on your web server to generate the Sitemaps automatically." These Sitemaps are automatically submitted to Google.

Each web page should include a menu at the bottom of the page in text form, in addition to the more graphic menus normally displayed across the top or left side of the page. This is especially important if Flash is used to create the primary menu. Search engine bots probably will not be able to read or understand the navigation information embedded in Flash-based menus.

Other tips include:

- Have a useful 404 page—the error page that a user gets when a link guides them to a non-existent page. You can customize your 404 page with suitable content and provide a link back to your root (home) page.
- Use better *anchor text* for links. The anchor text is the clickable text that users see as a result of a link (indicated by parentheses in the <a ref=> tag). Rather than use the text "click here," provide useful content in the anchor text. For example, instead of <a ref="http://www.bandname.com/tour-dates/gigs">Click Here</a>, replace the words Click Here with the band name and subject matter of the page, "Band Name Gigs."
- Optimize your use of images, including a suitable file name for the image; instead of image0013.gif, use "Bandname-at-Staples-Center" for the image

name. Also, use the "alt" attribute to provide information about the picture. Your photos are more likely to turn up in a Google image search by a user.
- Use heading tags to emphasize important text. Those are the <h1>, <h2>, etc. tags that create subject headings in the body of text on your page. There are six sizes, with <h1> being the most important. In addition to providing a visual cue for users (much like newspaper headlines), this text may end up in part of your search engine results.

OTHER OPTIMIZING TIPS

Share, "Tell a Friend" and "Bookmark This Page" Scripts

Word of mouth is unquestionably the best form of marketing communication. It carries a sense of credibility lacking in most other forms of marketing. Generally the person who is giving the recommendation has some knowledge of the recipient's interests. The recipient is more inclined to pay attention to the recommendation because of the credibility of the source. Money can't buy this type of marketing, although many record labels spend much time and effort to create street teams to give the appearance of word of mouth or street credibility. Well-designed web sites make it easy for visitors to spread the word and pass along information about the site. The newest generation of social networking music-oriented sites even has software that allows a visitor to pluck email addresses directly from their personal email accounts. One way that web sites make it easier to spread the word is through the "tell a friend" or "share this page" *JavaScript*—a bit of code embedded in one of the pages. Emarketer.com states, "A recent study showed that more than half (53%) of Internet users had visited web sites referred by friends or family members in the previous 30 days." Several web sites online will generate scripts for this feature. Some of the free versions add a viral message of their own, attached to the email.

All social networking sites and sites like Digg, Reddit, and StumbleUpon provide ready-made widgets that you can insert into your site, allowing visitors to easily and freely share content on your site among their friends within these networking services. For Facebook, go to the Facebook developers' area, core concepts, and social plug-ins. http://developers.facebook.com/docs/reference/plugins/like/. You can create a "like" button to embed in your web page. When users click the "like" button on your site, the story or content appears in that user's Friends feed, linking back to your web site.

"Share" Scripts
- www.plus2net.com/php_tutorial/tell_friend.php
- www.javascriptkit.com/script/script2/tellafriend.shtml
- http://developers.facebook.com/docs/reference/plugins/like/

Contests

Contests are a great way to generate traffic to the web site. They provide something of value to visitors and encourage them to return to the site. Contests should be

Step 1 - Get Like Button Code

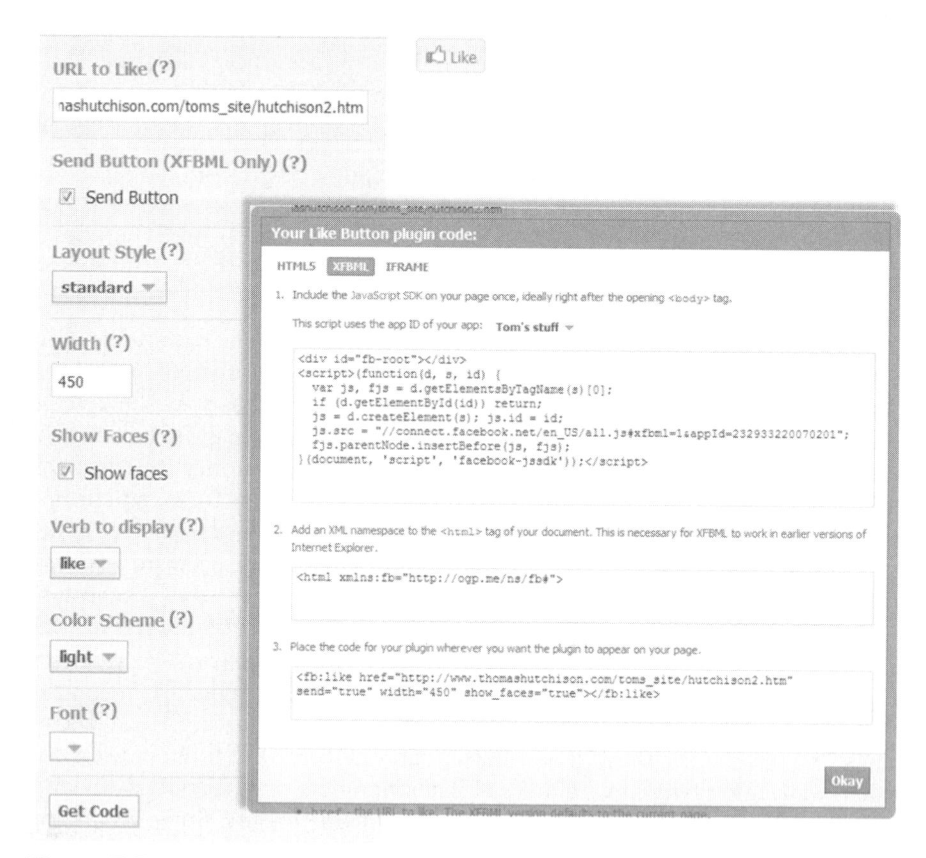

Figure 7.5
Facebook "Like" button for sharing

run on a weekly or monthly basis, with one or more prizes awarded at the end of each contest period. The contest should be designed to maximize its benefit to the site owner. Visitors should be required to provide their email address and other valuable marketing information to be entered in the contest. You can then use this information to create a mailing list, but be sure to alert visitors to this possibility and give them the opportunity to "opt out" of receiving future email correspondence. Contests can also be designed to encourage visitors to further explore the web site. Some contests pose trivia questions, with the answers found on other pages on the site—like a scavenger hunt. Repeating the contest weekly or monthly, with a new entry required for each contest period, will encourage visitors to return to the site, perhaps bookmarking the site in their browser. Contests also increase the amount of recommendations that visitors send to their friends. They also create great linking opportunities. There are many sites that link to any site offering contests.

Figure 7.6
Example of web site contest for an artist

General entry contests just require the visitor to "sign up for a drawing," but more creative contests can involve fans to a greater extent, thus making the contest more of a fun, interactive experience than just a game of chance. Fans can be asked to "pick the next" single after listening to several songs. Then a drawing can be held from among those who selected the song that winds up being the top pick. Another idea is to have fans send in their favorite digital photos from a recent performance of the artist. Contests have also been run where entrants create a new slogan, submit a song for consideration, create a *mashup*, come up with a new band name, or create a new band logo. If the purpose of the contest is to promote a new CD, it might not be best to use that CD as the contest prize. Fans may delay purchasing the album until the end of the contest. Offer something else of value: a T-shirt, an earlier release, or other swag.

Blogs

Blogs, or weblogs, have become popular on the Internet lately with the introduction of Web 2.0 and many sites and software programs that offer blogging opportunities. MarketingTerms.com defines a blog as

> a frequent, chronological publication of personal thoughts and Web links. A blog is often a mixture of what is happening in a person's life and what is happening on the Web, a kind of hybrid diary/guide site, although there are as many unique types of blogs as there are people.

Blogs are a good way to encourage repeat visits to your web site. Visitors know that with each visit to the artist's web site, they have a chance to read this fresh new material. The diary-like quality of blogs gives music fans the chance to feel closer to the artist, getting to know them better and establishing rapport. They bring a human element to an otherwise impersonal medium (Hurlbert 2004a, 2004b).

Blogs are also popular with search engines. The fresh, keyword-rich content of blogs is easy to find and catalog. Search engine spiders will crawl a site more frequently if it is updated regularly. Blogs frequently contain links to other information on the same site, luring the visitor deeper into the site. Blogs also commonly link to each other and these inbound links raise the popularity rating of the site. Blogging software and sites usually contain comments or feedback features, setting up a dialogue between readers and the blog author.

All artists and musicians should maintain their own blog, creating entries when the muse strikes, but on a regular or frequent basis. Artists generally blog about experiences on the road or in the studio, and the fans enjoy the behind-the-scenes aspect of the artist's blog. Artists have been known to pull out the

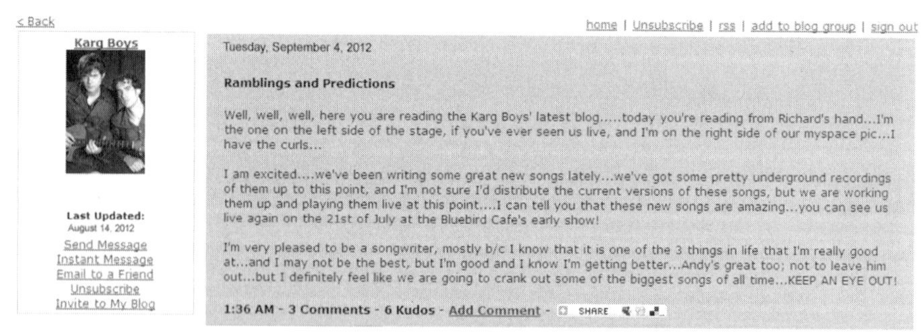

Figure 7.7
Artist blog entry

laptop on the bus right after the event and write their latest blog entry, thanking the fans for their attendance and support. The instant publication of blogs makes it attractive to readers. Bloggers also use cell phone cameras to enrich the blog entries with photos. Software companies and web sites make it easy for bloggers who may not otherwise know how to post material on the Internet, and social networking sites generally provide easy-to-use blogging opportunities. Other sites, such as Blogger.com, offer services either for free or for a small fee. These blogs can then be embedded into the artist's site and fed to other web sites and social networking sites through RSS (see Chapter 6).

VISITOR REGISTRATION

The most valuable piece of information that can be obtained from consumers who visit a web site is their email address and permission to add them to the "mailing list." A list of email addresses, and permission to contact the addressees, is necessary in this new age of *spamming*. Spamming is the activity of sending out unsolicited commercial emails. It is the online equivalent of telemarketing (more on spamming later). An effective, up-to-date email list is a valuable marketing tool and allows for e-newsletters to be sent to fans who have shown

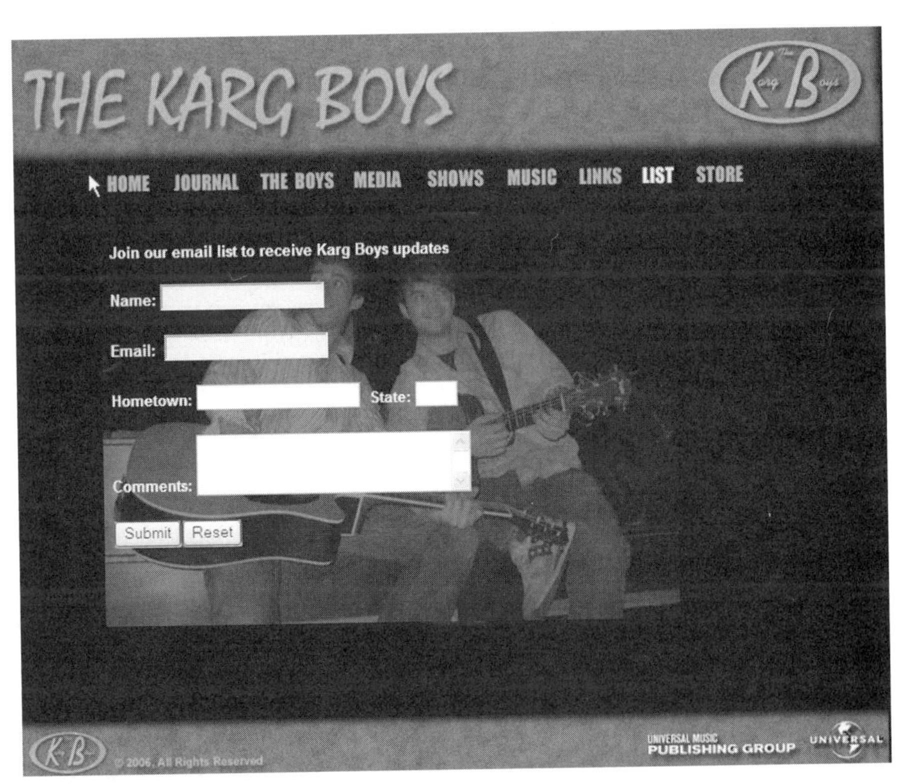

Figure 7.8
Example signup form

enough interest to sign up. Web sites should post their privacy policies to avoid any confusion or legal complications if visitors end up on a mailing list.

When recruiting visitors to sign up, it is more attractive to present this as either a guest book or free membership to the artist's fan club, rather than just signing up to receive emails. Visitors can be enticed into signing up to gain "membership" to restricted areas of the web site that may contain free music downloads and allow fans to post messages on the site's message board. It is also a way for webmasters to monitor the site for inappropriate message posting and restrict access of repeat offenders. Members can also receive benefits such as priority in purchasing concert tickets and prerelease music purchases.

Web sites can include a visitor sign-in, registration, or "join now" button leading to a short online form. When visitors are asked to sign a guest book or register to enter the site, there are two strategies for adding these visitors to the email list: "opt-in" and "opt-out." With opt-in, the visitor selects a blank check box to be added to the email list. With opt-out, the button default is in the checked position and the visitor must uncheck the box to avoid being included in the email list. "Opt-in" means that visitors choose to join a site's mailing list—one that is generally aimed at notifying the visitor of new developments. Some marketers set the default setting to the "opt-out" approach by automatically checking the "Yes, sign me up" box. By default, visitors overlook the box, thus "giving permission" to contact them. Marketers advise to use the opt-in method only. This avoids annoying visitors who did not intend to sign up for emails and simply overlooked the "uncheck" function. It also reduces the amount of follow-up work that the webmaster must do to honor unsubscribe requests.

The Client Help Desk (www.clienthelpdesk.com) reports that almost 70% of Internet users say they unknowingly signed up for email distribution lists. Almost 75% of those who received unsolicited email took action to be removed from the sender's list. With opt-outs, the web site can claim a larger number of subscribers (the willing and the unknowing), whereas with opt-ins, the site can get a better understanding of how many people want to receive the emails or e-newsletters sent out from the list.

FAN-GENERATED CONTENT

In this era of Web 2.0, the power of creating content for the Internet has been turned over to the users, with sites like Wikipedia and Yelp relying on the visitors to add meaningful content that attracts other visitors to the site. The Internet is full of such sites, and site visitors do not hesitate to pitch in their contributions. These creative participants are likely to encourage their friends to visit any sites where they have posted content—giving a viral aspect to the endeavor. So how can artist web sites take advantage of Web 2.0 user-generated features? Using Web 2.0, artist web sites can now allow fans to post messages or photo galleries to share with other fans. Some also make it easy for fans to post videos of them dancing or singing karaoke versions of the artist's music."

SHOW ME YOURS makes fans part of the show by allowing them to upload videos and photos from their mobile phone, desktop, or straight from their webcam to muchmusic.com, with the best clips airing on TV. Once uploaded, this content becomes viewable on muchmusic.com. With SHOW ME YOURS, anyone in Canada can be the star of their favourite program, and Canadians can see what other MuchMusic fans from across the country are posting.

MuchMusic Press Release, November 2006

ANALYTICS: MONITORING WEB TRAFFIC

An important aspect of optimizing a web site involves monitoring traffic to the site. Web traffic refers to the number of visitors to your web site and the number of pages visited. Oftentimes, it is measured to determine the popularity of a web site and its individual pages and elements. By including a bit of programming code on each page of the web site, the webmaster can learn a lot about the visitors to the site. This helps the webmaster and other marketing professionals understand which elements of a web site are considered valuable to its visitors and which are not.

What to Measure

Some of the most important factors that are measured include the following:

1. *The number of visitors.* This is represented by the number of different people who access your web site over a period of time. From this information, you can determine which times are most popular for visitors. You can determine if your traffic is influenced by any marketing campaigns that may be unfolding, the impact of promotional materials such as email blasts, and the impact of advertising. For example, you send out an email blast to members of the fan club announcing a new tour schedule and notice a jump in the number of visitors to the site and the tour schedule page for the next couple of days.
2. *Whether these visitors are new or returning.* The effects of advertising and other marketing efforts to expand the market can be measured by observing the number of new visitors to the site. The number of returning visitors indicates the success level of efforts designed to bring visitors back to the site, such as blogs, new material regularly posted to the site, and so on.
3. *The number of page views.* This is a measurement of how many pages each visitor looks at on the site. If the ratio is high, meaning that each visitor on average visits a fair number of pages, that is an indication of the "stickiness" of the site. Stickiness means that the site is so compelling that visitors are inclined to stick around and visit other sections. However, this could also indicate that they are not finding what they are looking for, so they keep going on to the next page hoping to find what they need. Determining which of these two factors is in play is determined by the next measurement.

4. *Time spent per page.* If visitors are spending a lot of time on particular pages, one could conclude that these pages contain something of interest to the visitor. If other pages are glossed over quickly, then perhaps they are not as meaningful to the visitor or the visitor has not yet found what they are looking for. If certain pages don't get much traffic, or visitors tend to spend little time on them, they should be reviewed to determine if the level of interest is appropriate (it may be a page designed for a subsection of visitors, such as journalists) or whether the page should be revamped or combined with another page.
5. *Time spent on the site.* Visitors who spend a long time on the site are probably the most dedicated customers or fans, especially if they are returning visitors. The average amount of time spent on the site indicates the worthiness of the site in providing something of interest.
6. *Date and time.* It is helpful to know the most popular viewing times and days to plan when updates will be made to the site and if traffic is seasonal.
7. *Where visitors reside.* This information is not always accurate, as some visitors may use an Internet service provider (ISP) that reflects the location of the main servers instead of the visitor's hometown. But for most systems, country of origin and city are listed in the visitor statistics. You can determine if there is more activity on the web site coming from areas where the artist is touring. Then by combining that with information on page hits, you can determine how important or useful the tour information page is to visitors.
8. *Where visitors are coming from and which page they enter the site through.* This information can help you to determine which outside URLs are providing most of the traffic, whether it's other sites that link to yours, search engine traffic, or direct request (the user types in your domain name).
9. *Exit page.* Which page do visitors commonly view last before leaving your site? Sometimes the page content will help determine the reason people leave the site: they found what they were looking for, they didn't find what they wanted, you directed them elsewhere, or they made the purchase.
10. *The technology that visitors use.* This function indicates the resolution of the monitor, connection type, browser type and operating system of each visitor. It is helpful in determining whether users have the technology to handle the latest bells and whistles before deciding to add those features to the site.

HOW TO USE THAT INFORMATION

Building an effective web site involves more than just aesthetics and design. The site has to offer some value to the visitor. Decisions on other aspects of marketing rely on input from customer feedback forms, surveys and other devices. Often this requires effort on the part of the consumer to provide this valuable information to market research experts. One of the great advantages of the Internet is that it offers marketing analysts a rich body of marketing information based on where web visitors go, what they click on, and how long they engage with the marketing message. In an article "Five Reasons to Track Web Site Traffic," author Monte Enbysk pointed out that too many "small businesses build web sites, invest time in online marketing campaigns and then devote little or no effort to analyzing the return on their investment." He stated that *web analytics* tools can help the marketer in the following ways:

1. *Evaluate the effectiveness of marketing efforts.* You can see the results of each aspect of promotion and how it affects traffic to the site. You can find out what the keywords are that your customers use to find you.
2. *Figure out where your traffic is coming from.* This is generally known as the *referrer* function on web statistics programs. This function will let you know if your search engine optimization is working, or if most of your visitors are coming from some other source that you can devote more of your marketing efforts toward. You can determine if your advertising is working.
3. *Learn what your users like and don't like about your web site.* Find out if it's time to replace or take down those underperforming pages where visitors tend to bail out. You can assess modifications of an underperforming page by changes in visitor activity.
4. *Learn of defects in your site.* By analyzing what computer systems most of your visitors are using, you can determine if they are getting the benefit of your site design or if design elements are displaced because of incompatible systems and browsers.
5. *Get to know your customers.* After studying the data coming in and making adjustments to the site, you can learn what your visitors like and what they respond to. Tracking them can tell you what they are looking for when they visit your site.

WHERE TO FIND ANALYTICS TOOLS AND HOW TO APPLY THEM

The most basic tool and first to appear on the scene is the simple hit counter, consisting of a bit of programming language that would place a component on the bottom of each page that counted the number of visitors. Marketing experts advise against the use of the simple hit counter, stating that this information should be kept private for use by the site webmaster and marketing team. No one is impressed with coming across a web site and finding out you are the "seventeenth visitor to this site"—period. The information is also limited compared to what is available on the market now.

Web analytics is defined as the use of data collected from a web site to determine which aspects of the web site work toward the business objectives. Many services on the Internet offer web analytics features. After asking a series of questions about how you want to track and compile information, the service will create the code to be inserted into every page of the web site. The code helps the service track activity on the site. The webmaster logs in to the service to view and download the statistics that the system has gathered.

Figure 7.9
Visible web counter

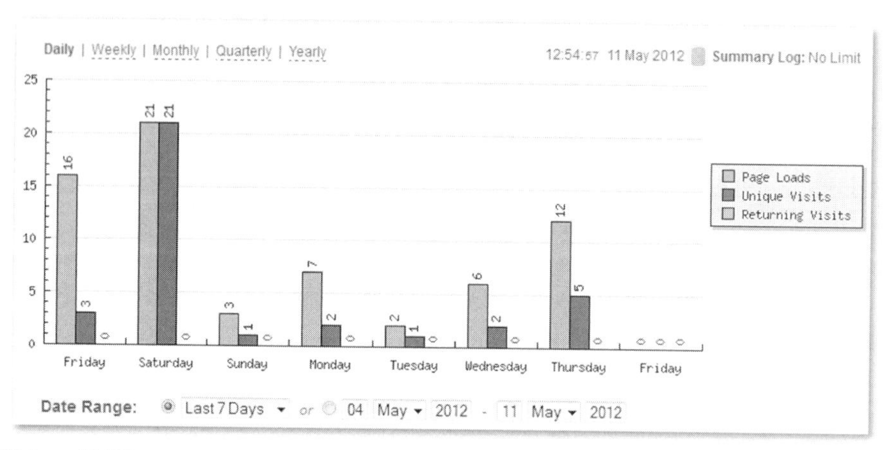

Figure 7.10
Visitor statistics

Google Analytics

In early 2005, Google bought the web analytics company Urchin. Later that year, Google revealed its repacked version of Urchin for free: Google Analytics. Google Analytics offers tracking of web visitors and provides the usual tracking statistics plus keyword reports and the ability to measure the effectiveness of AdWords programs. The process requires web owners to insert a small piece of Java code into the head tags of their pages. The statistical output can be viewed on the Google site with the assistance of a dashboard: customizable collection of report summaries.

Other Analytics Services

Alexa bills itself as a useful resource for people to discover information about web sites. Analytical information is more public in nature and can allow marketers to size up the competition. Alexa can reveal which search terms a site is using to successfully attract visitors, what kind of visitors they have, and where they are coming from. To access their pages, simply type the URL of any site into the Alexa Search box.

Statistics and Web Analytics Tools

- www.OneStatFree.com provides a free, simple stat counter service and more advanced fee-based services.
- www.Alexa.com monitors web traffic of people who use its toolbar.
- Nielsen Netratings is a commercial web monitoring service used by large sites.
- www.trafficestimate.com is a free tool to help estimate the volume of traffic that a web site gets.
- www.statcounter.com is a service that offers both free and paid statistics services.
- www.iwebtrack.com is another service with free and paid versions of the service.
- http://Analytics.google.com

Updated information available at www.focalpress.com/cw/Hutchison.

Quantcast is a team of web analytics experts building powerful statistical technology to understand internet audiences. Quantcast helps web publishers of all sizes understand the composition of their audiences and attract higher advertising rates, and helps advertisers and agencies find elusive online audiences wherever they might be. This web site allows anyone to view audience reports on hundreds of thousands of web sites. For quantified sites (those who are registered with Quantcast), demographic and psychographic information on web visitors is available. www.quantcast.com/

Compiling FAQs

What are FAQs?

FAQs, the abbreviation for Frequently Asked Questions, are a collection of the most commonly asked questions with the answers provided. This section may help a webmaster from having to personally answer emails asking the same questions time and again. It is also beneficial for placing information that will only be of interest to particular visitors.

How is FAQ pronounced?

Since the concept has its origins in text communication on the Internet, the pronunciation varies; it is pronounced as separate letters F.A.Q. or as "fak."

What does Wikipedia say about FAQs?

> Today 'FAQ' is more frequently used to refer to the list, and a text consisting of questions and their answers is often called a FAQ regardless of whether the questions are actually frequently asked (if asked at all). This is done to capitalize on the fact that the concept of a FAQ has become fairly familiar online.
>
> **Wikipedia http://en.wikipedia.org/wiki/FAQ**

What are the benefits of having FAQs on a web site?

Webmasters can learn a lot about what materials are missing or are obscure on the web site based upon visitor feedback and questions. From this, the list of FAQs is compiled and constantly revised to provide better service to the customer or site visitor.

Are FAQs commonly used on music web sites?

While it is not common to see a FAQ page on an artist web site, they are more common for record label sites. Some label sites have a FAQ page to let artists know how to submit material, or how candidates can apply for jobs or internships. A FAQ page is important whenever a contest is being held, to clarify entry policies and contest rules. Artists can also present an interview in FAQ form rather than standard interview form.

What are some example FAQs found on music web sites?

- How do I get in touch with the artists or set up an interview?
- How can I get fan club info?
- Can I use music and photos of Sony Music artists on my web site?
- Can I get permission to use lyrics and/or sheet music for a Sony Music artist's song?
- Where can I find sheet music?

From the Sony Music web site.

Trendrr is a service that tracks popular trends on the web by using input from sources like social networks, blogs, and video downloads, reported in real time as they are happening.

CONCLUSION

Building an attractive, enticing web site is very important. Adding these other elements will draw traffic to the site and encourage visitors to engage in word-of-mouth marketing. Search engine placement is an important aspect of helping people find your site. And once there, the elements of giveaways, contests, blogs, visitor registration, and fan-generated content encourage the visitor to return and tell others about your web site. Monitoring traffic to the site can be important for evaluating the effectiveness of the site and for guiding the webmaster when making changes and upgrades.

Glossary

404 page The error page that shows up on a web site when a link goes to a dead or non-existent page.

Anchor text The clickable text that users see as a result of a link.

Blog (Weblog) a frequent, chronological publication or journal of personal thoughts and web links posted on the Web.

Bots Software robots called "bots" are programs designed to perform an extensive range of automated tasks.

Breadcrumbs The part of web site navigation that shows you where you are, borrowed from the fairy tale "Hansel and Gretel." Breadcrumb trails are often found near the top of the web page and define both the current location within the site hierarchy as well as primary pages above the current page. They are hot links, allowing the user to quickly backtrack to a certain page in the hierarchy.

FAQs The abbreviation for Frequently Asked Questions, they are a collection of the most commonly asked questions with the answers provided.

Hot links A link that takes the web browser to another place upon clicking the link.

JavaScript A scripting language developed by Netscape and used to create interactive web sites.

Keyword A word (or phrase) that a search engine uses in its hunt for relevant web pages.

Mashup A mixture of content or elements.

SEO (search engine optimization) Various techniques that seek to improve the ranking of a web site in search engine results.

Spam Flooding the Internet with many copies of the same message in an attempt to force the message on people who would not otherwise choose to receive it. Most spam is commercial advertising.

Spider A program that automatically visits and catalogs web pages. Spiders are used to feed pages to search engines. Marketing companies also use them to gather information. The program is called a spider because it crawls over the Web. Another term for this type of program is webcrawler.

Web analytics The use of data collected from a web site to determine which aspects of the web site work toward the business objectives.

Bibliography

Clay, Bruce (2007). Search engine optimization (SEO). www.bruceclay.com/web_rank.htm.

Enbysk, Monte (2011). Five reasons to track web site traffic. websitetrafficsurge.com/website-traffic-counters/

Google Search Engine Optimization Starter Guide. http://static.googleusercontent.com/external_content/untrusted_dlcp/www.google.com/en/us/webmasters/docs/search-engine-optimization-starter-guide.pdf

Google Webmaster Tools Help: URL structure. http://support.google.com/webmasters/bin/answer.py?hl=en&answer=76329

Hurlbert, Wayne (2004a). Blogs as a website promotional tool. July 13. www.globalprblogweek.com/archives/blogs_as_a_website_p.php

Hurlbert, Wayne (2004b). Blogs as excellent public relations tools. September 15. www.seochat.com/c/a/Website-Promotion-Help/Blogs-as-Excellent-Public-Relations-Tools

Krishnan, Raj (2007). Improve snippets with a meta description makeover. http://googlewebmastercentral.blogspot.com/2007/09/improve-snippets-with-meta-description.html

Odden, Lee (2007, January 15). SEO benefits from blogs. www.toprankblog.com/2007/01/seo-benefits-from-blogs

SHOW ME YOURS on Muchmusic.com with new fan-generated content. (2006, November). Press release.

Sullivan, Danny (2007a, March 15). How search engines rank web pages. www.searchenginewatch.com/2167961/print

Sullivan, Danny (2007b, March 15). Search engine placement tips. www.searchenginewatch.com/2168021/print

www.searchenginewatch.com

www.webanalyticsassociation.org

Y-Times publication (2007). Are you using Google Analytics? Here's why you should be. www.ytimes.info/areyouusgoan.html

Chapter 8
Audio and Video for Your Web Site

It is essential for a music-oriented web site to provide audio and video samples of an artist's work in order to achieve the web site goals of engaging fans that are outlined in Chapter 5. Today, not only do web sites offer music to preview, but artists are finding ways to provide fans with access to their music through other sources such as social networks, either for download or streamed to the user's computer and mobile device. In his book *How to Promote Your Music Successfully on the Internet*, author David Nevue stated, "if you want people to buy your music online, you've got to give them a sample of the goods." The question becomes how much audio? Should you provide samples of songs, entire songs, or a few songs from each album? Nevue stated that some artists may not feel comfortable giving away their product, whereas others, including some major acts like Radiohead, are comfortable giving away their music if it increases their fan base and they can make up the money on concert tickets. The industry standard for online e-tailers is to offer 30–45-second samples of some or all songs on an album, but Nevue believes that songs samples should be longer, up to two minutes, "enough for your potential buyer to get into the groove of your music" (Nevue, 2005, p. 36).

AUDIO FILES ON THE INTERNET

Most audio files used by consumers for portable devices and Internet delivery are compressed. Compression is a means for reducing file size through discarding much of the duplicated data in the audio file. Consider the way "ditto" marks are used with text, and you get a basic analogy. The creation of the *MP3* compression format for audio opened up the possibility for the first time of transferring music over the Internet. MP3 stands for MPEG (Motion Pictures Expert Group) Audio Layer III, and it is a standard for audio compression that makes any music file smaller with little loss of sound quality (although that's debatable). Without this compression technique, one second of CD quality sound requires 1.4 million bits of data, so the average song in *WAV* format is extremely large, averaging around 50MB in file size. WAV is short for Waveform audio format, a Microsoft and IBM audio file format standard for storing audio on PCs. With MP3 compression, the file becomes one-twelfth its original size, at about 3 to 4 megabytes, depending on the bitrate.

MP3 was first introduced to consumers in 1997 with the launch of the popular Winamp player for computers and Diamond Multimedia's Rio

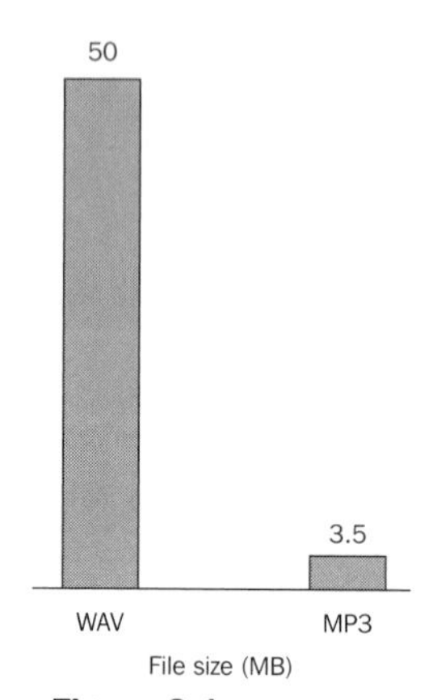

Figure 8.1
Comparison of file size for WAV file and MP3

portable MP3 player. The Recording Industry Association of America (RIAA) brought suit against Diamond, claiming that its Rio player did not fall under the protection of the *Audio Home Recording Act of 1992* because it did not employ a *Serial Copyright Management System* to protect the audio files from being exploited. The court rejected the argument, saying that because the obvious purpose of the player is for personal use, allowing consumers who owned a copy of the music to make a copy for their portable device constitutes fair use under copyright law.

MUSIC AND YOUR WEB SITE

A major issue is whether to offer music as downloads or streaming. Do you want visitors to be able to click on a song and start listening immediately (streaming), or do you want them to be able to download, store, and own your song or sample on their computer (or portable device) so they can listen at their convenience? A comparison of *streaming* and downloading will be discussed next, followed by an overview of the different types of formats for audio files with and without digital rights management technology.

Streaming Audio

PC Magazine describes streaming audio as "a one-way audio transmission over a data network. It is widely used on the Web to deliver audio on demand or an audio broadcast (Internet radio)." With streaming, the audio file is not transferred to the user's computer but portions are stored in the computer's buffer so that the content will play normally, without interruptions.[1] (The other option is to allow users to download files to play at their convenience. This may lead to file sharing and does not provide the level of protection found with streaming.) The Information Technology web site at Cornell University describes streaming audio this way:

> When audio or video is streamed, a small buffer space is created on the user's computer, and data starts downloading into it. As soon as the buffer is full (usually 10–30 seconds), the file starts to play. As the file plays, it uses up information in the buffer, but while it is playing, more data is being downloaded. As long as the data can be downloaded as fast as it is used up in playback, the file will play smoothly.
>
> **http://atc.cit.cornell.edu/course/streaming/index.cfm**

According to David Nevue, you should create an additional file called a "metafile" in order to stream MP3 files from your web site. This is done by creating a bit of HTML language with a link to that MP3 file and saving that language in a file with the extension .m3u. When the user clicks on the link with the m3u extension, the default Internet audio player should open the original MP3 file and begin streaming it.

According to John Haring of Nashville Independent Music:

This also can be accomplished in a cleaner manner by embedding a server-based player into the website that will pop up when called. This is usually done with custom Flash code, although there are a number of open source players available to embed into your web site. This way, native MP3 files can be played without having to convert them and without the user having to have a player installed.

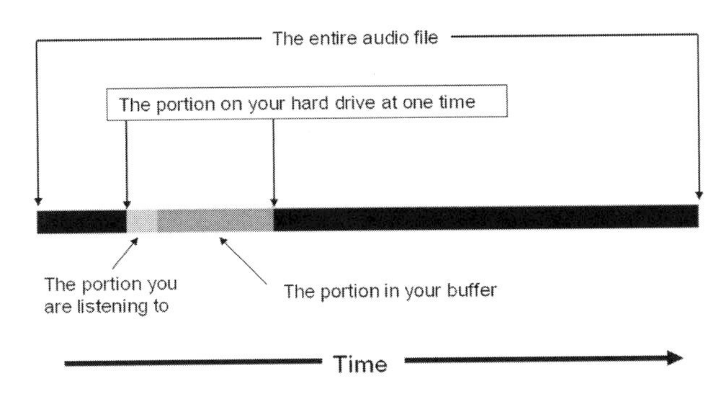

Figure 8.2
Illustration of streaming audio file

You can also use Windows Expression Encoder 4 to create streaming audio and video in the *Windows Media Audio* (WMA) format http://www.microsoft.com/expression/products/EncoderPro_Overview.aspx.

Web-based widget players. Recently, several companies have begun offering stand-alone streaming players to embed into your web site, so you can stream your music to fans. Others provide the storage vehicle for the music, incorporating it into widgets that allow for streaming from your web site and many of the social network platforms. One such player is the *Wimpy* MP3 player ($39). The Wimpy is a full feature player, with a playlist for your MP3 files but does not support other formats. Their more-versatile *Wasp* player supports three audio formats: MP3, AAC, and MP4, along with the most popular video formats.

SoundCloud describes their product as "a web-based tool for digital musicians to streamline the process of hosting and distributing audio … It allows users to effortlessly upload tracks and share them anywhere on the web or with anyone they choose, either privately or publicly." SoundCloud has a free, limited version, and several professional versions. With SoundCloud, you must upload your files to their server. *Podbean* offers a simple jukebox or single song player with simple instructions. Setup occurs on Podbean's web site, where you select the player style and then enter in information for song title and URL where the file is located. You can upload your files to their server, or use the URL address where the file is located. The player has nice features. Podbean has a free version and tiers of paid service for heavier use. *Alsacreations* is a French company offering a variety of players.

If you are a member of *ReverbNation* (as you should be) their widget page provides a nice variety of audio and video widgets that can be customized and implanted into your web page and/or the major social networks.

- Wimpy http://www.wimpyplayer.com/
- Podbean https://www.podbean.com/
- Alsacreations. http://www.alsacreations.fr/dewplayer-en.html
- SoundCloud http://soundcloud.com/
- Open Source player: http://flash-mp3-player.net/
- ReverbNation www.reverbnation.com.

Figure 8.3
ReverbNation streaming audio widget

Paste and play. You can create your own flash player with a bit of code and help from Google Reader. Copy and paste the following code within your web page (Digital Inspiration, 2008). Replace the string "MP3_FILE_URL" with the location of your MP3 file. A small player with controls is embedded for that one file. You will need to repeat this for each audio file.

```
<embed type="application/x-shockwave-flash"
flashvars="audioUrl=MP3_FILE_URL" src="http://www.google.com/
reader/ui/3523697345-audio-player.swf" width="400" height="27"
quality="best"></embed>
```

Yahoo offers a more advanced Flash-based player that will take all the links to MP3 files on a web page and insert them in their embedded jukebox player. Paste the following code into a web page with existing links to MP3 files and Yahoo will automatically load them into the player.

```
<script type="text/javascript" src="http://mediaplayer.yahoo.
com/js"></script>
```

CREATING A MUSIC FILE FOR DOWNLOADING

For downloading, it is necessary to offer the audio file in one of the more commonly used compression standards. Setting up a downloadable file is much easier than setting up streaming—just set up a link to the file on the server and when web visitors click on the link, they will be able to download the song. Add all music files to a folder on the server and set up a link to each one. As visitors click on the link, they will be prompted to download the file. The disadvantages of this method include the fact that the visitor cannot begin to listen to the song until it has finished downloading. Also, there is less protection of the music if it is given away to visitors to download to their own computers; there is the potential for file sharing. Artists often reserve the free downloads for acoustic or live versions of songs rather than give away their primary product. In fact, giving away a "bonus track" can be a great promotional tool and generate traffic to the web site.

AUDIO COMPRESSION FORMATS FOR DOWNLOADING

The MP3 format is the most popular and most flexible format, supported by most platforms including the iPod (Potts, 2002). As for computer software, Winamp, iTunes and Windows Media Player support the MP3 format. On the other hand, MP3 is the lowest-quality choice among the common formats.

Advanced Audio Coding (AAC) is iTunes' compression of choice because of the sound quality and, until 2009, the copy protection incorporated into their files. iTunes and other major download retailers adopted *digital rights management* (DRM) early on to copy-protect downloads so they could not be swapped in peer-to-peer networks. This was the only way the download retailers could convince the major record labels to license their products for downloading. AAC is in either MPEG2 Advanced Audio Coding or MPEG4 Advanced Audio Coding. MPEG2 AAC can produce better audio quality than MP3 using less physical space for the files. MPEG4 AAC can produce even better quality and smaller files than MPEG2 AAC. In 2007, EMI and iTunes announced they would be offering a DRM-free version of EMI's catalog available on iTunes in the AAC format for the premium price of $1.29 per download. AAC files offer better sound quality and are around 30% smaller than the MP3 equivalent. In 2009, iTunes dropped the use of DRM on most of its tracks.

Windows Media Audio (WMA) is another format widely used by many online retailers, but it does not offer much flexibility. It outperforms MP3 in terms of quality and compression, particularly at lower bitrates. Consequently, WMA is probably the format of choice for streaming at low bandwidths.

Free Lossless Audio Codec (FLAC) is an open-source audio compression format that does not degrade the audio quality like the others, but files are typically much larger (40% of original size).

Regardless of the format, any songs offered for download should be appropriately labeled with the song title, album title, artist, running time, and other important information. Be sure to include this information in any files created for download. Also, it may be wise to enter your track information in the universal Gracenote (CDDB) database. This is a web site that connects to audio software programs on computers and provides those computers with song identification data so that when the user loads up a CD in the computer, perhaps to rip the songs to a portable device, the software can download the track information from Gracenote. Any computer audio software program that works with Gracenote (such as iTunes or WinAmp) will enable you to enter in track information for your own CD and upload it to the system. Then when fans purchase your CD and play it in their computers, the correct track information will be available for them on their computer.

> The Gracenote Media Recognition Service is an Internet-based service that we license to software and hardware developers for use in their CD players, CD burners, MP3 players and encoders, cataloguers, jukeboxes, cell phones, car audio systems, and home media center applications (among others). The service allows these developers to display artist, title, tracklists, and other music-related information automatically and instantly in their applications.
>
> For example, when you insert a music CD in your computer, the software player application on your computer uses our service to first identify the CD, and then display the artist, title, tracklist, and other information. Most commercial music CDs do not contain any of this information on the CD itself.
>
> **Gracenote.com FAQ**

PROVIDING MUSIC SAMPLES

One safe way to preserve the value of an artist's music is to provide 30-second samples instead of letting visitors download or listen to the entire song. No one wants to make and distribute copies of 30-second segments of a song. And the sample gives potential customers an idea of whether they might like the song or not. If the artist's music is featured for sale on one of the major online download services, they usually provide 30-second samples, and it may not be necessary to create them for the artist's site. Visitors can be redirected to one of those e-tailers to preview the music.

To provide samples on the artist's web site, they must first be edited from the full song. The idea is to select a sample that best represents the aura of the song, not simply start at the song's introduction and take the first 30 seconds. One general rule to follow is to capture the end portion of a verse and most of the first chorus. At Nashville Independent Music, John Haring stated,

> We've found that offering 45 seconds of a song starting from the 20 second point forward captures most of a verse and chorus. We use this standard when creating clips in our automated process for nashvilleim.com. We also automatically create a one second fade-in and a four second fade-out for better listenability.

CREATING MUSIC SAMPLES FROM SONGS

The process of creating samples can be accomplished using any music editing software such as ProTools, Cakewalk, or one of the less expensive audio editing programs available on the Internet such as Audacity or Gold Wave. From the

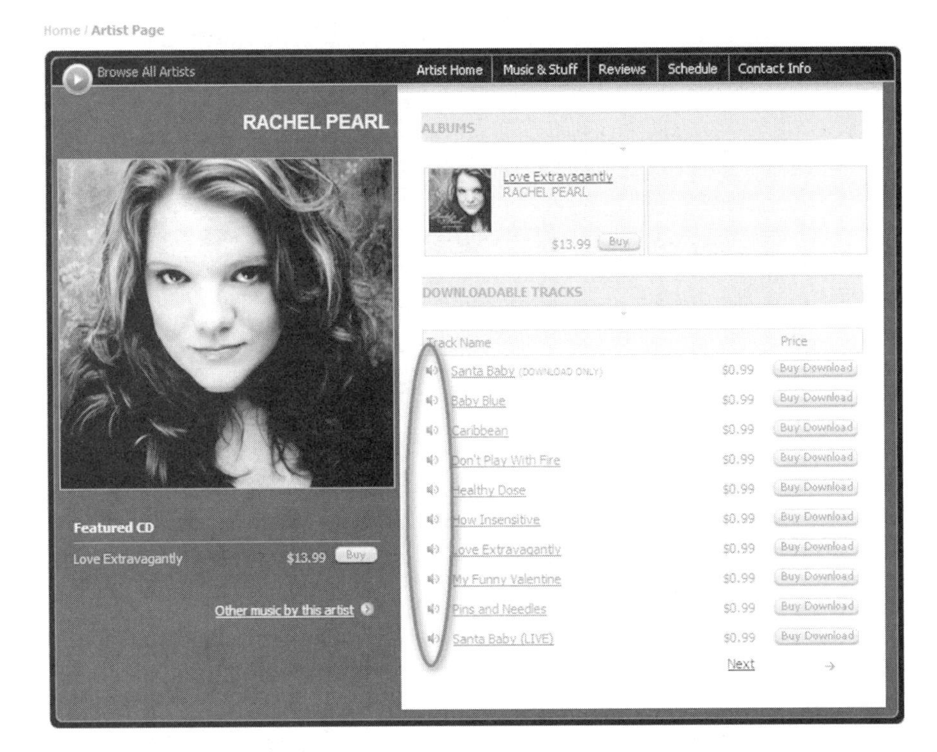

Figure 8.4
Audio samples indicated with the speaker icon (courtesy Nashville Independent Music)

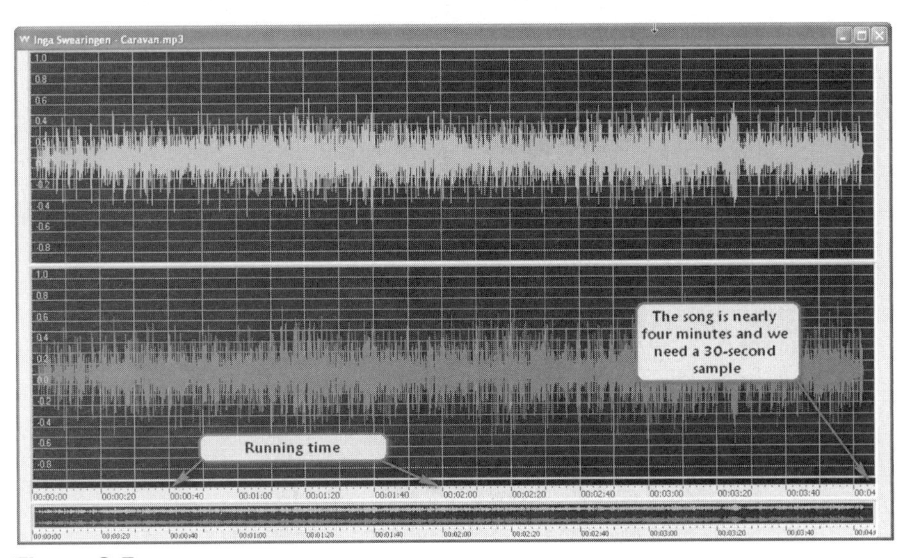

Figure 8.5
Graphical representation of song (with permission from Goldwave)

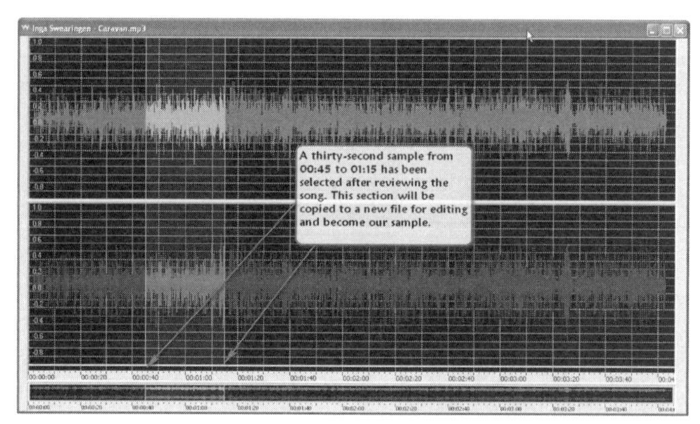

Figure 8.6
Highlighted 30-second song sample (with permission from Goldwave)

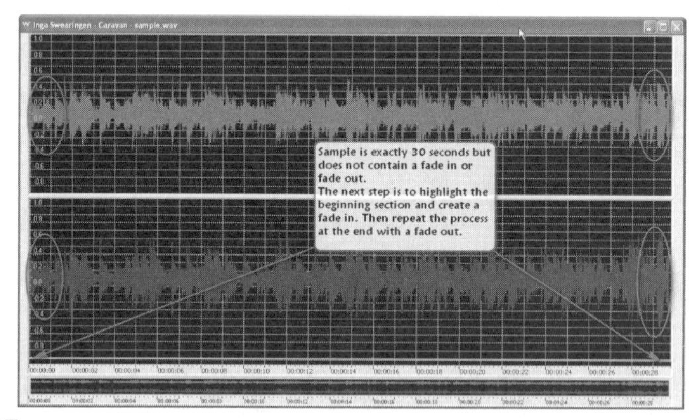

Figure 8.7
Sample expanded in a new file (with permission from Goldwave)

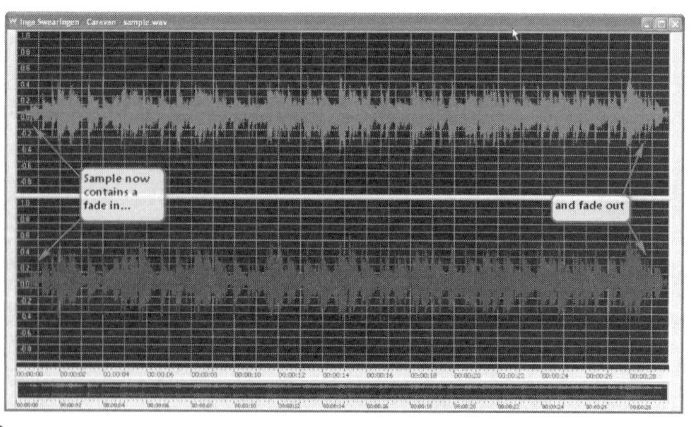

Figure 8.8
Song sample with fade in and fade out (with permission from Goldwave)

songs selected for sampling, simply open the song in an editing program and listen to various 30-second sections until you have found a section that best represents the overall song. Then follow these simple steps:

In most audio editing programs, the running time is listed either at the bottom or the top of the song file. The graphic representation is amplitude modulation, with loud parts of the song showing up with large bars and quieter sections showing shorter bars. With the highlighting tool, you can select 30 seconds and preview it to determine its suitability.

Once a 30-second sample has been selected that is a good representation of the song, highlight it and copy it to a new file for further editing.

The new file will contain the sample filling up the entire running time of the file—in this case, 30 seconds. At this time, you may want to preview the sample again to verify that it is the best possible representation of the song. If it meets those requirements, it still needs some editing.

To sound like a normal sample, it will need a fade in and fade out. These can be accomplished by highlighting first the beginning section of the song. This will be the area selected for a fade in, from silence to full modulation, so that by the end of the highlighted section, the song is playing at normal volume. The larger the highlighted section, the longer the fade in (see Figure 8.9). Select an appropriate size section for the fade in and use the fade-in tool to reshape the sound. First, highlight the section for the fade in. The appropriate fade-in rate may vary depending on the song and may take some experimentation. Repeat the process in reverse at the end of the sample so that it fades out to silence. This will

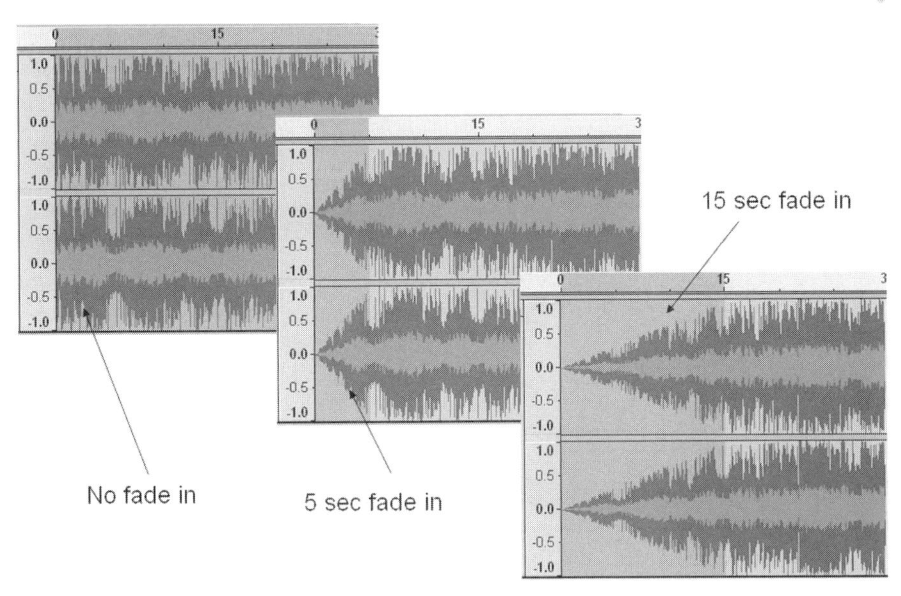

Figure 8.9
Illustration of fade-in options

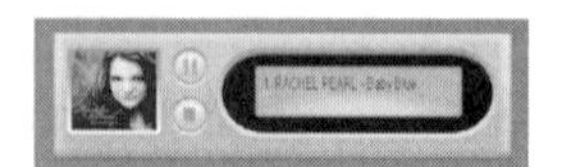

Figure 8.10
Embedded player user controls (courtesy of John Haring)

permanently alter the sample so that no further manipulation is necessary, and the listener will not be required to make any adjustments. Without this editing, the sample would have abrupt entry and exit points and not seem natural.

Then save the file as an MP3 file and upload it to the server. As each web visitor clicks on a link created to the MP3 file, the browser's player will open the file and play it on the visitor's computer. The alternative is to embed the player controls within the web page so that the visitor can click on them to access and listen to the sample. Most web design software programs include multimedia controls. This will allow the web designer to place more than one sample on a page. When setting up the music file on the web page, the options will generally include the following:

- Do you want this to play automatically when the page opens, or have the visitor select play?
- Do you want the song to play once, several times, or loop continuously?
- Do you want embedded user controls?

The advantage, or disadvantage of having the user's browser open the default media player is that the music will continue to play even if the user moves on to another web page. If the controls are embedded in the page, chances are that the music will quit when the user continues through the web site. However, if there are music files on several pages and all are set to open in the default media player when the page loads, several songs may play at the same time, confusing the visitor.[2]

When including several music samples on one page, it is best not to have any of them play automatically, so that visitors can select if and when they want to listen. A page that automatically plays a sound clip when it's opened may delay the loading process and cause the visitor to wait or give up.

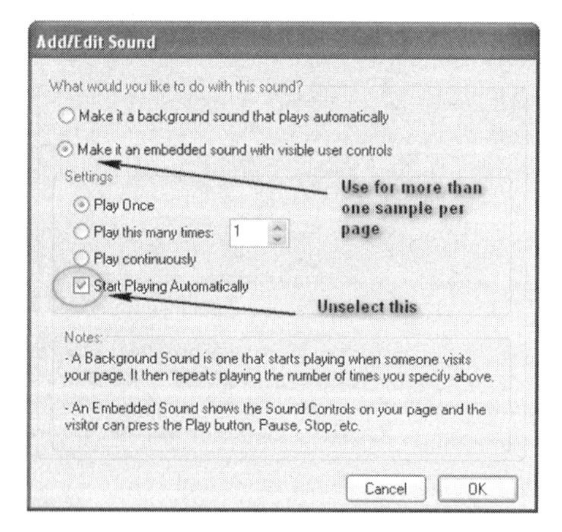

Figure 8.11
Example of embedding sound in a web page (courtesy WebStudio, www.webstudio.com)

Audio Editing Programs

- Goldwave audio editor, www.goldwave.com
- Audacity audio editor, http://audacity.sourceforge.net
- Audiobook Cutter, divides longer MP3 files into several smaller files; good for samples, http://www.audiobookcutter.com/
- Kristal Audio Engine, www.kreatives.org/kristal
- winLAME, converts audio files from one format to another (use in conjunction with an audio editor). http://winlame.sourceforge.net

Check www.focalpress.com/cw/Hutchison for updated lists.

EMBEDDING MUSIC IN FLASH PROGRAMS

Embedding media in web pages allows the delivery of an integrated multimedia experience that appears seamless to the user. Although this can be achieved with the above-mentioned streaming process, it is currently done with Flash by Adobe software. Flash is useful for building small player applications that present audio and video content. It will allow the user to peruse other content on the page while enjoying the audio file. The Flash program could also allow for, say, a slide show, to run while the music is playing. Apple mobile products and Android products do not support Flash. HTML5 will make the use of Flash unnecessary as more multimedia delivery programming is provided by the HTML standard, making Flash or other multimedia plug-ins obsolete. HTML5 accommodates a <video> tag.

Music Players for Your Web Site

- The Wimpy Player, www.wimpyplayer.com
- Secure-TS Player, fancy skinned secure Flash player for under $30, www.tsplayer.com
- XSPF Music Player, www.musicplayer.sourceforge.net.

MUSIC VIDEO CLIPS

Music videos made a big splash when they were first popularized by MTV back in 1981. Since then, they have been a staple in the marketing arsenal for any recording artist. Now, YouTube has done for Internet music videos what MTV did for established artists back then. The difference now is that anyone can participate with user-generated content—and it seems like everyone is. YouTube offers a way to share videos in a social networking environment and to create some sense of organization for the thousands of video clips contributed from members. Because video has now become an important marketing tool for musicians, it may be important to include some short video clips in the artist's online presence. You can accomplish this in one of two ways, either by embedding the videos into your artist's own web site or offering links to other video host sites such as YouTube so that the visitor can easily click to watch a video.

Editing Videos

It is unwise to upload raw footage of the artist's live performance without the benefit of editing. Most successful video clips are short in length. And a one- or two-minute video will need to feature the best of the best. Three minutes is tops for maintaining interest. Viewer attention span is short, and anything that drones on too long or takes too long to get to "the good stuff" is likely to fail. Viewers will tune it out before giving it a chance. Therefore, it is important to grab people up front. Once the raw footage has been shot and the content of the video is determined, video editing can be done either online, with editors such as eyespot (www.videotoolbox.com), or with an offline video editor.

MASHUPS

> Often as a pre-release promotion, record labels will distribute raw tracks in free loop bundles for people to download and use to create remixes. The resulting remixed versions of the songs are uploaded to a searchable archive that displays ratings, recommendations and comments from other remixers. More importantly, from a word-of-mouth marketing perspective, this personalized content gets shared through interpersonal networks of friends and family who are online and get remixed versions of an upcoming album release.
>
> **http://splinteredchannels.blogs.com/weblog/2005/04/mashup_video.html**

One popular Web 2.0 video editing hobby is creating *video mashups*. A mashup is the compounding ("mashing") of two or more pieces of complementing web functionalities to create a powerful web application. Mashups originated from the world of pop music where DJs would mix two or more songs together to present a new blended version (www.videomashups.ca). A video mashup is the combination of multiple sources of video—which usually have no relevance with each other—into a derivative work often lampooning its component sources. They are one of the latest genres of mashups and are gaining popularity (Wikipedia). If music videos introduced consumers to fast-paced video cuts and edits, mashups take this art form one step further by allowing the user to develop re-creations of previously released video content. It becomes the perfect new format for developing video projects to promote artists. Some major artists have even used contests to encourage fans to create their own mashups of the artist's materials. This user-generated content then has the potential for spreading virally as proud mashers share their creations with friends.

- Total Recut is one such social networking site that features video mashups. The site provides tools for creating video mashups, including online video

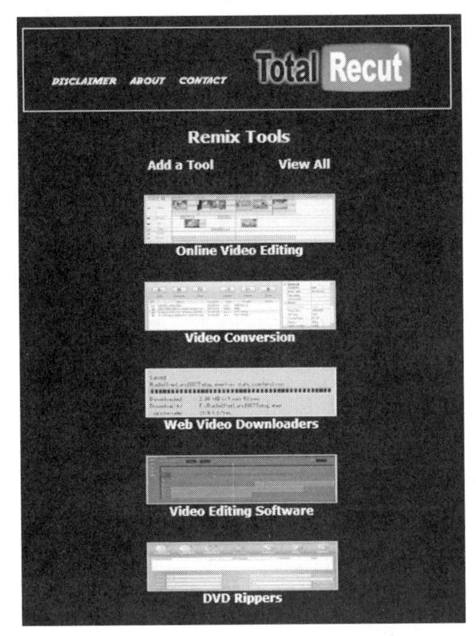

Figure 8.12
Video editing and mashup tools from Total Recut (courtesy of www.totalrecut.com)

editing, video conversion, and video editing software (www.totalrecut.com/totalrecutdosanddonts.php).

- Animoto is a service that lets you create a professional-looking music video based on your uploaded video, images, and music (http://animoto.com).

PLACING VIDEOS ON YOUR WEB SITE

Before deciding to upload videos to the artist site, consider the large amount of server space and bandwidth that videos consume. Bandwidth and file size is a function of the following:

1. Number of frames per second
2. Frame width in pixels
3. Height in pixels

For example, a video shot at 20 frames per second, with a pixel height of 450 and pixel width of 800, will use up 1,620 kilobits per second with moderate motion (http://sorenson-usa.com/vbe/index.html).

The next step is to make sure the video is in the correct format for the player that will be used and is one that is compatible with most user systems. Adobe Flash may be the best bet as it works with most browsers and allows for more control over the presentation and the player. If the original file is in some other format, such as Windows Movie Maker, a small conversion program such as Riva FLV Encoder, can be used to convert the video to Flash Video (FLV)

(Boutell). AnyVideoConverter is a free program that will convert video files from one format to another and upload them directly to your YouTube account. It supports these formats: AVI, FLV, MOV, MP4, MPG, M2TS, MTS, RMVB, AVCHD, MKV, WebM(V8), QT, WMV, VOB, 3GP, 3GPP2, and DivX. Adobe Flash program will allow you to embed the video in the web page with a simple Flash applet. An alternative, FlowPlayer provides many options to customize or "skin" your video frame. The free version includes the FlowPlayer branding, but allows for commercial use. A simple code generator for embedding a simple video player is available at FreeVideoCoding.com. Some web design programs also contain features for setting up video embedding.

Once you have the video edited, converted, and the appropriate audio track coupled with it, upload the file to the server. It may be wise to create a new folder on the server to hold the video file and the *applet* that runs the video. Then set up the player's attributes: size, look, controls, skins, and so on. It might be a good idea to feature a link on your web site for visitors to download a copy of the latest flash player.

Video Editing Tools

- Riva FLV encoder (free trial), www.rivavx.com/?encoder
- AnyVideoConverter, http://www.any-video-converter.com
- VTube Tools (skins for video), www.vtubetools.com
- Free Video Coding, http://www.freevideocoding.com/
- FlowPlayer, http://sourceforge.net/projects/flowplayer or flowplayer.org/
- Tutorial on how to add video to a web site, www.2createawebsite.com/enhance/adding-video.html
- Flash player, http://www.adobe.com/devnet/dreamweaver/articles/add_video.html

Updated information available at www.focalpress.com/cw/Hutchison.

POINTING VISITORS TO A VIDEO HOSTING SERVICE

An easy alternative to hosting your own videos is to use one of the video hosting services such as YouTube or Vimeo to store your videos. YouTube offers two ways to promote a YouTube video on your web site: a link to the page with the video or HTML text that embeds the YouTube video player in your web page. YouTube offers several options for embedding your video into other sites and modifying the size of the player. There are third-party companies that provide skins to customize your embedded videos to better match the look and feel of your site. VTube tools program is one of many.

The field of video hosting services has been reduced as many of the startups available several years ago have either been absorbed or folded. Vimeo and Blip.TV are two video hosting services that, like YouTube, do not charge for basic hosting (500 MB per week uploading for Vimeo). Vimeo offers a premium service, Vimeo Plus at less than $10 per month for more professional applications

. Chapter 11 has a section on promoting your videos on YouTube and other video hosting services. Blip TV touts that it's a video distribution service more than a social platform. Videos uploaded to the service are syndicated to AOL, Yahoo and MSN.

CONCLUSION

Today's Web 2.0 atmosphere offers many opportunities to display and share audio and video. Online consumers have come to expect a multimedia experience, and every music and entertainment web site should offer them one. The wise music marketer will take advantage of all options, including adding audio and video to the primary web site and creating pages on social networking sites that include audio and video of the artist. Music samples should be provided in easy-to-find and well-traveled places. Mashups are becoming popular now and can showcase musical talent in new and exciting ways. With today's ADD (attention deficit disorder) generation (Nalts, 2007), videos need to be short, and the mashup is a perfect format for that.

Glossary

Advanced Audio Coding (AAC) A standardized encoding scheme for digital audio, promoted as the successor to the MP3 format. AAC generally achieves better sound quality than MP3 at the same bitrate. Used by Apple's entertainment products and originally with DRM in place.

Applet A special type of Java program that can be included in an HTML page. Web browsers, which are usually equipped with Java virtual machines, can run the applets to perform interactive graphics, games, calculators, and so on.

Audio Home Recording Act of 1992 Established a number of important precedents in U.S. copyright law that defined the debate between audio and video device makers and the content industries, requiring all digital audio recording devices sold, manufactured, or imported in the United States (excluding professional audio equipment) to include the Serial Copy Management System.

Digital rights management (DRM) A systematic approach to copyright protection for digital media whose purpose is to prevent illegal distribution of paid content over the Internet (Bitpipe.com).

Embedded player (media) A term used to describe animation, video, audio or other types of media that are displayed within a web page. Embedded media makes it possible for web page users to have a seemingly integrated multimedia experience.

Gracenote Primarily a music recognition service that works with your computer's software player application to identify an audio CD and display the artist, title, tracklist, and other information.

Mashup A mixture of content or elements; the compounding ("mashing") of two or more pieces of complementing web functionalities (audio or video) to create a powerful web application.

MP3 file Stands for MPEG (Motion Pictures Expert Group) Audio Layer III, and it is a standard for audio compression that makes any music file smaller with little loss of sound quality.

Serial Copyright Management System Created in response to the digital audiotape (DAT) invention, in order to prevent DAT recorders from making second-generation or serial copies. SCMS sets a "copy" bit in all copies, which prevents anyone from making further copies of those first copies. It does not, however, limit the number of first-generation copies made from a master. SCMS was an early form of digital rights management (DRM). It was also included in consumer MiniDisc and DCC players and recorders.

Streaming audio "A one-way audio transmission over a data network. It is widely used on the Web as well as private intranets to deliver audio on demand or an audio broadcast (Internet radio).

Unlike sound files (WAV, MP3, etc.) that are played after they are downloaded, streaming audio is played within a few seconds of requesting it, and the data is not stored permanently in the computer" (ZDNet).

Video mashup The combination of multiple sources of video—which usually have no relevance with each other—into a derivative work often lampooning its component sources.

WAV file Short for waveform audio format, a Microsoft and IBM audio file format standard for storing audio on PCs.

Windows Media Audio An audio data compression technology developed by Microsoft.

Notes

1 Depending upon the speed of the Internet connection. Streaming audio performs very poorly on dial-up systems. The portion in the buffer will play fine, but there will be significant delays as the player catches up to the buffer. The player will eat up data much faster than the buffer can refresh it.

2 Adding a pop-up player allows the music to play after leaving the page. However, the down side is that, if you click on more than one song, you may have all of them playing at the same time in different players.

Bibliography

Cornell University Information Technology. http://atc.cit.cornell.edu/course/streaming/index.cfm

Digital Inspiration (2008). How to embed MP3 audio files in web pages. http://www.labnol.org/internet/design/html-embed-mp3-songs-podcasts-music-in-blogs-websites/2232/

Haring, John (2008). Personal interview.

Hinchcliffe, Dion (2007). More results on use of Web 2.0 in business emerge. ZNet. http://blogs.zdnet.com/Hinchcliffe/?p=103 (April 3).

http://www.webmonkey.com/webmonkey/05/46/index2a_page5.html?tw=multimedia.

Nalty, Kevin (2007). How to promote your videos on YouTube. http://willvideoforfood.com/2007/06/01/how-to-promote-your-videos-on-youtube/

Nevue, David (2005). *How to Promote Your Music Successfully on the Internet*, The Music Business Academy. www.musicbusinessacademy.com

PC Magazine online (2012). Definition of: streaming audio. www.pcmag.com/encyclopedia_term/0,2542,t=streaming+audio&i=52132,00.asp

Potts, Daniel (2002). Audio compression formats compared. Australian PC World. http://www.pcworld.idg.com.au/index.php/id;1090733979;fp;2;fpid;206 (July 31).

Recording Industry Association of America v. Diamond Multimedia Systems, Inc. June 15, 1999. http://www.internetlibrary.com/cases/lib_case13.cfm

www.videomashups.ca.

Chapter 9

E-commerce and Financial Transactions

For most artists and record labels, the ultimate goal is selling a product, especially recordings and concert tickets. To do this, it is necessary to get into e-commerce: having a retail presence on the Web. The first decision involves whether to "set up shop" or leave it to the experts. This chapter will cover the transactional options for selling products both through your web site and via third-party retailers. Still, as of 2011, two-thirds of all albums sold were in physical form, which means even the novice musician needs to sell CDs, and not just digital downloads.

There are many online retail stores that handle either physical product, digital downloads, or both. Should you decide to do it yourself, there are several online services that handle the complex portions of self-distribution, including financial transactions, setting up the web site storefront, and inventory management and handling. Some of these services basically offer software to interface with your artist's web site and use a database to manage shipping information, or they simply deal in financial transactions (see Table 9.1).

DOING IT YOURSELF

Doing it yourself requires several components in the process of engaging in commercial transactions: processing orders online, providing financial security, handling the financial transaction, inventory management, and shipping out orders.

Fulfillment

Fulfillment is defined as order processing that includes documenting when an order was received, when and how it was shipped, and when and how it was paid for. Record labels, artists, and their managers need to weigh the options when deciding whether to handle their own fulfillment. It requires persistent attention to the web site and prompt follow up on all orders received. If an artist is on the road touring, fulfillment should be left to a third party to handle. Advances in computer technology have streamlined the fulfillment process. With the right software program in place, much of the order processing can be automated, from keeping tabs on inventory levels to actually printing out shipping labels and bar codes for the delivery services.

Table 9.1 Options for e-commerce: comparison of using fulfillment services versus DIY

	Advantages	Disadvantages
Doing it yourself	Keep more of the profit per unit No up-front costs or monthly fees Good for those with more time than money	Requires constant commitment Requires knowledge of web development and security systems Must handle storage and shipping Customers may be reluctant to provide credit card numbers
Using shopping cart software or services	Reliable and secure transactions Inventory management is more organized Suitable for artists with limited Internet skills but who are available to handle fulfillment	Up-front costs to set up shopping cart Monthly or transaction fees cut in to profit Still must handle storage and shipping
Using full-service fulfillment	They are responsible for fulfillment, order processing, inventory management Suitable for artists who are busy or on the road Can provide marketing services	Take a much larger percentage of sales, much like a retail store

FINANCIAL TRANSACTIONS

Credit Cards

To set up shop, it is necessary to process transactions and collect money from your customers. Credit cards are the most popular form of financial exchange online because of their convenience and speedy processing. Years ago, mail-order businesses mostly requested money orders or cashiers' checks; some would take personal checks but would then wait until the check had cleared before sending out the product. Back then it was common to see the disclaimer "allow four to six weeks for delivery." In today's immediate gratification society, four to six weeks is not an acceptable timeframe for most customers. Credit cards increase impulse buys. In *How to Promote Your Music Successfully on the Internet,* David Nevue stated, "To run a successful business on the Internet, credit card acceptance is an absolute must." However, credit card processing is not without costs and requires an elaborate setup. There are web service companies now that provide the credit card processing services along with the software to integrate into your web site. To set up a site to accept credit cards, your company must have a merchant bank account, security and encryption measures in place (*secure socket layer* server or SSL), credit card verification services (also called a payment gateway), a shopping cart page, and the software to process and track orders and shipments.

There are many credit card processing companies set up to cater to small businesses. Selecting from one of these options may take some time, but is worth the effort to do your comparative shopping. There are several factors to consider, including startup fees, transactional fees, cost per month, customer service, and web features.

- **Monthly fees**. Most vendors charge a monthly fee and/or have a monthly minimum in sales volume if they rely on transactional charges for their fee for services. Among the monthly fees are:
  - *Statement fees*: The amount charged each month for compiling and providing the itemized statement of transactions. This can range from about $7.50 to $10.
  - *Gateway fees*: A monthly charge for the service. These are put into place by the vendor to ensure they have a monthly income from slower accounts, since transactional fees may not cover their costs. There can be an inverse relationship between gateway fees and transaction fees. Many services for small businesses charge up to $15/month.
  - *Monthly minimum*: Some companies rely more on transaction fees and may require a monthly minimum in sales volume. If that volume is not reached, the gateway fees kick in.
- **Transaction fees**. These are the percentage fees or flat fees charged for each transaction by credit card. They include:
  - *Discount rate*: The percentage of your sale that the credit card processing company keeps for themselves. This is generally about 2– 2.5%.
  - *Transaction fee*: A required charge per transaction; regardless of whether the transaction is $0.50 or $500, this fee is always the same. Generally, the fee is about 20 cents per transaction.
  - *Address verification fee*: Each time a credit card address is verified, this small fee is added to the cost. Some companies do not charge this fee, while others charge about five cents.
- **Setup fees**. This fee is charged for setting up a new account.

When selecting a credit card service company, it is best to estimate or anticipate how much sales volume you expect through credit card transactions. If your sales volume is low, you may prefer a service that has lower fixed charges (monthly fees) but higher transactional fees. However, if you expect a large sales volume, it's best to keep the discount as low as possible while being less concerned about monthly fees. If your large volume in sales comes from an abundance of small purchases, such as digital tracks, then it is best to stay away from large transaction fees. If you are selling a digital track for 99 cents, a 20 cent transaction fee, along with a 5 cent address verification fee and a 2.5% discount rate can reduce profits substantially.

A comparison chart is available at http://credit-card-processing-review.toptenreviews.com/v2/

An important consideration to acknowledge is that many customers are apprehensive about giving out their credit card numbers to an unknown vendor on a small web site and prefer to use a more reputable retailer or more secure service such as PayPal.

PayPal

A report from Market Platform Dynamics states that online payment platforms such as Paypal and Google checkout were developed to provide a much-needed service for the small sellers and buyers on web sites such as eBay who needed a cost-effective, safe, reliable way to transfer money.

PayPal launched in 1998. It is an online service that allows registered users to transfer funds to and from bank accounts set up as their PayPal accounts. It also allows nonregistered users to make a payment to a registered user via a major credit card. The payee will feel more secure providing his or her credit card number to PayPal than the small, unknown vendor who is selling items on the Internet. PayPal then credits the payment to the vendor's account. The vendor pays a service fee of 30 cents plus a small percentage of each transaction (2.9% for a standard business account or 2.2 for a merchant account). The service also offers online shopping cart services through third-party vendors (see Table 9.2).

Google Checkout

Google Checkout is a service offered by Google. When combined with Google's AdWords program (see Chapter 11), the costs are greatly reduced, with merchants waiving the monthly transaction fees on $10 in sales value for every dollar spent on advertising with Google. Google charges 2% and 20 cents per transaction. The main advantage of a Google Checkout account is its integration with Google AdWords; fees are reduced when using AdWords and income from AdSense is used to pay the fees.

Keep in mind that doing it yourself involves a commitment to maintain accurate accounts of inventory on the web site and to promptly respond to each transaction with a confirmation email and shipping information. It also requires being ready to pack up and ship products out the door on a frequent and consistent basis. For musicians who are on the road, this responsibility is best left to the experts.

For handling e-commerce on the web site, it is necessary to follow these recommendations:

1. Provide thorough product descriptions, including graphics.
2. Prominently display the product name and price. If several formats are available, clearly identify the format.
3. Make it easy for the customer to purchase; the fewer clicks, the better. Make the "buy" button obvious.
4. Make sure the customer knows when the order is completed.
5. Once the order is placed, send an immediate email confirmation.
6. Make sure the orders go out as quickly as possible.

7. Make sure the customer knows how long it will take to receive the order.
8. Make sure you have the inventory to fill the orders. If you are out of stock, modify the storefront page immediately to reflect this fact.

Shopping Carts

The *online shopping cart* offers customers a convenient and straightforward way to purchase items from a web vendor. Webopedia describes the shopping cart as

> a piece of software [or service] that acts as an online store's catalog and ordering process. Typically, a shopping cart is the interface between a company's Web site and its deeper infrastructure, allowing consumers to select merchandise; review what they have selected; make necessary modifications or additions; and purchase the merchandise.

The shopping cart software also offers the vendor seamless management of inventory, revenue and tax collection, and shipping information. Many shopping cart software vendors and services have the capability to design the shopping cart page so that it is integrated into the look and feel of the rest of the web site.

There are two options for integrating a shopping cart into the artist's web site: (1) stand-alone software programs that allow the web designer to specify parameters and create the cart or (2) online services that offer inventory management features and the ability to customize their shopping cart to conform to the design of your web site. The web site "Top Ten Reviews" lists the following leading software providers: Volution, Shopify, Ashop Commerce, BigCommerce, 3D Cart, ShopSite, and a few others. Most of these are under $100. The advantage of using a software program is that generally there are no commission fees to pay, only the initial cost of the software, however there are still transaction fees associated with payment. This arrangement may be preferable for companies that do a great deal of business and can dedicate an employee or two to monitor and maintain the interface. Arnie Kuenn (2008), in his article "Choosing the best online shopping cart software" states that "Finding the right software will be crucial to making sure that your online storefront thrives." His favorites include 1ShoppingCart, MerchandizerPro, and Network Solutions. CubeCart is another popular program. If you change from one system to another, a popular migration software program called Cart2Cart can help.

The second option, hiring an online service to handle and host the shopping cart services, can reduce the possibility of complications as most services have security and transaction management procedures covered. These services also provide database management and keep records of transactions for the vendor. PayPal is the most popular service for providing transactional services, but also offers a shopping cart feature. PayPal has moved from a single source to offering integrated third-party services, depending upon the merchant's needs. AShopCart is another online service and is setup to sell digital music in a variety of formats. http://ashopcart.com/.

The web site EarlyImpact.com outlines the features one must examine when selecting which shopping cart service to adopt. The first is to focus on features

Table 9.2 Shopping cart sites offered in conjunction with PayPal

Beginner
GoDaddy: $9.99/month includes free hosting; up to 20 products.
Shopsite Starter: $6.95/month including hosting; up to 15 products.
ASecureCart: $15.95/month with no setup or transaction fees.
Shopify Basic plan: $29/month for hosted plans; up to 100 products. 2% transaction fee.
Intermediate
Café Commerce: $30/month; no setup or transaction fees; no long-term contracts.
E-Junkie: No setup fee, no transaction fee, and no bandwidth fee. Starting at $5, monthly plans vary by catalog size.
X Cart: One-time licensing fee of $149 for X-Cart GOLD. Free evaluation. No periodic fees.
Shop Factory: One-time licensing fee starting at $89, no monthly fees. Or rent a plan from $29/month.
For higher volume merchants
Pinnacle Cart: Free templates; free migration from other carts, packages start at $29.95 per month.
Miva Merchant: $49.95/month. Includes: support, e-template, domain name, dedicated SSL certificate.
Magnento Go: $15–$65/month; 100–1000 SKUs; no transaction fees; no setup fees.
Yahoo! Small Business: One-time licensing fee starting at $89, no monthly fees. Or rent a plan from $29/month.

that are important to your needs and eliminate programs and services that don't match those needs. Second, visit live stores that use each of the services or programs you are considering. Walk through the customer experience by shopping and adding things to the shopping cart. Continue exploring by manipulating the items in the cart. Is it easy to remove, add, change quantities, and so on? The third aspect of comparing options is to determine what the total cost will be for each of the possibilities. Some offer a lower setup fee, or monthly rate, but charge more commission per transaction. This may be suitable for low-volume business. But at higher sales volumes, it may be more cost effective to pay a higher up-front setup fee or higher monthly fees and make up the savings with lower commission rates.

Next, review store management tools, which can be done through product demonstrations. Note how easy or complicated it will be to perform inventory management and shipping functions such as processing orders, authorizing funds, sending out automated emails to the customer, manually changing orders, batch processing multiple orders, and so forth. It is also a good idea to look up published reviews of the product and evaluations of product support.

Top 12 Problems with Shopping Cart Design

In 2001, Barbara Chaparro conducted research on various shopping cart programs and published the "Top ten mistakes of shopping cart design." The study was repeated in 2007 (Naidu and Chaparro, 2007) to determine if any of the problems had been addressed. The following summarizes the findings of the problems:

1. Calling it something other than a shopping cart. Whereas the earlier study found that U.S. consumers are accustomed to the term *shopping cart,* the later study revealed that the term was used more literally in some other countries to mean only the physical shopping carts in bricks-and-mortar stores.
2. Requiring users to click the BUY button in order to add something to their cart. As in physical stores, the shopping cart should be a place to put items that may or may not be purchased upon checkout, without requiring customers to commit to each item at the time they place the item in the cart. An "add to cart" button is more appropriate. Some programs now offer a "wish list" function for customers who prefer to purchase at a later time.
3. Giving little to no visual feedback that an item has actually been added to the cart. With some more ambiguous programs, the customer may not be sure their item has been placed in the cart, only to discover later that they have three or four of the same object added to the shopping cart list when they prepare to checkout. The author states that as of 2007, about two-thirds of sites take the user to a shopping cart page when an item is added.
4. Forcing the shopper to view the cart every time an item is added to the cart. This takes the shopper away from shopping mode and may decrease overall sales. If the cart information is included on the shopping page, such as with Amazon.com, it is not necessary to take the customer to the cart page after each item is added.
5. Asking the user to buy other related items before adding an item to the cart. Again, disrupting the flow of shopping at this point in the shopping experience is not a good idea. It is better to show related products either just after an item is placed in the cart, or better yet, just before the customer checks out.
6. Requiring users to register before adding an item to the cart. Customers report that they do not like to provide personal information until they are ready to check out. In addition to the intrusion, it also disrupts the shopping flow.
7. Requiring a user to change the quantity to zero to remove something from the cart. It is much better to provide a remove or delete button next to each item in the shopping cart, yet the "change quantity to zero" function still persists on 15% of sites tested.

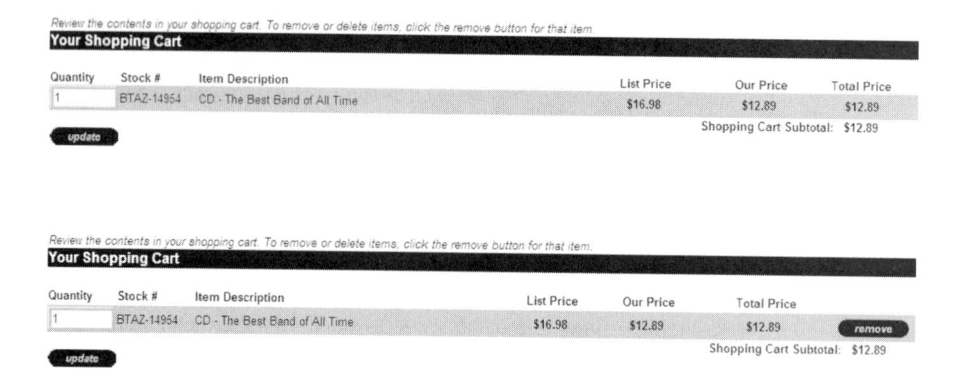

Figure 9.1
Portion of shopping cart shown without and with the remove button

8. Including written instructions to update the items in the cart. Today's online shoppers do not bother reading instructions. Using the shopping cart should be intuitive enough that they can focus on shopping instead of managing the cart.
9. Requiring a user to scroll down to find important buttons on the cart page. The update cart button, as well as the checkout button should probably be located at the top *and* the bottom of the shopping cart page so that customers with a lengthy list of products will not have to scroll to find them.
10. Requiring a user to enter shipping, billing and personal information before knowing the final costs, including shipping and tax. Some of the newer sites open a separate window to compute shipping and tax, dependent on the

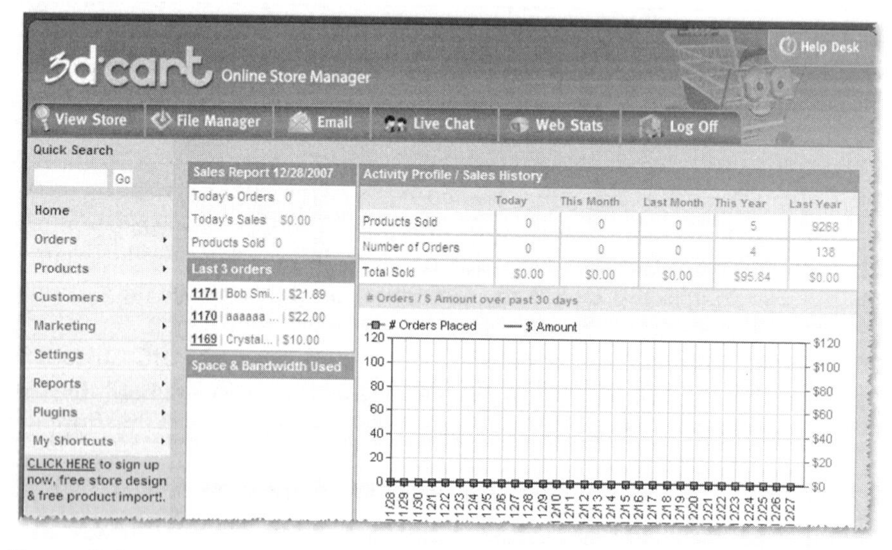

Figure 9.2
Example of what information the vendor has access to (courtesy of www.3Dcart.com, by permission)

customer's shipping address. But nearly half of the sites tested still required most of this information before showing the final total.

The 2007 study included some new issues not covered in the original study.

11. Security is an issue with consumers, so verified secured sites are necessary.
12. Out-of-stock items were a problem in that some consumers were not notified that their items were out of stock until they reached the checkout page.

Digital Download Sales Widgets

For those artists who do not want or need a storefront and who are perhaps only selling digital tracks through their web site, there are several companies that offer widgets and services for selling downloads. One such service is Website Music Player. They offer widgets with features such as color customizing, setting up preview samples, tracking downloads, and integration with social networks. Monthly fees start at $8 and there is no transaction fee.

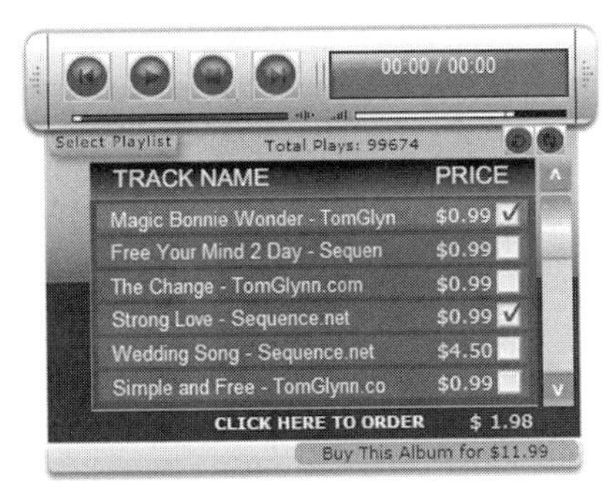

Figure 9.3
Website Music Player (used with permission)

CD Baby offers sales widgets for members and will handle fulfillment for physical CD orders. All the information on the widget is populated by your CD Baby account, so when you make changes to your CD Baby dashboard (such as a price change) those changes are automatically updated to your widgets. The widgets can be used on any site that supports Flash and uses HTML and social networking sites. It should be noted that the widget does not work on personal Facebook pages but does work on your Facebook Fan Page.

Nimbit is a direct marketing web-based service that provides a widget for digital sales and a host of other distribution and fulfillment services including CDs, electronic tickets, and T-shirts. Widgets are available for interfacing with Facebook and other networks. ReverbNation set up their Reverb Store in 2009 and began offering standard e-commerce tools for selling physical merchandise and digital downloads, however, as of early 2012, it was still in beta testing. You can set your own retail price, and ReverbNation charges a flat 30 cent fee per download. It is recommended, however, that sales through mobile devices be left up to specific apps designed for e-commerce, iTunes and others who support e-commerce on all mobile platforms.

Chat Online: Providing an Extra Level of Service

Sometimes customers may have a question about an item for sale and would like a quick answer before continuing with their purchase. Providing an instant chat feature may be one way to avoid a lost sale. Several companies provide widgets that make an instant chat feature available for your shopping cart page. Zopim is one such company that offers an instant chat widget and a dashboard for less than $10 per month. A small widget is installed on your web page. When the customer clicks on the widget, they have access to a chat

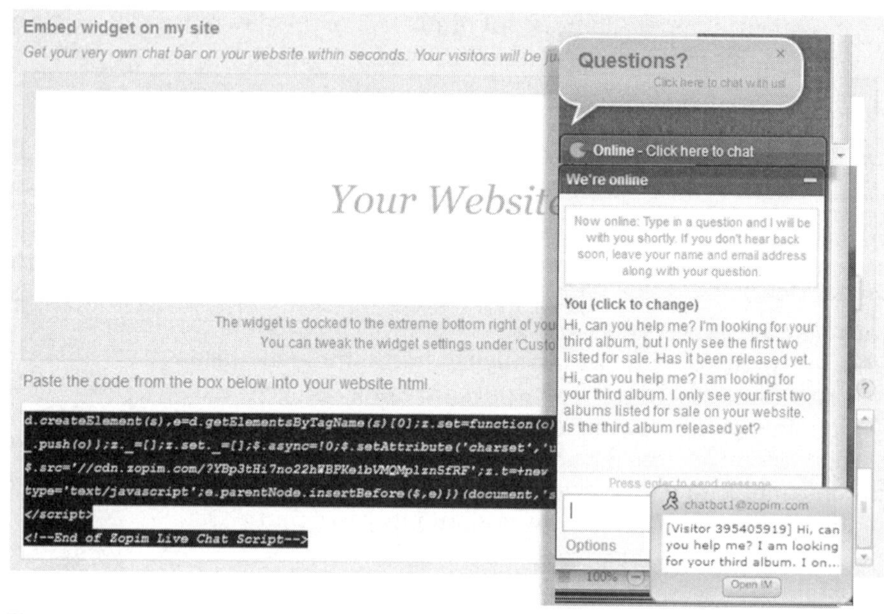

Figure 9.4
Online chat feature from Zopim (used with permission)

window without leaving the shopping cart page—an important feature in converting shoppers into buyers. The inquiry is delivered to the artist or web manager via a variety of options, selected by the seller—even to mobile devices using SMS text and allowing for direct responses from the seller's cell phone back to the shopper. www.zopim.com. A dashboard allows the web manager to view chat histories. Boldchat (recently purchased by LogMeIn) is another company specializing in chat features for a low price. The free version allows up to three simultaneous chats, with one operator. The basic paid version is less than $10 per month.

Online retailers

There are advantages to signing on with an online retailer. The major decision is whether you want to handle the money and product fulfillment yourself, or whether it is worth sharing some of the profit with an online retailer and letting the retailer manage transactions and shipment. Online retailers can provide benefits such as marketing, reputation, and traffic—much like a mall offers shopping traffic to its tenants. New to the scene are the business models created by streaming music services like Rhapsody and Spotify. Generating income from those services is covered in Chapter 12.

Online Distributors and Retail Stores

There are many online retailers that offer services to small vendors (musicians and indie labels), including product fulfillment, collection of revenue, and some online promotions. The most popular online retailers for music are iTunes, CD Baby, and Amazon.com.

- *CD Baby.* CD Baby is quite popular with independent and developing artists. David Nevue (2011) stated, "CD Baby is the largest seller of independent music." The setup cost per title is $35 to cover setting up the account and web page. CD Baby's web pages include song samples, artwork, artist bio, customer reviews, and a link to the artist's web site. CD Baby does not charge a monthly fee, only the $35 setup fee and $4 per unit sold. They ask for five CDs from unproven artists to start the process. As those are sold, the artist is notified and more CDs are requested. CD Baby also features artists, provides bar codes, and reports sales data to SoundScan.
- *Amazon.com.* Amazon is a bit more expensive but is very good for raising your profile. Music is submitted for review, which means there is some quality control. Labels should set up an Advantage account, which provides an online storefront for the label's catalog. The advantage of Amazon is the extremely high traffic generated by the site. The disadvantages include the fact that Amazon takes a large portion of the profit from each sale and charges an annual fee to participate. Amazon has also launched a digital download service, available to independent musicians.
- *Independent Online Distribution Alliance (IODA).* IODA is a distributor of digital music. Applications are subject to review and not automatically accepted. IODA distributes to all the major digital retailers and streaming music services, including Amazon, Spotify, iTunes, Rhapsody, eMusic, Vevo and MOG.
- *TuneCore.* Tunecore is an online distributor that caters to the independent artist by offering a low-cost distribution arrangement. Upfront costs are low and the service allows sellers more control over what is sold and where.
- *The Orchard.* The Orchard is a distributor of physical CDs and digital music, but not an online retailer. The company distributes to iTunes, Napster, Rhapsody, and others. Independent labels can apply for an account and receive worldwide digital distribution if their music is accepted. The Orchard is probably not the way to go for independent artists. http://www.theorchard.com/work-with-us
- *iTunes.* The best way to get placement on iTunes is through one of the distributors. However, if you are distributed by more than one distributor, only designate one as the official wholesaler representing you to iTunes.

No longer in operation:

- *CD Now.* Founded in 1994, CD Now was one of the first online CD mail order companies. Financial difficulties forced a sale to Bertelsmann to combine with their direct mail division. The CD Now brand was then purchased by Amazon in 2001 and absorbed into the parent company.
- *AmieStreet.* Amie Street was an experimental business model that built upon fan participation to drive sales of new artists. As songs became more popular, the retail price would rise, with some of the additional income

distributed to earlier customers who recruited their friends to buy the track. The model did not succeed and the assets have been absorbed into Amazon.

- *Snocap.* Launched in 2004 by Napster founder Shawn Fanning, Snocap was an early player in the music retailing business, offering independent artists a way to sell digital downloads through their web site, MySpace and the now defunct imeem. Snocap was plagued by problems and all but disappeared in 2009 (Buskirk, 2009).

MOBILE PAYMENTS AT THE GIG

Wireless technology and advances in payment processing have made it easier to sell CDs and other items at shows and accept credit cards and PayPal. Until just a few years ago, the cost of renting a credit card processing machine (swiper), and the transaction fees that accompanied it, made cash-only transactions standard procedure for selling merchandise at live shows. Now, in the age of point-of-sale (POS) mobile applications, the ease of using PayPal, mobile card and near field communications, cash is no longer the only form of currency for selling goods at shows.

Square is a mobile swiping service that provides the credit card reader for free and charges 2.75% per transaction, with no minimum (www.squareup.com).

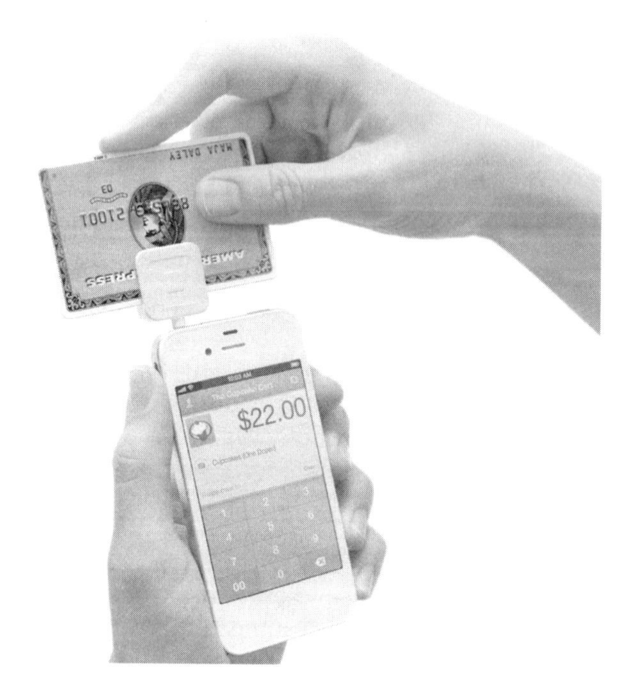

Figure 9.5
Square mobile credit card swiping (courtesy of Square)

Concert Ticket Sales Online

The opportunity for small business owners and DIY musicians to sell tickets online to upcoming events is new, as several services have developed in the past few years. The most popular services are Eventbrite and Eventbee. Eventbrite combines an event invitation or announcement with ticketing and transaction features. Quinn and Andrei, in their article "A few good online event-registration tools" state "It lets you cap the number of attendees, give simple surveys, and easily post your events to other sites like Eventful or Google Calendar. It also supports discount codes and ticket levels and provides limited ability to tailor a registration form to match your site and your needs" (2011). They charge 2.5% of the ticket fee, with a minimum of $.99 and a maximum of $9.95 per ticket. The service allows you to create a customized event page, offer professional ticketing with discount codes and the ability to dispense bar-coded tickets, to offer easy online registration for events, and to create widgets to embed in your own web site. It uses PayPal and Google Checkout. Eventbee offers a flat fee of $1 per ticket and provides widgets. Both offer mobile applications and bar codes or QR codes for mobile devices. TicketLeap is another online ticketing service, offering features to manage reserve seating and 24/7 customer service. The cost is a $1.97 flat fee for any ticket over $10, and $.97 for any ticket under $10. TicketLeap also offers to rent hardware for on-site ticket sales.

Raising Money Online for Projects

Raising money online or sending an email asking fans to financially support your projects presents some problems. The first concern is that some recipients may think your request is not valid, but is the work of hackers who have accessed your email account. So credibility is important. Another issue is transactional: how will your fans get their money to you? You could offer to collect through PayPal.

Perhaps a better way to resolve the issue is through Kickstarter.com. This site is designed to allow "investors" or benefactors to contribute to a project. The site is dedicated to projects in the creative arts and does not cater to fundraising for charities. Kickstarter must approve your project proposal, including your fundraising goal. The company encourages project creators to offer rewards to donors (tickets to show, free signed copy of CD, name mentioned in album credits, etc.). Donation requests can be set at various financial levels, with corresponding rewards for each.

If the goal is not reached within the time frame specified, no money changes hands, and "would be" donors are not charged. This is done to protect benefactors and project creators alike. If $5,000 is needed to record a new album, and you only raise $2,000, the project can't move forward and those who pledged money initially are not charged and will not expect the outcome promised by the project creator. There are no startup fees for Kickstarter and no financial risk to the project creator. If the project meets it goals and is funded, the creator is charged 5% of the money raised.

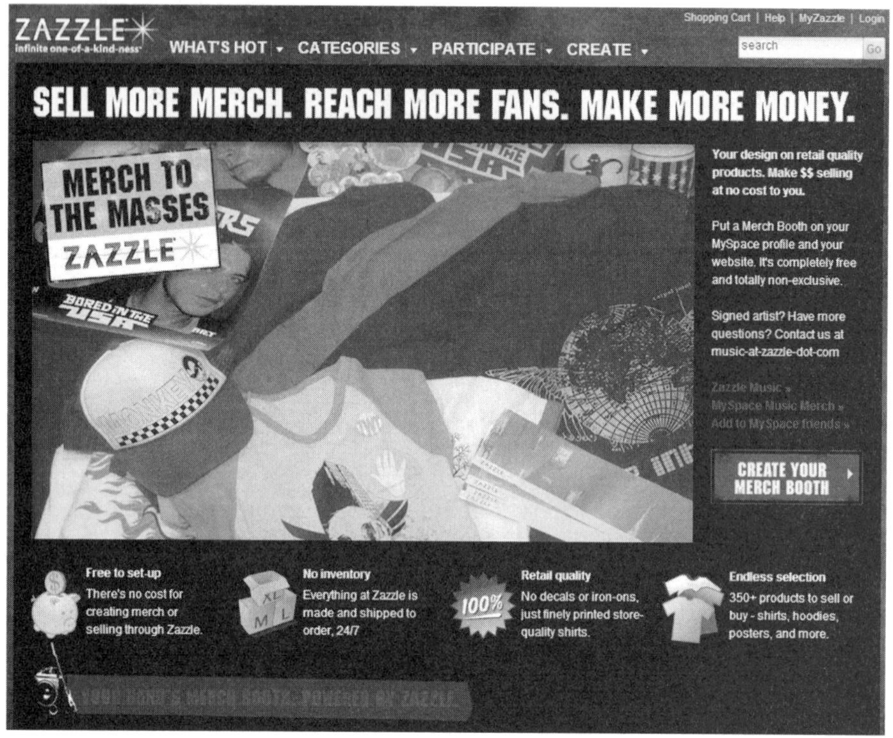

Figure 9.6
Online T-shirt and swag merchandising store

Other Artist-Related Products

In addition to selling recorded music, artists often sell branded T-shirts, hats, coffee mugs, bumper stickers, and other items commonly referred to as *swag* (stuff we all get). Although these items are often sold at live performances, they can also be offered for sale online. For artists who do not want to personally handle the manufacturing and fulfillment of these items, there are companies available online who will handle swag e-commerce. Café Press is ideal for independent artists who don't want to invest in a large inventory of products but who want to provide options for fans. There are no startup costs for a basic account, and the company prints on demand with no minimum order, but the base price is high, leaving little room for markup. Another online site, Zazzle, offers a feature for setting up your own shop and posting it on MySpace and Facebook. Zazzle also prints to order and specializes in band merchandise. For more serious merchandising, PrintMojo charges a much smaller fee per item but does require a minimum order of 25 shirts (www.printmojo.com).

The Last Word On E-commerce

While online stores are starting to become more consumer-friendly, the online shopping process is still in the evolutionary stage of development—unlike traditional retail stores that have been researched for decades. According to Innotrac survey in February and April, 2012:

- Only 7% of merchants surveyed offered one-page checkout as an option.
- 96% of merchants were able to confirm inventory of a purchased item.
- 65% could ship packages to customers in three days or less.
- 38% utilized some sort of branded packaging.
- Only one of the eleven DR marketers surveyed offered free shipping, whereas roughly two-thirds of those in the Q4 2011 study did.
- The average shipping cost for orders was $11.33.

Despite ongoing research and analysis of metrics in e-commerce, conversion rates remain low. A December 2011 article from "Level 343" found:

- The average web site only has a 2% to 3% conversion rate.
- 7 out of every 10 shopping carts never make it to checkout.
- Approximately $18 billion is lost in sales every year.
- Shopping cart abandonment is up to 75%.
- 88% of online consumers have abandoned a shopping cart at least once.

The bottom line is that customers expect the shopping experience to be pleasant, without hassles, confusion, or security concerns. We are now living in the era of convenience—bringing the product to the customer. All efforts at e-commerce should focus on meeting this goal.

Glossary

Fulfillment Processes necessary to receive, service, and track orders sold via direct marketing. "The primary functions of fulfillment systems are (1) to respond quickly and correctly to an order by delivering the item ordered, (2) to maintain customer records, (3) to send invoices and to record payments, (4) to respond to customer inquiries and complaints and resolve problems, and (5) to produce purchase and payment information on an individual customer basis." (Answers.com)

PayPal PayPal is an online service that allows registered users to transfer funds to and from bank accounts and credit cards that they set up for their PayPal accounts. It also allows nonregistered users to make a payment to a registered user via a major credit card.

Product fulfillment The gathering of orders from a sales transaction and the process of completing the order through delivery of the ordered merchandise.

Secure socket layer (SSL) A protocol developed by Netscape for encrypting private documents for secure transmission via the Internet.

Shopping cart A piece of software that acts as an online store's catalog and ordering process, allowing consumers to select merchandise, review what they have selected, make necessary modifications or additions, and purchase the merchandise.

Swag Souvenir or promotional items associated with a product or brand, or short for "stuff we all get."

Bibliography

Answers.com.

Buskirk, E. V. (2009). MySpace/Imeem deal leaves thousands of artists unpaid. *Wired Magazine* online. http://www.wired.com/epicenter/2009/12/myspace-imeem-deal/

Chaparro, Barbara (2002). Top ten mistakes of shopping cart design. http://www.surl.org/usabilitynews/92/shoppingcart.asp

Early Impact. Review and compare shopping carts: how to choose the best ecommerce software. Navigating software reviews and making a good decision, www.earlyimpact.com.

eMarketer.com (2007). Many web retailers miss the basics. www.emarketer.com/Article.aspx?id=1005666 (December 3).

Innotrac (2012, February 28). Innotrac shares findings of direct response operations benchmarking. Innotrac press release. http://www.innotrac.com/news/view/innotrac-shares-findings-of-direct-response-operations-benchmarking

Innotrac (2012, April 17). Innotrac shares findings of February eCommerce benchmarking study. The Business Journals. http://www.prnewswire.com/news-releases/innotrac-shares-findings-of-february-ecommerce-benchmarking-study-147761885.html

Kuenn, Arnie (2008). Choosing the best online shopping cart software. *Small Business Newz*. http://www.smallbusinessnewz.com/expertarticles/2008/06/03/choosing-the-best-online-shopping-cart-software

Level 343 Team (2011, December 12). E-commerce marketing: why are your shopping carts being abandoned? Level343. http://level343.com/article_archive/2011/12/12/why-are-your-shopping-carts-being-abandoned/

Market Platform Dynamics (2006). Payment industry futures: The global adoption and evolution of eWallets. http://www.marketplatforms.com/MPD/Uploads/2_The%20MPD%20Dialogue%20Series_Payment%20Card%20Industry%20Futures_The%20Global%20Adoption%20and%20Evolution%20of%20eWallets.pdf

Naidu, S., & Chaparro, B. (2007). Top ten mistakes of shopping cart design revisited: a survey of 500 top e-commerce web sites. http://psychology.wichita.edu/surl/usabilitynews/92/shoppingcart.html

Nevue, David (2011). *How to Promote Your Music Business Successfully on the Interne*t. The Music Business Academy. www.musicbizacademy.com.

Top Ten Reviews (2012). Shopping Cart reviews. http://shopping-cart-review.toptenreviews.com/

Quinn, L., & Andrei, K. (2011). A few good online event-registration tools. TechSoup.org. http://www.techsoup.org/learningcenter/software/page7510.cfm

Chapter 10
Finding Your Market

Perhaps the most important aspect of marketing involves finding and getting to know your market. On the most basic level, markets can be segmented into three sections: (1) fans and current users, (2) potential fans and users, and (3) those people who are not considered part of the target market. Perhaps this third group includes people who cannot or will not consume your products. For music, that may mean people who do not particularly care for the genre that your artist represents. It may include people who do not consume music, people who are unwilling to pay, and those without access to become consumers. For example, there have been midlevel and star-level recording artists who have enjoyed a fan base that includes very young children, for whom regular concert attendance would be impractical. Perhaps the concert is too late, too loud and rowdy, or is restricted to those of legal drinking age. Artists who are aware that they have a following of small children have been known to add a special matinee performance to the schedule to accommodate families with small children. This is an example of why it is important to learn as much as possible about your market so that you can adjust your marketing strategy to increase the likelihood of success. This chapter focuses on the first two market sections: current fans and potential fans.

MARKET SEGMENTATION

The basic goal of *market segmentation* (subdividing a market) is to determine the target market. Because some markets are so complex and composed of people with different needs and preferences, markets are typically subdivided so that promotional efforts can be customized—tailored to fit the particular submarket or segment. For most products, the total potential market is too diverse or heterogeneous to be treated as a single market. To solve this problem, markets are divided into submarkets called *market segments*. Market segmentation is defined as the process of dividing a large market into smaller segments of consumers who have similar characteristics, behaviors, wants, or needs. The resulting segments are homogenous with respect to characteristics that are most vital to the marketing efforts. That means that members of the segment have enough in common with each other that customized messages can be more effective. This segmentation may be made based on gender, age group, purchase occasion, or benefits sought. Or they may be segmented strictly according to their needs or preferences for particular products. The Internet has revolutionized the way markets are segmented because so much more data are available on consumers' interests and purchase behavior.

To be successful, segmentation must meet these criteria:

1. *Substantiality.* The segments must be large enough to justify the costs of marketing to that particular segment. Because the Internet allows for more effective targeting of markets and the cost of reaching each member is lower, this is starting to change so that smaller and smaller "segments" are worth individual attention.
2. *Measurability.* Marketers must be able to analyze the segment and develop an understanding of their characteristics. The result is that marketing decisions are made based on knowledge gained from analyzing the segment. The Internet has allowed for data mining and *behavioral targeting*, creating market segments based on what Internet users purchase online and what types of sites they visit. For example, based on weather reports and restaurant listings online, a search engine company can determine where someone lives. And based on searches they have conducted on the Web and what keywords they have used, the company can determine what products that person might be interested in receiving information about (Jesdanun, 2007).
3. *Accessibility.* The segment must be reachable through existing channels of communication and distribution. With web marketing, accessibility is no longer the problem it once was. The only limitations are with the technology—some users still have dialup Internet service and care should be taken not to overload them with slow-loading files.
4. *Responsiveness.* The segment must have the potential to respond to the marketing efforts in a positive way, by purchasing the product. Internet users are becoming more comfortable with purchasing products online. Services such as PayPal have increased the willingness of consumers to deal with small, unknown web vendors with confidence and ease. However, there are still some segments of the market whose members are reluctant to order products online for security reasons, because they do not possess a credit card, or because they are unwilling to provide the credit card number to the vendor.

MARKET SEGMENTS

The process of segmenting markets is done in stages. In the first step, segmentation variables are selected and the market is separated along those partitions. The most appropriate variables for segmentation will vary from product to product. The appropriateness of each segmentation factor is determined by its relevance to the situation. For example, age may be a significant factor for products that are age related, such as acne medicine or denture adhesive. For other products, age may not be as significant. After the salient market characteristics are determined and the market is segmented, each segment is then profiled to determine its distinctive demographic and behavioral characteristics. Then the segment is analyzed to determine its potential for sales. The company's target markets are chosen from among the segments determined at this stage.

There is no single correct way to segment markets. Segmentation must be done in a way that maximizes marketing potential. This is done by successfully targeting each market segment with a uniquely tailored plan—one that addresses the particular needs of the segment. Markets are most commonly segmented based on a combination of geographics, demographics, personality or psychographics, and actual purchase behavior. Traditional marketing has relied on these types of segments or combinations of them (using some demographics combined with psychographics), but it has been evolving to include more purchase behavior as technology provides a means for measurement. Behavior segmentation is a more effective way to segment markets, because it is more closely aligned with propensity to consume the product of interest.

WE'VE GOT OUR SEGMENTS, NOW WHAT?

Target marketing involves identifying a market segment to "go after" with some sort of marketing campaign. The decisions about what that campaign should involve are made based on information gathered about the market segment. Table 10.1 lists some of the information you might want to know about your target market.

Once you know these characteristics about your target market, you can decide what to say and where to say it.

MARKETING RESEARCH COMPANIES

There are several marketing research firms that collect and provide data and analyses on consumer groups. Many of these specialize in Internet consumers, and several also cover the recording industry and technology. Generally, the reports are for sale, and it may be worth purchasing one or two before starting out on a new venture for an artist or a label. These companies are also contracted by the various industry associations to conduct specialized research that is then made available to association members.

Forrester Research is one of the major market research firms focused on the Internet and technology; the company conducts research for the recording industry on all aspects of music and the Internet. Forrester also offers custom research and consulting services to its clients. *Edison Media Research* is a leader in political, radio, and music industry research with clients that include major labels and broadcast groups. *Music Forecasting* does custom research projects on artist imaging and positioning.

The *NPD Group* provides marketing research services through a combination of point-of-sales data and information derived from a consumer panel. NPD's research covers music, movies, software, technologies, video games, and many other product groups. *ComScore* offers consulting and research services to clients in the entertainment and technology industries and conducts audience measurements on web site usage. Free information is available through their press releases, blogs and complimentary reports. *Taylor Nelson Sofres* (TNS), a

Table 10.1 Application of marketing research to strategy

	What we know about our target market	Marketing factor based on that knowledge
1	Age, gender, ethnicity, hometown. To know more about the demographic and geographic makeup of our market.	Where can we find our market? What categories do they fall in to?
2	How did you hear about this CD? What influenced you to buy this CD (e.g., radio, TV, in-store, concert, magazine, etc.)?	Which of our marketing efforts are most successful in reaching our customers and which were wasted money?
3	How many other albums do you own by this same artist?	Loyalty: how loyal are you to this artist?
4	What do you read? Watch? Listen to?	How can we best reach you with future advertising and media placement?
5	How much do you buy online? Mail order? Catalog? Toll-free phone numbers?	To what extent should we be using alternative distribution methods?
6	What is your favorite beverage? Sneaker company? Car or truck? Music store?	What companies should we team up with for promotional tie-ins?
7	How much time do you spend online? Shopping? Out to dinner? At the mall?	Where can we find you to target you with our marketing efforts?
8	Do you own a DVD or video game player, computer, cell phone? Do you go to the movies, clubs?	What is competing with music for your entertainment dollar?
9	Do you own or use any of the following consumer electronic devices: DVD, MP3 player, CD-R burner, computer music software, multiuse cell phone, etc.?	What configurations should we use to deliver our recorded music and marketing messages?
10	What are your hobbies and interests?	Where else can we find you when you're not listening to our music or doing any of the above activities?

UK firm, provides both syndicated and custom research of media usage and consumer behavior. Based in France, *IPSOS* is a global group of researchers providing survey-based research on consumer behavior. *BigChampagne* (owned by LiveNation, previously by Clear Channel) tracks online P2P usage and reports, among other things, the most popular songs on P2P networks. *Juniper Research* also provides valuable information on Internet and mobile customer trends.

Conducting Your Own Marketing Research

Never pass up an opportunity to learn more about your market. Online forms and companies such as Survey Monkey have made it easier to ask questions of web site visitors. Survey Monkey has a free version that will allow up to 10 questions and 100 responses. The survey can be embedded into your own site or you can send a link via email. Freedback is another service designed to develop survey forms and manage the data. Subscription rates start at less than $10 per month for one survey of unlimited questions. Google Forms provides a free alternative through Google Docs. Anyone with a Google account who has Google Docs set up can click the create button to create a form. The form can then be customized and imbedded into a web page. Data collected from submissions is available in spreadsheet form in your Google Docs account.

On a more traditional basis, record labels have been known to use response cards inserted into jewel boxes. Entering and analyzing the data is time consuming and should only be undertaken if there is a real need for more information about the market.

TRACKING CONSUMER BEHAVIOR ON THE WEB

Before the emergence of the Internet, marketing researchers used a variety of techniques to learn more about consumer behavior. Many of these studies were not comprehensive, meaning that shopping behavior may be measured on one group of consumers while advertising exposure is measured on another. It was difficult to conduct a comprehensive measurement program without being intrusive. One company attempted to measure media consumption and consumer purchases in the same household. Generally, participants had to subject themselves to extensive monitoring and extraordinary procedures to collect the data. In some ways, that made them unlike the general marketplace to which the results would be generalized. So the measuring had a tendency to get in the way of the natural consumer behavior. With the Internet, data collection is more transparent—web users are not really aware that their movements through the Web are being recorded and analyzed.

Cookies

The Internet has made it easy to track what consumers do, where they go, and what interests they have. One way of keeping track of that data is through the use of *cookies*. Webopedia defines cookies as a "message given to a Web browser by a Web server. The browser stores the message in a text file. The message is then sent back to the server each time the browser requests a page from the server." David Whalen made this analogy on www.cookiecentral.com:

> You drop something off [at the dry cleaners], and get a ticket. When you return with the ticket, you get your clothes back. If you don't have the ticket, then the [dry cleaner] man doesn't know which clothes are yours. In fact, he won't be able to tell whether you are there to pick up clothes,

> or a brand new customer. As such, the ticket is critical to maintaining state between you and the laundry man.

Webopedia goes on to explain:

> The main purpose of cookies is to identify users and possibly prepare customized Web pages for them. When you enter a Web site using cookies, you may be asked to fill out a form providing such information as your name and interests. This information is packaged into a cookie and sent to your web browser which stores it for later use.

So when you return to that same web site, your browser will send the cookie to the web server letting it know who you are—it's your ID card or your frequent shopper card. Then, the server can use this information to load up personalized web pages that may include content that interests you, based on information the site collected the last time you visited. So, for example, instead of seeing just a generic welcome page you might see a welcome page with your name and features on it. The use of cookies is frowned upon by privacy advocates but hailed by marketers and webmasters alike in its ability to offer customized information to visitors.

WEB ANALYTICS RESOURCES

There are several web sites that compile information on consumer profiles of visitors to well-known sites. Some of this information is available publicly.

- *Quantcast* is a web analytics service that compiles web traffic information with consumer surveys to identify market characteristics for web sites. Quantcast identifies demographic characteristics of visitors and lifestyle information such as their "audience also likes" section, which is based on other web sites these visitors frequent. One suggestion is to look up well-known artists' sites and use their demographic data to extrapolate information on your target market. This web site allows anyone to view audience reports on hundreds of thousands of web sites.
- *Alexa* is a web analytics and information company that tracks web traffic and provides public data. By looking up popular sites, one can find data on web traffic stats, search analytics, audience demographics, and clickstream–information on which web sites visitors were on prior to visiting that site (upstream) and where they visit upon leaving (downstream). By registering for an account, you can begin monitoring this information for your own site. To access these pages, simply type the URL of any site into the Alexa Search box.
- *Compete.com* offers traffic statistics for major web sites and provides information on similar sites, "also visited" sites, keyword density and related topics. They help you find "the most relevant similar sites based on related content, web site structure, link analysis algorithms, detailed user surfing behaviors and a large community of user rankings" (www.compete.com).

COLLABORATIVE FILTERS

Collaborative filtering software examines a user's past preferences and compares them with other users who have similar interests. When that user's interests are found to match another group of users, the system starts making suggestions of other things this group likes. In the article "Collaborative filtering," author Francis Heylighen (2001) stated,

> The main idea is to automate the process of "word-of-mouth" by which people recommend products or services to one another. If you need to choose between a variety of options with which you do not have any experience, you will often rely on the opinions of others who do have such experience.

Suppose you normally never listened to jazz, but you liked artists A, B, and C a lot. If numerous other people who don't normally listen to jazz also like A, B, and C, but also like band D, the system might suggest D to you and be relatively confident that you'll like it. Amazon.com uses the technology to recommend other products with its "people who bought this product also purchased these other products" feature. Unlike previous music sorting procedures that required judgments based on personal tastes and opinions, web-based collaborative filtering is usually a process developed through the input of consumers. So we now have consumers dictating which songs belong in which category and should sit beside each other on playlists. Web 2.0 has increased the user-generated process by allowing users to create and share their playlists with other members of the social networking world. Through services such as iTunes' Genius and Pandora, data is gathered when members group songs together into playlists. That cumulative information combined with an individual user's profile and activities, helps the system make intelligent recommendations of additional music each user is likely to enjoy.

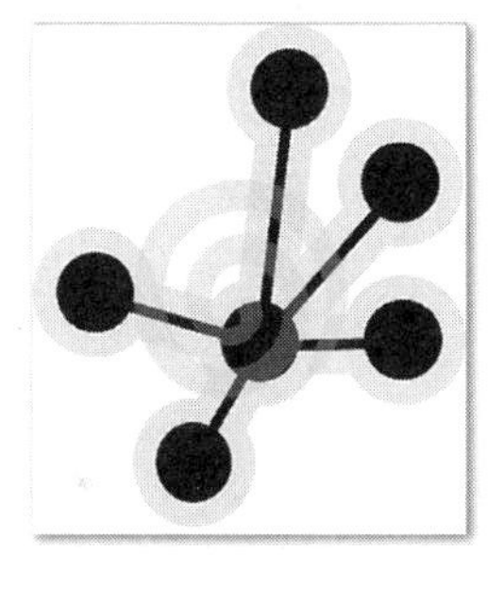

Collaborative filtering can help artists and their marketing analysts determine where an artist fits in relative to other artists, but only if the artist has a large enough following to be included on one or more collaborative filtering web sites (see the section on determining the target market for your artist).

GENERAL INFORMATION ON FINDING YOUR MARKET ONLINE

Author Frances Vincent, in her 2007 book *MySpace for Musicians*, brought up some good points for identifying your target market. Among those, Vincent suggested asking your fans directly; researching contemporaries and competitors; becoming a student of pop culture by listening, watching, reading, and going places; and networking.

One marketing article by Donna Gunter (2006) put it in these terms: "Are you fishing where the fish are?" Gunter went on to describe several research tools available online to help research the target market. Here is an expanded list of those sources:

1. *Professional associations.* They generally have information related to the market based on commissioned research studies. For the music business, that would include, but is not limited to, the National Academy of Recording Arts and Sciences (NARAS), the National Association of Recording Merchandisers (NARM), the American Association of Independent Music (A2IM), various genre-specific trade organizations, and the International Federation of Phonographic Industries (IFPI).
2. *Professional conferences.* Many of the trade associations hold annual conferences with panels and presentation on the latest research in consumer trends. Several notable conferences worth attending are South by Southwest, NARM, and MIDEM.
3. *Trade and consumer publications.* Read up on the market and the industry by subscribing to the top publications. For the music business that would include, but is not limited to, *Billboard, CMJ Network, Pollstar, Hollywood Reporter, Rolling Stone* and *Variety*. Check out Billboard.biz. There are also genre-specific magazines such as *The Source, Vibe, Weekly, Downbeat,* and *Alternative Press*.
4. *Online discussion forums/groups.* Go for the industry-oriented forums rather than fan-oriented sites. Entering terms such as "music business forums" into a search engine should return a plethora of sites. For example, www.starpolish.com and www.getsigned.com are good industry reference sites. ArtistHouseMusic.org is a good resource for industry information and advice. Also try AllMusic.com and music.AOL.com.
5. *Online networking.* Find out where other industry folks hang out online and get to know them. Or visit their web site and look for a "contact us" link. Drop them an email to encourage online discussion.

Table 10.2 Music industry trade associations

Music industry trade associations
National Association of Recording Merchandisers (NARM): www.narm.com
National Academy of Recording Arts and Sciences (NARAS): wwww.grammy.org/recording-academy
American Association of Independent Music (A2IM): www.a2im.org
International Federation of Phonographic Industries (IFPI): www.ifpi.org
Songwriters Guild of America (SGA): www.songwritersguild.com
American Federation of Musicians: www.afm.org
Audio Engineering Society: www.aes.org
National Association of Recording Industry Professionals (NARIP): www.narip.com
National Association of Broadcasters: www.nab.org
International Alliance for Women in Music: www.iawm.org

Source: http://musicnewsdaily.com/org.html.

Table 10.3 Examples of industry conferences
Billboard Magazine sponsored events: The music industry's most powerful business-to-business events, including Digital Music Live, Mobile Entertainment Live, Billboard Music and Money Symposium, Billboard Touring Conference and Awards, R&B Hip Hop Conference and Awards, Latin Music Conference and Awards, and more. www.billboardevents.com/billboardevents/index.jsp
South by Southwest SXSW Music & Media Conference: Austin, Texas Showcases hundreds of musical acts from around the globe on more than 50 stages in downtown Austin. By day, conference registrants do business in the SXSW Trade Show in the Austin Convention Center and partake of a full agenda of informative, provocative panel discussions featuring hundreds of speakers of international stature. www.sxsw.com
MIDEM: Palais de Festivals, Cannes, France Nearly 10,000 music and technology professionals from more than 90 different countries, including delegates from the recording, publishing, live, digital, mobile, and branding sectors gather to do deals, network, learn, and check out new talent. www.midem.com
Americana Music Association Convention: Held in Nashville each fall, this conference brings together independent musicians, record labels, and other industry executives for panels, workshops and showcases. http://americanamusic.org/about-conference
Millennium Music Conference: For over a decade, the Millennium Music Conference has educated emerging talent, showcased new music and entertained the community. Music industry professionals attend as panelists, speakers, mentors, exhibitors, and talent scouts network, do business, and share their experience with musicians, registrants, and attendees. The community joins in the celebration of emerging talent, independent artists, and new music. http://musicconference.net/mmc12
Audio Engineering Society Conventions are held annually in the United States and Europe. Each convention has valuable educational opportunities, including a full program of technical papers, seminars, and workshops covering current research and new concepts and applications. An integral part of each convention is a comprehensive exhibit of professional equipment. www.aes.org/events
National Association of Recording Merchandisers: The NARM Annual Convention & Marketplace is considered the industry's premier venue to learn, discuss, and meet. Industry players come together to make business deals, hear live music, get the latest research, see the most up-to-date technology, showcase new product lines, hammer out solutions to industry issues, and network to find new business partners. http://www.narm.com/events/2011-event-recap/

6. *Blogs.* Leaving comments for other bloggers in the industry can help raise your profile in addition to fostering intellectual conversation and learning from the bloggers. Don't hesitate to leave questions in your blog comments—they may be answered by other readers as well as the blogger. Check out AOLMusicNewsBlog.com and Billboard.blogs.com.
7. *E-zines.* The use of e-zines to promote an artist is covered in the chapter on Internet promotion, but these e-zines can also be a good place to gain an understanding of the target market. If the publication is professional and successful, the publishers probably already know a great deal about their readership. Although they may not share that information willingly or without cost, the results of their knowledge and understanding should be evident in the subject matter of the e-zine. (For example, if the e-zine has an article on the quality of MP3 files compared to other formats, then perhaps the readership is technology oriented.) Gunter noted that reading these e-zines will "give you a great overview of typical issues and problems faced by your target market." Check out www.ezine-dir.com/Music.
8. *The radio industry.* Radio has a long history of studying music listening trends. Radio stations routinely conduct marketing research. Arbitron.com has a section on ratings that lists the top radio markets and some format and demographic details on each. Arbitron also supplies the annual Radio Today report as well as others on their "Free Studies and Reports" page. http://www.arbitron.com/home/studies.htm.

MAKE IT EASIER FOR YOUR MARKET TO FIND YOU

In the early 1990s, Internet researchers alluded to the coming age of broad*catch*, moving away from the days of broad*cast*, where multitudes of people consumed a common message or programming simultaneously. Broad*catch* on the other hand, refers to the concept of media consumers seeking out programming, entertainment, or information. Under this scenario, the role of marketing and advertising shifts from that of "carnival barker" to that of information provider—being there with the product information when the consumer is in the market and looking. With that comes the shifting role of marketing to making it easier for the market to find you rather than you finding your market. Part of that effort is outlined in the chapter on optimizing your web site—the idea of being at the top of search results for the product category. For this effort to be successful, it is necessary to think like the consumers and gain an understanding of the process *they* go through to find *you*. What search terms are they likely to use? Do they always correctly spell the words? What if your band is an alternative spelling of a commonly known word, such as Boyz instead of Boys. Will your potential fans know to spell Boyz with a "z"? If not, all keyword-based advertising and search engine optimization should include spellings and keywords most likely to be used *by members of the target market*. That is when marketers regret it if they have to use a domain name with a hyphen, or an extension other than dot-com, because most potential customers may not remember the subtle differences.

THINK VIRTUALLY, ACT LOCALLY

Be sure to find and post notices at every possible local site where your target market is known to hang out—just like sniping (putting up posters) in the nonvirtual world. If there are local web sites that post upcoming events and concerts, make sure to find these sites for each geographic market on the tour schedule and post a notice of the event, complete with viral components directing the visitor to the artist's web site. Potential fans may be perusing these sites looking for things to do locally and may have no prior awareness of your artist. Again, this requires getting into the mind of the potential consumer and imagining the process they may go through in discovering your artist.

DETERMINING THE TARGET MARKET FOR YOUR ARTIST

Established artists with a consumer base have the advantage when it comes to identifying the artist's target market. Generally these fans have been through some type of marketing research process as they were buying tickets or music, attending concerts, or just using the artist's web site. All of those fan interactions provide the opportunity for marketers to gather information about the fans to improve the marketing efforts. Established artists are more likely to appear on collaborative filtering web sites such as imeem, where a bit of research can determine the extent of overlap in fan base with other artists. Go through all the music-oriented sites that use collaborative filtering to recommend music and search for mentions of your artist. When they appear, go to the profile pages for these fans and get to know them. See what other music they are listening to, and make some inferences from that feedback.

It's generally more difficult to identify the target market for a new artist. Without a sales track record, the marketing department of a record label must make some assumptions about a new artist. Marketers would be wise to examine market characteristics that relate to other artists who are perceived as similar or who would appeal to a similar market. They may look at consumer information for more established acts and decide that this is the market they should go for. And at times, the product itself, in this case the artist, may undergo some modifications to successfully appeal to the "target market." It is also possible to survey visitors online or at live shows regardless of the size of an artist's fan base.

This is also an area where online marketing and traditional marketing need to overlap. Data gathered in the real world can be applied to the virtual world and vice versa. At clubs and venues where the artist appears regularly, analyze the other musical acts that perform there to determine if there is any audience overlap. (It is the same principle as collaborative filtering, only performed in a more casual way.)

SPEAK TO YOUR MARKET

One of the most important aspects of successful marketing is in knowing *how* to communicate with your market. It is not enough to know where they are and how to find them, but you must also understand how to *reach* them with your marketing message. Certain market segments, particularly teenagers and young adults, will reject marketing messages from companies they perceive as not understanding them. So the message itself becomes important, and not just the words, but the look and feel of the web site, the logo, the images, and so on. These design elements and messages must be fine-tuned for the market, in the vernacular that the target market is accustomed to hearing.

In his book *How to Promote Your Music Successfully on the Internet,* author and musician David Nevue introduced a concept he calls "targeting by site." The idea is that a certain segment of *your market* may be looking for or respond to something in particular or may respond to something of a more general nature. For example, if your artist is a blues slide guitar player from the Panhandle of Florida, perhaps you should design a web site around North Florida Blues in an attempt to attract anyone who is interested in that genre in that area of the country.

Landing Pages

Nevue even suggested creating particular topic-oriented web pages and submitting them separately to search engines. For record labels that represent more than one genre of music, the label would do well to feature more than one web site. The label would also want to create specific pages that feature each of the artists who are currently the top priority in marketing. These pages are referred to as *landing pages*. Landing pages are specifically designed pages that are intended to be the page that the web visitor "lands on" when looking for something in particular and is being directed to your site. This page would be more finely tuned to appeal to the segment of your market that is looking for something specifically and has found you through the use of a particular set of keywords that indicate their interest. In an article titled "Creating effective landing pages" on TamingtheBeast.net, author Michael Bloch described landing pages this way:

> In marketing terms, it's a specialized page that visitors are directed to once they've clicked on a link, usually from an outside source such as a Pay Per Click ad. The page is usually tightly focused on a particular product or service with the aim of getting the visitor to buy or take some form of action rapidly that will ultimately lead to a sale.
>
> **www.tamingthebeast.net/articles5/landing-pages.htm**

Bloch explained that too many web site marketers labor under the assumption that the product purchase scenario goes like this: the customer arrives at the company's home page, the customer then selects an option from a menu or from an offer on the page, the customer clicks to the page with the product he or she is interested in, and then the customer buys. But it often does not work that way.

The landing page should be the first page that customers are directed to when they have found your product and your web site through search engine keywords, through online advertising, or in response to your email touting the particular product. At that point, the customers know what they want, the seller knows what the customers want, and the seller should deliver the customers to the product as quickly as possible. One analogy is that if a customer wanted to buy a pair of tennis shoes and the retailer had the ability to magically transport the customer to the correct shoe rack in the correct store, why would the retailer drop the customer off at the front entrance of the mall?

Use different landing pages to test different offers and creative treatments. In his article "How to write an effective landing page," Ivan Levison suggested, "You can test variables by sending prospects to unique landing pages. Just measure the click through rate and you'll find out fast what works best." The Web offers companies a unique opportunity to target customers by creating alter ego web pages and web sites that appeal to each segment of the market. By using landing pages, a particular customer may never be aware of the variety of other markets this company serves.

CONCLUSIONS

The first step in developing a marketing plan is identifying and learning about your target market. The more you learn about that market, the more effective the marketing plan can be. It is important to reach out to customers with a communication strategy that speaks to them and makes them feel like the company or brand understands them. On the Internet, much of that involves being easy to find. To do that, you must get inside the consumer's head and go through the scenario the consumer will enact when shopping for and deciding on your product. How do customers go about finding you in the vastness of the Internet? What techniques do they use? How does your fan base find out about new music and new artists? How loyal is your fan base? Is it necessary to constantly find new customers, or do previous customers keep coming back for more?

Then, once the customers have found you, it is important to give them the right information to help them make that purchase decision. What motivates them to buy? Are they driven by costs, quality, uniqueness, a bond with the product? Major corporations spend millions of dollars on research to better understand their market and can modify not only the marketing messages but sometimes the product itself, and certainly the web site.

Glossary

Behavioral targeting Creating market segments based on what Internet users purchase online and what types of sites they visit.

Broadcatch The concept of media consumers seeking out and controlling their consumption of programming, entertainment, or information rather than being passive consumers.

Clickstream The sequence of links that are clicked on while browsing a web site or series of web sites.

Collaborative filtering Examines a user's past preferences and compares them with other users who have similar interests. When that user's interests are found to match another group of users, the system starts making suggestions of other things this person may like.

Cookies Parcels of text sent by a server to a web browser and then sent back unchanged by the browser each time it accesses that server. The main purpose of cookies is to identify users and possibly prepare customized web pages for them.

Landing pages Also known as jump pages. It is the particular page a customer is directed to by a link based on what that customer is looking for. The page is usually tightly focused on a particular product or service with the aim of getting the visitor to buy.

Market segmentation Breaking a total market down into groups of customers or potential customers who have something significant in common in terms of their needs and wants or characteristics.

Target market A segment of a specific market that your company has identified as your customers or clients.

Bibliography

Bloch, Michael (2011). Creating effective landing pages. www.tamingthebeast.net/articles5/landing-pages.htm

Gunter, Donna (2006). How to find your target market online. www.amazines.com/article_detail.cfm/124420?articleid=124420

Heylighen, F. (2001, January 31). Collaborative filtering. *Principia Cybernetica*. http://pespmc1.vub.ac.be/COLLFILT.html

Jesdanun, Anick (2007, December 1). Ad targeting grows as sites amass data on web surfing habits. The *Tennessean* daily newspaper.

Levison, Ivan (2012). How to write an effective landing page. http://www.levison.com/august_2012_cc.html

Los Angeles Times (2011, December 14). Live nation acquires BigChampagne. http://latimesblogs.latimes.com/entertainmentnewsbuzz/2011/12/live-nation-acquires-big-champagne.html

Vincent, Frances (2007). *MySpace for Musicians.* Thomson Course Technology, Boston, MA.

Whalen, David (2002, June 8). FAQs about cookies. www.cookiecentral.com.

www.webopedia.com

Chapter 11

Successful Promotion on the Web

INTRODUCTION

A musician's career is like any other business. Creating the perfect product does not guarantee success. In fact, marketing expenses often can outweigh production expenses. From a business standpoint, it wouldn't make any sense to open a restaurant—to put all that effort into designing and building a fine establishment—without a strategy to promote the restaurant to potential customers. But that is what people do when they spend their entire Internet budget and effort on building a perfect web site with no plans to promote it.

In the restaurant analogy, management must make certain obvious marketing efforts to become successful, including advertising, getting reviews, creating word of mouth, and offering coupons, to name a few. There is a common expression: "No one wants to eat at a restaurant with no cars parked out front." The presence or absence of a full parking lot indicates a restaurant's popularity and level of success. Online, search engine ranking achieves the same purpose: a high search engine ranking denotes that your web site is one of the more popular sites in its particular keyword-driven category, and as a result, more Internet users are likely to check it out. But how does one achieve this high ranking? Restaurants have been known to have employees park their cars out front on slow nights to give the appearance of a popular spot. Web sites use search engine optimization (see the section on search engines) to achieve this same result.

The restaurant would also need to advertise to create awareness, establish the brand, and bring customers in the door. Likewise, a web site developer will need to buy banner ads and use Google's or Yahoo!'s advertising programs. The restaurant would actively seek out restaurant reviewers, travel guides, and other opinion leaders in hopes of spreading the word through media outlets. Ditto for the artist's web site, where the Internet marketer will attempt to get bloggers to write about the artist and the site and will try to get online magazines to publish features about the artist.

The successful restaurant will employ techniques to reward customer loyalty and repeat visits, including frequent customer promotions (buy 10, get the next one free), drawings for free meals, and other promotional incentives for repeat customers. The successful web site should offer similar incentives to encourage visitors to return to the site periodically. And certainly, word of mouth is one of the best forms of promotion and one that successful marketers foster by using subtle techniques to encourage satisfied customers to spread the word.

This chapter and the next will examine the online promotional tools necessary to turn a good web site into a popular, successful, and profitable web site.

BASICS FOR INTERNET PROMOTION

The Internet has opened new opportunities for musicians and small businesses to easily and economically promote their products on a global scale. Web marketing should be a part of every recording artist's marketing plan, but it should not be the only aspect of the plan. Even though the Internet has become a great tool for selling music, the traditional methods of live performance, radio airplay, advertising, and publicity are still significant aspects of marketing and should not be neglected.

Rule 1: Don't make the Internet your entire marketing strategy. Internet marketing should not be a substitute for traditional promotion. The two strategies should work together, creating synergy.

In the music business, it is necessary to build brand awareness. Whether it's with the artist, a record label, or a studio, you need to create a sense of familiarity in the consumer's mind. This can be done through many tactics available both on and off the Internet, which should be designed to lure customers to the artist's web site or drive them into the stores to buy the artist's records. Those efforts should always be coordinated. Although the Web is a great tool to reach a large number of people with minimum expense, it is so vast that unless potential customers are looking for a particular artist, they are unlikely to stumble across an artist's site by chance. Also, many customers are more responsive to traditional marketing methods such as radio airplay and retail store displays. An artist's web site is a great place for customers to learn more about the artist's products—live shows and recordings—but it is not necessarily the best way to introduce new customers to the artist's products.

Rule 2: Build a good web site, but don't expect customers to automatically find it on their own.

A solid marketing plan incorporates the company web site into every aspect of marketing and promotion. And while the web site will be the cornerstone of Internet marketing efforts, it is just the start. Building a good web site is crucial to marketing success, but the old adage "If you build it, they will come" does not apply to the Internet.

Every marketing angle should tie in to the web presence. Internet marketing is more effective when it is conducted in conjunction with other aspects of the plan. Every piece of promotional material should contain the web site address to direct the potential customer to the web site. Any posters, flyers, postcards, or press releases should have the artist's web site address prominently displayed. And it goes without saying that CD tray cards should also contain the address. One band had hand stamps printed up at the local office supply store. The stamps had the band's web address on them, and the band requested that the bouncers use them to stamp customers' hands as they entered the venue where the act was performing.

Rule 3: Incorporate the artist's web site address into everything you do online and offline.

Now that we have established the relative function of Internet promotion, we will next examine the various structures for promoting one's brand and products on the Web.

SEARCH ENGINE SUBMISSION

Search engines are online directories of web sites and web pages that web visitors use to find a topic of interest or a specific site. They help organize the Web so that visitors can make some sense out of the vastness of the Internet (Dawes and Sweeney, 2000). One of the goals when setting up your web site is to ensure its prominence in search engine results so that potential customers can find it.

It is gratifying to type your artist's name into a search engine and have their official web site listed as the first result. That is likely to happen for well-established artists, but it is not as common with lesser-known acts, especially if the band or artist's name is a common one or is associated with another product. Even if the site does come up as one of the search results, the web visitor will need to have knowledge of the artist and the intent to seek out that particular artist online. What about potential fans who are not yet familiar with the artist? How do you reach them? And how do you improve your standing in search engine results when the person doing the search is using more vague terms, such as "blues music" or "female blues vocalists in Atlanta."

As of press time for this book, the most popular search engines are those shown in Figure 11.1.

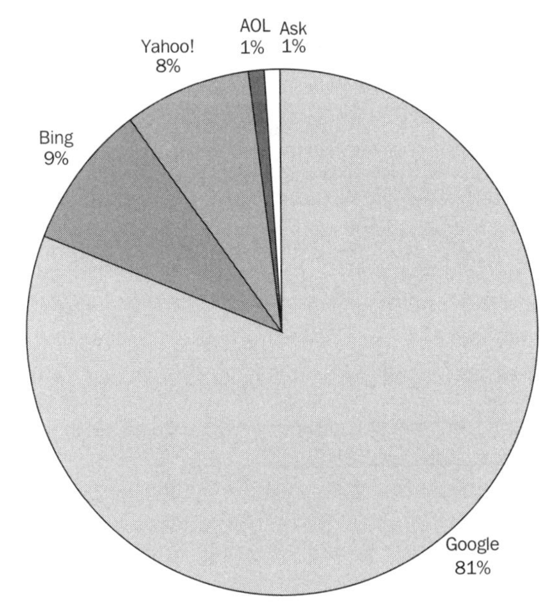

Figure 11.1
Search engine market share (source: StatOwl.com)

From 2007 to 2010, Google solidified their lead in market share from 56% to nearly 86% in 2010. Yahoo! had almost one-quarter of the market in 2007, but had slipped to just 6.4% in 2011. Meanwhile, Microsoft rebranded its search engine as Bing and began to turn around its market share which shrank from over 11% in 2007 to 3.3% in 2010. In December 2011, Bing moved ahead of Yahoo! into second place behind Google. By 2011, it was up to 13%, with Google holding 67%.

The various popular search engines each have their own way of categorizing and prioritizing their listings and search results. It is wise to understand how each finds and categorizes new web pages and then proceed with establishing your presence on each. Even though search engines are likely to find your site and list it, you should still submit a request to be listed and be prepared to list the relevant keywords and subject categories that apply to your site. Some of these search engines use spiders, software programs that scour the Internet and analyze web pages. Others use humans, or a combination of both.

There are three ways to get a web site listed in a search engine: (1) submit the site directly to the search engine via an URL submission form, (2) wait for the search engine to find the site, (3) pay the search engine to index the site. Sometimes the submission page is buried deep in the site; for Google, you must click the link "about Google" and then select "submit your content to Google" from a menu. Yahoo! has a link at the bottom of the page that says, "Suggest a site." More information is available on search engine optimization in Chapter 7.

To submit a site for Google, they provide a URL submission page via their webmaster tools at http://support.google.com/webmasters/?hl=en.

Google posts the following instructions at http://support.google.com/webmasters/bin/answer.py?hl=en&answer=1352276.

- On the Webmaster Tools Home page, click the site you want.
- On the Dashboard, under Diagnostics, click Fetch as Googlebot.
- In the text box, type the path to the page you want to check.
- In the dropdown list, select Web. (You can select another type of page, but currently we only accept submissions for our Web Search index.)
- Click Fetch. Google will fetch the URL you requested. It may take up to 10 or 15 minutes for the Fetch status to be updated.
- Once you see a Fetch status of "Successful," click Submit to Index, and then click one of the following:

To submit the individual URL to Google's index, select URL and click Submit. You can submit up to 500 URLs a week in this way.

To submit the URL and all pages linked from it, click URL and all linked pages. You can submit up to 10 of these requests a month.

Bing posts information on their site for getting listed. The easiest way to submit the URL to Bing is to use the anonymous listing tool at https://ssl.bing.com/webmaster/SubmitSitePage.aspx. The best way to have your web site listed with Bing is to sign up for a Bing Webmaster Tools account.

> You will receive in-depth and valuable data about how your website is seen by Bing. You should take advantage of the Webmaster Tools "Submit URL" feature for faster URL submission. Bing Webmaster Tools Accounts are free, and require a simple verification process be completed.
>
> **Bing web site**

The best way to submit to Yahoo! is through the directory listing submission page at https://ecom.yahoo.com/dir/submit/intro/. Ask.com does not offer a way to directly submit your URL for free. You can indirectly increase the chances of being listed by signing up at DMOZ (the Open Directory Project) at http://dmoz.org/. DMOZ is a volunteer-built guide to the Web (Search Engine Watch, 2007). ODP also reports to AOL's search engine. It is suggested that you first search the site dmoz.org to make sure the site is not already listed.

Numerous commercial services offer search engine optimization. Webopedia defines search engine optimization as

> the process of increasing the amount of visitors to a web site by ranking high in the search results of a search engine. The higher a web site ranks in the results of a search, the greater the chance that that site will be visited by a user.
>
> **www.webopedia.com**

Most search engines rank their results based on several factors. By paying attention to these factors, you can improve the chances of getting a top listing:

1. *Keywords.* Choose keywords that are targeted and represent the web site's topic accurately. These keywords should appear in the page text in addition to being listed in the meta tag keyword section. Many search engines rely on page text to confirm that the site's topic is accurate and corresponds to the keywords. If certain words have alternative spellings or are commonly misspelled, the variations should be listed in the meta tag section because these misspellings will not be visible to the web visitor. Keywords should also appear high on the page, within the first 100 characters. When ranking results, search engines give higher priority to keywords that appear high on the page, in the title, in the description, in the URL, in headings, in the ALT attribute of the IMG tag for graphics, and in the link text for inbound links (Monash, 2004). See Chapter 6 for more information on tags.
2. *Content.* Make sure the content of the site is represented accurately. Search engines monitor traffic to your site. If web visitors quickly leave your site because it is not what they expected, search engines will know that the subject matter is not what the visitor is looking for based on the search terms they entered (Walker, 2006). The search engine may then lower the ranking if many search engine users fail to click through and explore your site.
3. *Links.* The number and quality of inbound links helps the search engine spiders determine the validity and popularity of your site. Link popularity is one factor the search engines gauge when determining ranking. If many

people link to your site, then it must be one of the more popular sites. Danny Sullivan on SearchEngineWatch.com has suggested the following:

> Go to the major search engines. Search for your target keywords. Look at the pages that appear in the top results. Now visit those pages and ask the site owners if they will link to you. Not everyone will, especially sites that are extremely competitive with yours. However, there will be noncompetitive sites that will link to you—especially if you offer to link back.
>
> **Sullivan, 2007a**

One alternative to paying a company to list your site is to use http://freewebsubmission.com/. Free Web Submission also has tools such as link popularity checker, page rank checker, and keyword popularity checker.

E-ZINES AND BLOGS

Electronic or online magazines offer a good way to introduce your target market to your act. MarketingTerms.com defines e-zines as an electronic magazine, whether posted via a web site or sent as an email newsletter. (The term is loosely used to describe email newsletters, albeit without the magazine formatting; see section on using email for promotions). Short for electronic magazine or fanzine, some are electronic versions of existing print magazines, whereas others exist only in the digital format. The web-posted versions usually contain a stylized mixture of content including photos, articles, ads, links, and headlines, formatted much like a print equivalent. Most e-zines are advertiser-supported, but a few charge a subscription.

Many established music e-zines are genre specific or have particular subject areas dedicated to genres. They may feature music news, concert and album reviews, interviews, blogs, photos, tour information, and release dates. As a result, their readers are predisposed to be receptive to new and unfamiliar artists and their music, provided that the artist is within the genre that the e-zine represents. A study on readers of the Americana music magazine *No Depression* found that 90% of their readers found out about new music from an article, one published either in a print magazine or in an online version.

A blog is defined as a Web site that contains an online personal journal with reflections, comments, and often hyperlinks provided by the writer. Blogs may or may not be appropriate for adding to your mailing list, so ask permission first. But bloggers who write about music may want to receive periodic updates on news items and new recordings. The more followers a blogger has, the more potential there is for a feature about your music to translate into sales.

WHAT TO SEND

E-zines and bloggers are mostly interested in feature articles and press releases pertaining to some newsworthy item (such as an album release or a tour schedule announcement). A feature story should include biographical information as well as the newsworthy information—in other words, it is part bio and part press release. It helps if you write the story as if it were going to appear unedited in the online publication, in the inverted pyramid style. Always include a publicity photo or two along with the article for submission. And always include plenty of links to specific pages on your web site that pertain to the featured topics (tour page for touring news, product page for record release news, etc.).

Do not send out a press release if you have nothing that is considered newsworthy. A redesign of your web site might be interesting to fans who have signed up for emails, but to the casual reader who was probably unfamiliar with your old web site, it's not news and doesn't warrant a press release. In his book *How to Promote Your Music Successfully on the Internet*, David Nevue emphasized that before you send out a press release announcing your new CD, ask yourself "who cares?" and let the answer to that decision determine if, and to whom, you should distribute that information.

It might be best to first make an electronic press kit (EPK) consisting of music, photos, videos, bio, discography, tour dates, contact information, and other elements found in a standard press kit. Sonicbids is a web site that creates and stores EPKs for musicians. The site states, "Your EPK™ does not replace your web site. It replaces the costly physical press kits you put together to get booked at festivals and venues or to secure other performance and licensing gigs" (www.sonicbids.com). But do not send an unsolicited EPK. Get permission before sending attachments to e-zines and bloggers.

Blogs that write about music can be found by searching the Web. Chris Bracco (2011), in his publication "How to really get your music on blogs," states that "for artists, this has become an extremely effective medium to garner positive and sometimes career-changing exposure." He warns that getting exposure in popular blogs is a "meticulous and time-consuming process." Bracco recommends starting by writing down keyword descriptions of the band's music, lifestyle, and fans. The next step is doing the research to find suitable blogs and to get involved in the music communication networks. The blogs to target are those that fit the same niches as your music, fans, and lifestyle.

Bracco recommends considering non-music blogs also, as long as the market is the same. He suggests crafting a killer pitch letter that talks about them first, and you last. And he recommends providing links to download your music from Dropbox or SoundCloud, and to end with a soft pitch to feature the music. As with any other publicity campaign, it helps to develop a spreadsheet list of targeted bloggers and to use that list for contact information and to keep up with where you are in each stage of the process with each blogger.

In an ASCAP blog, Heather Browne (2010) suggests that before you submit anything to a blogger, you follow these suggestions:

- Familiarize yourself first with the blogger's work. Get to know the blogger's taste in music to see if yours is a good match. The last thing you want is a bad review.
- Cultivate relationships with your favorite bloggers. A few bloggers who are avid fans are better than many bloggers who are only peripherally interested.

Blog Name	Author's Name	pitch letter	response	music sent	response	will feature	has featured	follow up	Notes
The Big Fictitious Music Blog	Jane Smith	yes	yes	yes	yes	yes	yes	yes	want to come to show
The Fake Name Music Blog	John Pearson	yes	yes	yes	no				
Music Blog #3	Maggie Jones	yes	no	no	no				send follow up in 2 weeks
Made Up Blog Name	Bill Johnson	yes	yes	yes	yes	maybe			Send more tracks
Just Another Blog	Angie Stewart	yes	yes	yes	yes	yes			March issue

Figure 11.2
Example spreadsheet to track blog pitches

- Tell the story behind your music. Browne states "Music blogs' personal relationship with the music is what draws most of the readers to the best-loved sites."
- Most of the time, a digital submission is preferred, but ask first before sending attachments.
- Be clear in your email subject line. Also, get to the point in your email without being long winded.
- Don't get impatient if you don't get an immediate response to your inquiry. Bloggers may be busy answering a stack of emails.
- When you find bloggers who write about you, follow up. Keep them in the loop for future exposure.
- If you are touring in their area, invite them to the show.

WHERE TO SEND IT

The Ezine Directory has a listing of many of the better-known music e-zines, along with descriptions and ratings of each (www.ezine-dir.com/Music). The goal is to find those with the correct target market and submit articles, music, and photographs to the editor, encouraging him or her to include a link to the artist's web site. Some e-zines have submission forms available on their web site, whereas others are not as specific about their submission policy. But don't give up; Dawes and Sweeney (2000) suggest sending an email to the editor to inquire about the potential to have your artist featured in the publication. Provide editors with writeups to make their job easier. When these reviews and articles appear, place a link on the artist's site directed to the specific page on the e-zine site that includes the article. To find a list of e-zines, try one of the directories, such as the ultimate band list or www.ezine-dir.com.

The Music Industry News Network (www.mi2n.com) features articles and news of various independent artists and will accept submissions for news items. On the home page, click "submit your news" from the menu bar at the top and then fill out the form. The submission form allows you to include your web site address, a link to sound files, and a link to a graphic file. For a fee (currently $19.95 for the PR Express), their PR Syndicate service will send out your press release to major music news sites, including the appropriate "groups" at Google and Yahoo! and their related Music Dish Network sections of major social networking sites.

BeatWire.com is a web site dedicated to press release distribution for independent musicians and record labels. The cost is $149 and includes distribution to all the major music publications and media outlets, including

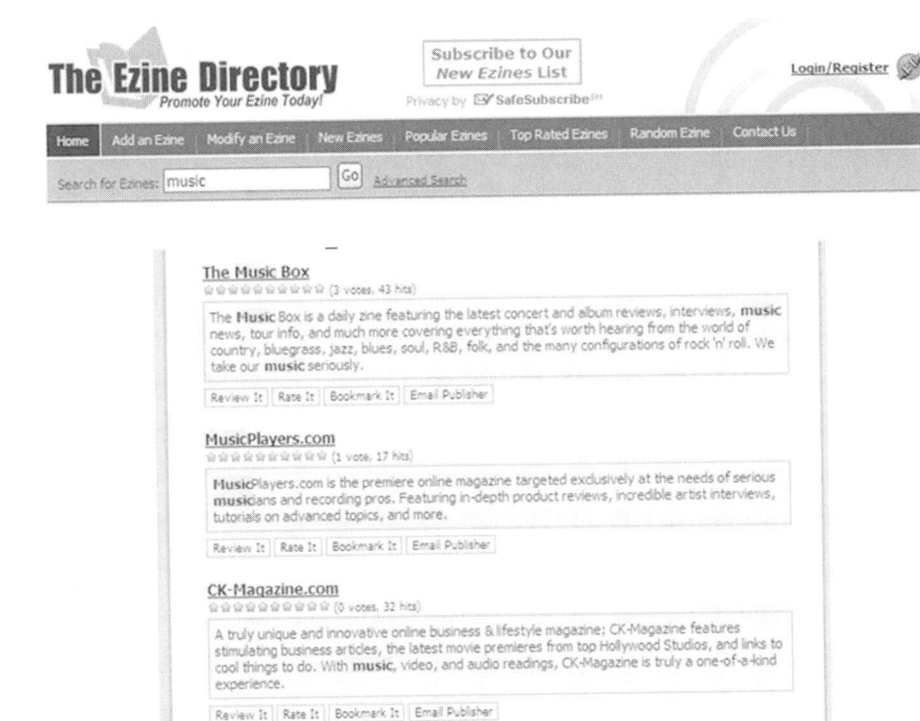

Figure 11.3
The Ezine Directory (courtesy of The Ezine Directory, www.ezine-dir.com)

monthly music magazines, college radio stations, and weekly and daily newspapers in hundreds of markets. This would be a good idea for an artist who is ready to move to the next level and has a compelling news release that is likely to be picked up by these national and regional publications.

To find lists of appropriate music bloggers, Bracco recommends Google Blog Search, Elbo.ws, Captain Crawl, The Hype Machine, and Shuffler.fm.

- Google Blog Search. http://blogsearch.google.com
- Elbo.ws. http://elbo.ws
- Captain Crawl. http://captaincrawl.com
- The Hype Machine. http://hypem.com
- Shuffler.fm. http://shuffler.fm

Lists of music bloggers can also be found at:

- Best of the Web. http://blogs.botw.org/Arts/Music/
- Music Blog Wiki. http://musicblog.wikia.com/wiki/Music_Blog_Directory
- Hype Machine. http://hypem.com/#!/blogs/alpha/all

Resources for E-zines and Distribution of Press Releases

- The Ultimate Band List: www.ubl.com
- CD Baby: www.cdbaby.org
- BeatWire: www.beatwire.com
- The Ezine Directory: www.ezine-dir.com
- Music Industry News Network: www.mi2n.com
- MusicDish: www.musicdish.com (try the open review and "submit your article")
- Dmusic: http://news.dmusic.com/submit
- PRWeb: www.prweb.com

Updated information available at www.focalpress.com/cw/Hutchison.

USING EMAIL FOR PROMOTION

Perhaps one of the most successful marketing strategies brought about by the Internet is the use of email to effectively target and promote commercial enterprises. The use of periodic mass emailing allows an artist to announce new recordings and live appearances, contact the press, disseminate news and accomplishments, and keep fans coming back to the web site. The effective use of email requires that you develop a list of interested fans, keep the list updated, and correctly target portions of the list for appropriate email announcements. Research has shown that email is very effective in driving traffic to the web site, but only if the email contains something of interest for the consumer.

Building and Managing Your Fan List

The best possible way to develop a quality email list is by requesting that fans and web visitors provide their email address so that they can receive valuable updates, news, and perhaps savings opportunities. It is important to provide a signup form in a prominent location on the web site. One way to motivate fans to provide their address is by offering something extra to those who become fan club "members" by providing their address. This can be anything from access to "restricted" areas of the web site, where they can receive free items such as audio tracks, to offering contest prizes to those who sign up.

While it is important to provide web site visitors with the opportunity to submit their email addresses when they visit the web site, it is equally important to provide signup opportunities at gigs and in retail settings. Always provide an email signup sheet at the bar or the front door, or have a worker pass around a signup sheet. It also helps to make an announcement from the stage asking fans to provide their email addresses so they can receive valuable updates and information on future performances.

G-Lock software advises not to "generate email addresses using special tools you may find on the Internet; don't harvest the email addresses from web sites; and don't purchase a ready-made list." The best quality lists come from grassroots efforts to sign up those people who are interested in the artist and willing to accept email newsletters.

It is important to inform people that by providing an email address, they will be receiving periodic updates—in the form of an email—about the artist. In this era of spamming, it is important to provide recipients of the email newsletter with the opportunity to unsubscribe to the email list. Spamming is the activity of sending out unsolicited commercial emails. It is the online equivalent of telemarketing. Fans who provide their email addresses should be told that by providing this information, they will be receiving correspondence—the specifics of which should be spelled out. You should never sell or offer your email lists to third parties. The addresses should be kept confidential and care should be taken when sending out bulk emails that the addresses of other recipients are not visible and available to any of the recipients. In the absence of specialized software for managing email lists, you can always send the email correspondence to yourself and blind copy (BCC) the message to those on the list. By using this commonly found email function, the addresses of the other recipients are not visible in the message heading.

Some of the important factors in email list management include the following:

1. Categorizing the list to send targeted emails only to appropriate groups. Touring information is only important to fans in the area of the performances. Leave the comprehensive touring schedule to the web site and use email for only those fans within driving distance to the venue.
2. Keeping the list current by purging addresses that are no longer valid.
3. Promptly removing subscribers who asked to be removed.
4. Providing automated ways for new subscribers to be added to the list (there are many software programs that will provide this feature).

The available software programs, both web-based and computer-based, help manage email lists and enable you to create subcategories. Some of the features to look for in email management software include managing the list, mail merge, verifying emails, removing dead addresses, managing unsubscribe requests, and email tracking. It is recommended that a webmaster invest in one of these programs if there is any volume of email that cannot be handled promptly at all times. Mail merge is important for personalizing messages with the recipient's name in the salutation. An automatic responder is helpful for sending a confirmation to people who have responded to your email or purchased your product.

THE EMAIL NEWSLETTER

The email newsletter, sent to willing recipients who have signed up to receive the newsletter, can be an extremely effective and inexpensive way to promote an artist. In his article "7 Reasons to Start an Email Newsletter Today," Rich Brooks extolled the virtues of utilizing this marketing technique, for the following reasons:

1. They complement your web site—a newsletter is the counterpart, the outreach portion of your online marketing plan, or, as Brooks stated, it

is "like white wine to fish (not that your web site stinks like fish)." Most people spend a great deal of time reading and responding to emails, so this puts you right in front of their faces.

2. They are more cost effective than print newsletters. Forget about all those printing costs and postage!
3. They are interactive. You need to goad people into turning on their computers, typing in your web address (correctly, you hope). The fans read, they see a link of interest, they click on it.
4. You can track their effectiveness. "You can track which links in your newsletter are being clicked on and which are being ignored," stated Brooks.
5. They are viral. Word of mouth is easy because readers can forward the email to friends they think will find it interesting.
6. You are preaching to the choir. Newsletters only go to fans who have signed up to receive them. These fans want to know where the next gig is or when the next recording is coming out.
7. You can start to build your fan base and subscriber base now to take advantage of the next generation of communication technologies.

Managing the Email List

Once an artist's career starts to take off, managing the email list can become a daunting task. It is important to be able to send the appropriate emails to the appropriate group of people without difficulty. Email management software and/or services are generally required once the list becomes more diverse and less manageable. Fanbridge is a fan management platform for musicians that will manage email distribution and notices to social networks. Fanbridge is used by small independent artists and major stars such as Lady Gaga and Linkin Park (CrunchBase, 2012). The free version allows up to 400 emails per month. Subscriptions for premium service start at $5.99 per month (www.fanbridge.com). A nice alternative is Mail Chimp (www.mailchimp.com), which offers a free (MailChimp branded) service for up to 2,000 subscribers and up to 12,000 emails per month. ReverbNation also provides email management services. When selecting a mail management system, take into account what your needs will be several years down the road. The last thing you want to do is have to migrate a large email database from one service to another if you decide to change providers later.

Content and When to Send Email Newsletters

Most news emails take a form adapted from print newsletters, with headlines and a brief overview of each topic, but with links to the full story on the web site for the added multimedia and graphic content. They should entice the reader to click on a link to get "the rest of the story" or view the item of interest. Be sure that each newsletter link directs the reader to the appropriate page of the main web site rather than the home page. Have a great opening, use short sentences, and focus on the recipient's self-interest. Newsletters can also contain fun items of interest, including stories, recipes, jokes, and links to sites of significance. The

G-lock EasyMail web site advises users to not use all capital letters, not to write too much, correct misspellings and typos, and send a text version of the message along with the HTML part. The site also advises users to "excite curiosity in the subject line" to motivate the recipient to open and read the email, to write as if you are conversing with another person, to keep it simple in format, and to place links at various intervals throughout the email. The best email newsletter is one that contains subject headlines with a teaser paragraph on each subject, providing enough information to get the point across but with links to more in-depth information on the artist's web site.

Skellie (2007) posts some tips for managing email effectively, including creating useful labels and folders, processing emails in batches, answering each email as you read it rather than returning to it later, keep it short and sweet, proof-reading before hitting reply, and to use a "bridge" email when necessary. A bridge email is a general response acknowledging receipt of their email and that you can't respond at the present, but will at a later time. A good signature and a good salutation or sign-off is also important.

Much like a standard press release, email newsletters should only be sent out when there is something new to report such as new music releases, announcements of upcoming touring, and award nominations. Newsletters that are sent out on a scheduled basis sometimes don't contain enough news to hold the reader's attention, and these senders often find that many of the recipients eventually request to be removed from the list.

Don't send out large files full of graphics and attachments. Instead, rely on the web site to provide the images. Although you may be tempted to send newsletters in the form of a PDF file, some recipients prefer HTML or text files and may find it difficult to download PDF files. The idea is to keep the email correspondence as a small, easy-to-download file so that it is not filling up inboxes or being filtered by spam filters because of the active content or file attachments. Spyware, adware, and computer viruses (see Chapter 13) are often spread through the use of attachments, and many email programs automatically filter them out.[1]

You may find it necessary to ask your recipients to unblock your email address so that your newsletters don't automatically end up caught in the spam filter. Ask your recipients to go to their spam folder, find your email, and unblock the address—but, of course, this will be difficult to do if your only method of contact is the recipient's email address. You can always use your personal email address to contact members of your newsletter list who have inadvertently blocked your emails and notify them. It also might be a good idea to provide a notice when visitors sign up for your newsletter that will remind them to unblock your address.

Note that in the example of the Sauce Boss Gazette, the formatting is simplified for quick downloading of a small file—one that is less likely to fill up the recipient's mailbox. The use of bold caps for headlines and hot links is important for easy reading and to encourage readers to follow the links for more information on the official and other related web sites. Make sure that each link directs the reader to the specific page of the web site containing the material

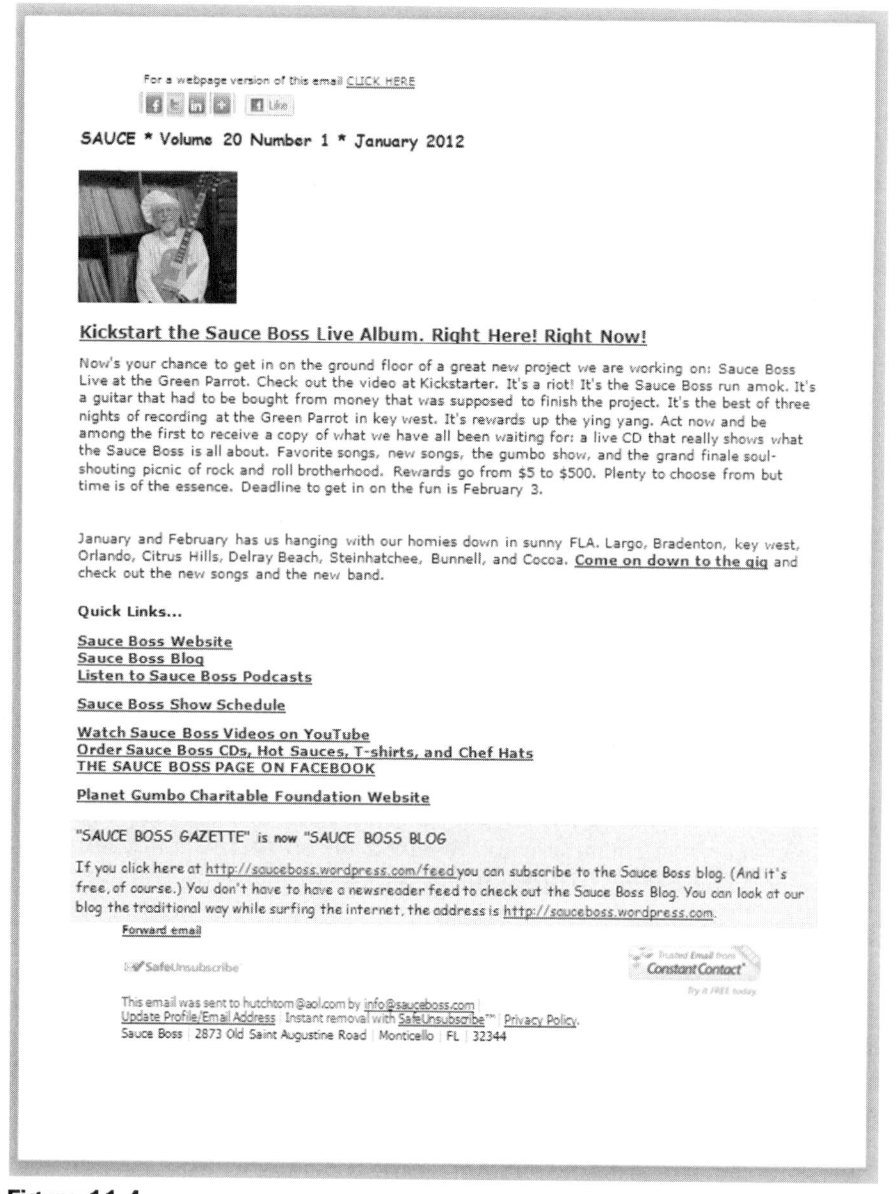

For a webpage version of this email CLICK HERE

Like

SAUCE * Volume 20 Number 1 * January 2012

Kickstart the Sauce Boss Live Album. Right Here! Right Now!

Now's your chance to get in on the ground floor of a great new project we are working on: Sauce Boss Live at the Green Parrot. Check out the video at Kickstarter. It's a riot! It's the Sauce Boss run amok. It's a guitar that had to be bought from money that was supposed to finish the project. It's the best of three nights of recording at the Green Parrot in key west. It's rewards up the ying yang. Act now and be among the first to receive a copy of what we have all been waiting for: a live CD that really shows what the Sauce Boss is all about. Favorite songs, new songs, the gumbo show, and the grand finale soul-shouting picnic of rock and roll brotherhood. Rewards go from $5 to $500. Plenty to choose from but time is of the essence. Deadline to get in on the fun is February 3.

January and February has us hanging with our homies down in sunny FLA. Largo, Bradenton, key west, Orlando, Citrus Hills, Delray Beach, Steinhatchee, Bunnell, and Cocoa. **Come on down to the gig** and check out the new songs and the new band.

Quick Links...

Sauce Boss Website
Sauce Boss Blog
Listen to Sauce Boss Podcasts

Sauce Boss Show Schedule

Watch Sauce Boss Videos on YouTube
Order Sauce Boss CDs, Hot Sauces, T-shirts, and Chef Hats
THE SAUCE BOSS PAGE ON FACEBOOK

Planet Gumbo Charitable Foundation Website

"SAUCE BOSS GAZETTE" is now "SAUCE BOSS BLOG

If you click here at http://sauceboss.wordpress.com/feed you can subscribe to the Sauce Boss blog. (And it's free, of course.) You don't have to have a newsreader feed to check out the Sauce Boss Blog. You can look at our blog the traditional way while surfing the internet, the address is http://sauceboss.wordpress.com.

Forward email

SafeUnsubscribe

Trusted Email from Constant Contact
Try it FREE today

This email was sent to hutchtom@aol.com by info@sauceboss.com
Update Profile/Email Address | Instant removal with SafeUnsubscribe™ | Privacy Policy.
Sauce Boss | 2873 Old Saint Augustine Road | Monticello | FL | 32344

Figure 11.4
Example of a Sauce Boss newsletter (used with permission)

for that topic, rather than using a link to the home page (see landing pages in Chapter 10).

The email newsletter, if used correctly and cautiously, can be an effective and inexpensive marketing tool that drives traffic to the web site, fans to the concerts, and increases recorded music sales. Rich Brooks, president of Flyte New Media stated, "an email newsletter is the sharpest tool in your Web Marketing toolbox.

If you're not sending one out you're missing a great opportunity to connect with your customers and prospects."

INTERNET RADIO AND STREAMING MUSIC SERVICES

Internet radio—often referred to as webcasting and sometimes as streaming audio or radio—is the concept of broadcasting much like a radio station, but via the Internet in lieu of a terrestrial signal. The obvious advantage over terrestrial radio is the absence of geographical limitations. Although the success of Internet radio has risen and fallen with the expenses associated with payments of performing rights royalties, it has consistently been a successful outlet for indie labels and emerging recording artists who do not always expect to be compensated for airplay of their music. Terrestrial radio airplay has been the key to selling recorded music since the days of early radio. It is still the most commonly used form of marketing new music to audiences and potential customers. But it is impossible for independent artists to gain much exposure through radio airplay on traditional terrestrial radio stations. Satellite radio, with its broader playlists and multitude of stations, has offered some opportunities for niche music, but Internet radio is the bastion of hope for frustrated emerging artists seeking exposure for their music. There are literally thousands of independent radio stations and Web 2.0 has created opportunities for anyone to create and manage their own online radio station with the playlist of their choice.

> Internet radio is growing at a rate of about 100,000 average active sessions per month (AAS is similar to Arbitron's Average Quarter Hour and represents the average number of active sessions at any point during the day). As of early 2012, Internet radio had some 1.6 million AAS and is projected to reach 3 million AAS by 2015—representing 10% of radio's total audience.
>
> **Paul Maloney (2012). Radio and Internet Newsletter**

Music podcasts are simply prerecorded shows, often in a radio-type format with narration and commentary intermingled with music. The major difference between podcasting and web radio is the time-shifting factor: podcasts can be downloaded, transferred to portable devices, and enjoyed at the convenience of the listener. Consideration must be given to the fact that portions of a podcast, including copyrighted music, may be copied and redistributed, whereas streaming radio is less likely to be captured and distributed illegally. This section concentrates on Internet streaming radio and getting airplay on web stations.

Streaming on-demand services are gaining popularity as mobile devices become more music-oriented. Rhapsody has been in business since 2001, and is now joined by Spotify. Pandora uses a slightly different programming model to create radio stations around the user's requested song or artists. The playlist reflecting that type of music is then created by Pandora. It is important for the artist to get music to each of these and other services. Equally important is registering with SoundExchange. SoundExchange is an independent non-profit royalty collection organization that collects royalties for music played on Internet and satellite radio, cable TV music channels, and other streaming platforms. Registration online is easy (www.SoundExchange.com), but you must provide tax information and fill out a W-9 IRS form. There is no cost to sign up, but SX does collect a small administrative fee from royalties paid.

POPULARITY OF INTERNET RADIO

An Arbitron report titled *The Infinite Dial 2010: Digital Platforms and the Future of Radio* estimated 27% of the U.S. population 12 years of age and older had listened to online radio within the past month, and nearly half of people in the 18–24 age demo and one-third of people under 50. Arbitron stated that about 70 million had listened within the past month with 43 million weekly listeners in the U.S. The 2012 Arbitron study revealed that the weekly audience for Internet radio had increased 30% in one year, and now comprised about 76 million U.S. listeners.

Among the top online radio services mentioned in the Arbitron survey, Pandora was the clear leader. As Pandora gears up to provide wi-fi audio streaming in automobiles and other services begin to take advantage of mobile and TV-top wireless devices, the potential for streaming radio is enormous. Twenty-seven percent of radio listeners in the Arbitron survey said they were very interested in a device that would allow Internet radio listening in thieir car. In 2013 certain Chevrolet, Nissan, and Hyundai models will include web access to Internet radio stations like Pandora. And, Apple is planning to enter the Internet radio business which will mean more offerings to users.

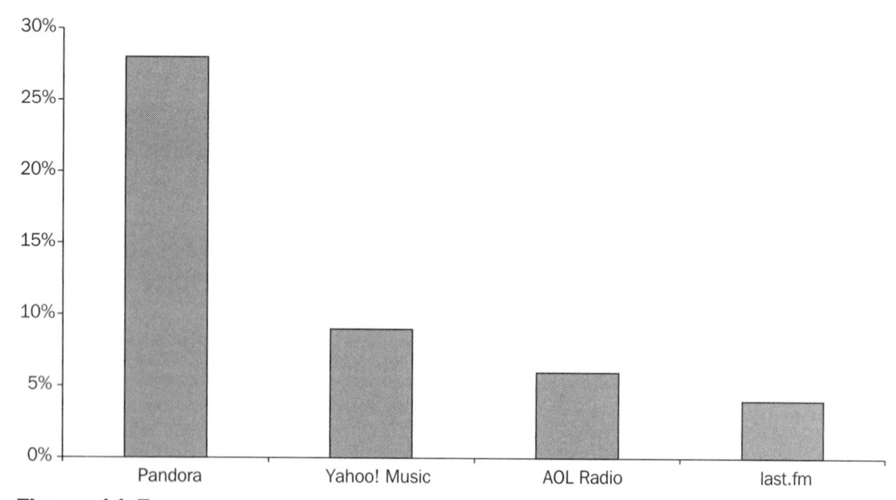

Figure 11.5
Top of mind awareness of Internet radio services (source: Arbitron, 2010)

GETTING AIRPLAY ON INTERNET RADIO STATIONS

There are three levels of webcasting stations on the Internet: (1) commercial (mostly FM) stations that simulcast their programming, (2) large "pureplay" commercial webcasters (such as Pandora) that provide bundled, syndicated or subscription streaming audio services, and (3) the plethora of small hobbyist stations including those on services such as Live365 and SHOUTcast that allow music fans to operate their own stations.

Simulcast of Commercial Media Stations

Columnist John Taglieri commented that you can "literally travel around the globe … [and] still listen to your favorite radio station from home." Taglieri went on to state that while this is advantageous for listeners, it has not helped independent music because the emerging artist still cannot break into the playlist for the Internet version of these commercial stations because the programming is not unique to the Internet broadcast, but is the same program broadcast commercially in the station's geographic area. The advantage for major artists is that exposure is no longer limited to the geographic area of the station, and airplay on these stations could have an impact globally. The 2010 Arbitron report found that 31% of FM radio listeners have visited a station's web site. Of those, more than half said it was to listen to the station online.

In his article "Radio Airplay 101: Traditional Radio vs. the Web," radio promoter Bryan Farrish stated that in the future, it will be just as difficult to get airplay on a big web radio station as it is to get airplay on a popular terrestrial station now. Farrish stated that the competition will heat up for slots on the most popular web stations, with artists and labels competing for the coveted airplay slots as they do now on FM radio. He added that as more sales come through downloads, the money saved on manufacturing will shift to promotion. Farrish concluded that "the amount of work it takes to get your songs heard will always be directly proportional to how many listeners you are trying to reach." A 2009 court ruling opened the way for terrestrial stations to simulcast without paying additional royalty payments. In March of 2011, the U.S. Copyright Office published the final decision of the Copyright Royalty Board on the statutory rates for Internet radio royalties.

Large Pure Play Commercial Webcasters

Pure Play stations are popular Internet radio stations that are online only, and are not a simulcast of an FM or AM signal. Pandora and AOL radio are among the top "web radio" services that have not evolved from traditional FM radio stations. These entities use a new business model that takes advantage of innovations in computer and mobile technology. Pandora accepts submissions from independent artists, with no guarantee of airplay.

Submission of Your Music to Pandora
To submit music to Pandora, you'll need these items:

1. A CD of your music.
2. A unique UPC code for that CD (There is a link on the web site to obtain your own UPC code at a low cost. If you already have a UPC code for this particular CD—for instance, through your record label—use that one.)
3. This CD to be available through Amazon as a physical CD (and not just as MP3s)—and for the name of each track to be listed in the track listing section on the page for that CD.
4. The legal rights to your music/comedy.
5. A standard free Pandora account, based on a valid email address, that will be associated with this submission.
6. MP3 files for exactly two tracks from your CD.

Note: It is only possible to submit one CD at a time to Pandora

Pure Play Webcasters per Performance Royalty Rates

- 2011—$.00102 per performance
- 2012—$.00110 per performance
- 2013—$.00120 per performance
- 2014—$.00130 per performance
- 2015—$.00140 per performance

An example of a successful commercial streaming radio station is from pop icon Jimmy Buffet, who sponsors a streaming radio station Margaritaville Radio at www.radiomargaritaville.com. The station started as webcast only in 1998 but was picked up by Sirius in 2005 and is now simulcast (Deitz, 2005).

For both media-sponsored webcasts and podcasts, getting on the playlist requires having some connection to the sponsoring organization, whether it's through membership or just targeting the right webcast sites and submitting appropriate material. David Nevue suggested identifying stations that take outside material and match the genre, and then contacting them with a request to submit songs for consideration. Nevue stated, "some broadcasts receive a number of CDs to review, so it may take several weeks for them to get to yours." Some of the more popular online radio services are Pandora, finetune, Grooveshark, Jango, Deezer, Slacker, and Songza.

Plethora of Small Hobbyist Stations

The smaller, less popular web stations are easier to break into if they are targeted correctly. It is necessary to identify those stations for which your music is appropriate. In his article "Web Radio Stations and Getting Played," Bobby Borg (2005) wrote that the submission policy for these small stations is usually simple and involves emailing a music file to the webcaster. These small stations usually lack the exposure of larger, more popular stations, but getting airplay on these stations is inexpensive, and as more stations pick up the songs, the audience is cumulative. Borg stated, "SOME exposure is better than NO exposure—especially if it leads to a listener buying your record or coming out to one of your shows." Many of these amateur stations can be found on the services Live365.com and SHOUTcast. SHOUTcast provides listenership data for all stations for a 30-day period under the "stats" menu item.

To find appropriate stations on Live365, select the genre from the list, select the appropriate stations (a broadcaster profile and broadcast schedule is provided), and contact the moderator of each. A free membership account is required to gain email access. If your inquiry receives a favorable response, email an MP3 file of the song you would like considered for broadcast. By all means, keep track of the stations on which you are receiving airplay. It helps to then promote the stations that support your music by including a link to them on your artist web site. To create your own station on Live365, you must pay a fee of $9.95 per month, which includes 150 MB of file storage and allows for up to 25 simultaneous listeners. SHOUTcast is free but requires that you use your own server and download the appropriate software. To request airplay, you must select the appropriate genre and scroll through the stations. A link to each station will send you to that station's site. Submission policies may vary.

LICENSING AND THE FUTURE OF WEBCASTING

In the article "The Day the (Web) Music Died," Robert Atkinson stated that mandatory broadcast licensing fees threaten to end the enormous popularity of web radio. He stated that "Internet radio is a classic Web 2.0 application, offering diverse programming that caters to a range of specific groups." Whereas terrestrial broadcasters are required to pay royalties to performing rights organizations to cover licensing of the musical compositions for the songs they play, they are exempt from paying royalties for the actual sound recordings—money that would go to the label and artists. But for webcasting and satellite broadcasting, this additional royalty must be paid, usually to SoundExchange. SoundExchange's web site describes the organization as follows:

> SoundExchange is an independent, nonprofit performance rights organization that is designated by the U.S. Copyright Office to collect and distribute digital performance royalties for featured recording artists and sound recording copyright owners (usually a record label) when their sound recordings are performed on digital cable and satellite television music, Internet and satellite radio (such as XM and Sirius). SoundExchange currently represents over 3,000 record labels and over 20,000 artists and whose members include both signed and unsigned recording artists; small, medium and large independent record companies; and major label groups and artist-owned labels.
>
> **www.soundexchange.com**

Before 1995, there was no performance right for copyright holders of sound recordings in the United States. Other countries have been collecting royalties for sound recordings for years, providing revenue to recording artists and record labels for commercial airplay. SoundExchange goes on to state:

> The Digital Performance in Sound Recordings Act of 1995 and the Digital Millennium Copyright Act of 1998 changed that by granting a performance right in sound recordings. As a result, copyright law now requires that users of music pay the copyright owner of the sound recording for the public performance of that music via certain digital transmissions.
>
> **www.soundexchange.com**

Subsequently, there was much debate over what the royalty rate should be based on and what is deemed fair to all parties involved. Rates were set to increase on June 15, 2007, to $0.0008 per performance: defined as streaming one song to one listener. Small broadcasters claimed that the new rate was too cost prohibitive; they banded together to protest the new rate, which is also set to increase 30% per year. In response, in August of 2007 SoundExchange offered a new rate to small broadcasters, defined as those earning $1.25 million or less in total revenues. Small broadcasters would pay a royalty rate of 10% or 12% of revenue, but only for SoundExchange member labels, and only if the small broadcasters signed on before September 14 of that year.

In 2009, the issue was finally resolved when an agreement was reached that would set royalty rates at either a per-song basis, or a percentage of income from the broadcaster. The rates were further delineated based upon size of the station's market. According to Rob Pegoraro of the *Washington Post*,

> Large Webcasters—defined as those who rake in more than $1.25 million in annual revenues—will pay either 25 percent of those revenues or a per-performance rate in the neighborhood of a tenth of a cent, whichever is greater. Small Webcasters—those raking in less than $1.25 million a year in revenues and fewer than 8 to 10 million listener-hours a month of broadcasts, depending on the year—can choose between paying a percentage of their revenues (12 percent of the first $250,000, then 14 percent of everything on top of that) or their expenses (7 percent).

Artists are paid through SoundExchange.

Resources for Internet Radio Promotion

- Rhapsody: www.Rhapsody.com
- SomaFM: http://somafm.com
- Pandora: www.pandora.com
- Live365: www.live365.com
- SHOUTcast: www.shoutcast.com
- Digitally Imported: http://www.di.fm
- Radio and Internet Newsletter: www.kurthanson.com
- New!: httpwww.clearchannelmusic.com/cc-common/artist_submission/index.htm?gen=2

GRASSROOTS

Grassroots marketing is defined as unconventional street-level marketing using word-of-mouth influence and that of opinion leaders to disseminate a marketing message among potential customers. Eschewing the top-down approach, grassroots marketing bypasses the traditional marketing methods that rely on expensive, slick advertising messages delivered through traditional media vehicles. Guerilla marketing is another term closely associated with grassroots marketing. Whereas grassroots marketing refers to a street-level, door-to-door campaign of peer-to-peer (P2P) selling, guerilla marketing refers to low-budget, under-the-radar niche marketing, using both P2P and any inexpensive top-down methods. Grassroots marketing for large companies may indeed involve a large budget to pay street teams. On the Internet, grassroots marketing includes placing conversational presentations of your product or artist in discussion forums and message boards, placing self-produced videos and music videos online, and encouraging bloggers to write about your artist. This can be accomplished by recruiting fans to be your online street team or e-team.

Larger corporations have joined the grassroots movement by soliciting media content from customers through contests to "create an ad," "name the product," or "write the slogan." The company Current TV runs contests for companies such as Sony, Toyota, and L'Oreal that encourage customers to submit their home-made 30-second spots. Winners receive a cash award and get their video promoted through mainstream media outlets (Mills, 2007). Large companies also hire members of the target market to spread the word to peers about the artist or product. On the Internet, these online street teams seek out web locations where the target market tends to congregate and infiltrate these areas to introduce the marketing message. Online street team members have been known to engage in the following activities to promote recording artists:

1. *Posting on message boards.* Street team members who are actually part of the target market visit relevant sites that allow for message posting and casually encourage readers to "check out my new favorite recording at …" These sites usually have specific rules about the posting of commercial messages and harvesting email addresses (which is universally frowned on by group

Table 11.1 **Comparison of grassroots versus guerrilla marketing**

Grassroots	Guerrilla
Bottom up/uses people in the market	Low budget
Works on word of mouth	Niche marketing
Avoids mainstream media	Nontraditional
Has street credibility	Avoids costly mainstream media
Used by both small and large companies	Usually used by small companies

moderators and system operators). Many groups, however, welcome brief messages from industry insiders notifying interested members of new products. Sometimes street team members do not announce their affiliation with the artist; instead, they engage in communication as if they were just enthusiastic fans. The message postings often are embedded with hot links to the site actually promoting or selling the product. Some user groups require that you register before being allowed to post messages.

2. *Visiting and participating in chat rooms.* Chat rooms allow for real-time interaction between members, whereas bulletin boards allow individuals to post messages for others to read and respond to. Because these members or users have a mutual interest in the site topic, user groups offer an excellent way for you to locate members of your target market. Marketing professionals as well as web surfers often engage in "lurking" behavior when first introduced to a new user group. Lurking involves observing quietly—invisibly watching and reading before actually participating and making yourself known. Often, user groups have their own style and "netiquette" (Internet etiquette), and it's best to learn these rules before jumping in.
3. *Blogging or submitting materials to bloggers.* Successful bloggers (those with a substantial audience) are opinion leaders, and their message can influence the target market. Online street team members will seek out blogs that discuss music and reach out to the bloggers to check out their artist and write about the music, much the same way a publicist will seek out album reviews in traditional media. Blogging, however, has the potential for street cred(ibility) that is sometimes absent in mainstream media.
4. *Pitching/promoting to online media.* This is much like pitching to bloggers only more formal. Under these conditions, professional media materials must be supplied to the publication, whether it's an online-only publication or one that also has a traditional media presence. The publication will most likely want a photograph. It's best to supply completed articles and press releases to busy journalists who don't have time to conduct their own research and whose decision to run your article may be influenced by convenience factors.
5. *Visiting social networking sites and posting materials including music, artwork, and videos.* Major record labels are now using interns and young entry-level employees to maintain their artists' presence on social networking sites. These young marketers are responsible for setting up the artist's page on each of these sites, fielding requests from fans to be added to the social group or "friends network," providing updated materials (music, news, photos, videos, etc.), and visiting related pages to engage in street team promotion. Some labels have teamed up with these sites to conduct promotions in the form of contests and giveaways such as encouraging fans to create their own YouTube music video for an artist.
6. *Finding fan-based web sites and asking the site owners to promote the artist.* We address this strategy in the section on fan-based sites. It is often up to online street team members to seek out and identify these fan-based sites. Sometimes they engage in dialog with the site owner, but often the list of fan-based sites is compiled by the street team member and then passed

along to a more senior member of the marketing team. Then these sites are managed by more experienced web marketers who can supply RSS feeds and other technical web assets to the site owner.

7. *Acting as a "clipping service" by scouring the Internet for mentions of the artist.* Google and Yahoo! news alerts have made this task easier, by allowing anyone to set up news alerts based on selected keywords. Any mentions of the artist in third-party online sites should be identified and decisions made about how to incorporate them into the overall web marketing plan. Should you create a link to the article? Should you request permission to reprint? Should you rebut negative publicity?
8. *Finding sites that attract the target market and then working with those sites* (see reciprocal links section). The following section on turning your competition into partners outlines many of the street team tasks performed in this area. The foremost task is finding and identifying where your target market hangs out on the Web—what web sites do they frequently visit? Then you can make marketing decisions about how to work with those sites, whether through advertising on those sites or setting up reciprocal links to generate cross traffic.
9. *Submitting the site to search engines and music directories* (see the section on search engines).
10. *Writing reviews of the artist or album on sites that post fan reviews.* Many retail sites such as CD Baby allow customers to post product reviews, including music reviews. It is not uncommon for artists to ask their fans to post reviews on these sites or for a label to have street team members post favorable reviews.

TURN YOUR COMPETITION INTO PARTNERS

One way to maximize grassroots marketing is by turning your competition into partners. Because music consumers generally buy several albums per month and are fans of more than one artist, sharing your fan base with similar artists makes sense.

It may be difficult at first locating appropriate partners. The main criterion is that the partners have identical markets of people who would easily cross over from one artist to the next. Cooperative efforts are generally only successful among artists who are all at the same level in their career. A superstar artist's web site is not generally prone to support and feature unknown artists unless the superstar has a personal interest in them. To find artists at the developing level, turn to places like CD Baby and scroll through the artists listed in your artist's genre. For more established artists, Amazon is a good example of a place to find potential co-op partners. On Amazon, it's easy to see the common thread by looking at the section titled "customers who bought this item also bought …" Then visit those artists' Amazon page and look at their section on "customers who bought …" After two layers of this, the accuracy of located artists in the same market is somewhat diminished. It is recommended that you only go two layers deep, but that should yield some good prospects to start with, although

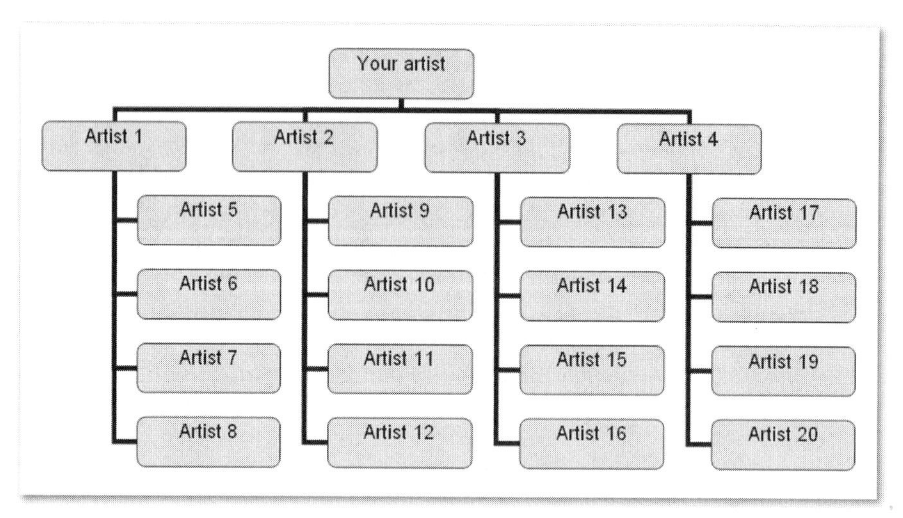

Figure 11.6
Creating a network for networking and reciprocal links

it will probably yield some overlap. The chart in Figure 11.6 indicates how four connections on the artist's Amazon page can ultimately yield 20 potential partners, going only two layers deep.

After potential business partners have been identified, the next step is to determine the nature of the partnership and contact each potential partner. The possibilities are described next.

RECIPROCAL LINKS

You may just be requesting to trade links, meaning that you will place a link to another's web site on yours, if the potential business partner will reciprocate. This can also be done with banner ads, but that limits the number of partners involved in the promotion. If six people join forces to swap links, visitors to one site will be exposed to the other five artists, thus essentially pooling the customer base. Reciprocal links, also known as link exchanges and link swaps, benefit the site in another way. Search engine ranking is partially based on inbound links to your site, indicating popularity and the importance other web sites put on having people visit your site. To find potential link-swap partners, visit their web sites and examine their link pages. Determine if a link to your site would be appropriate for them.

In the article "How to get reciprocal links," Allan Gardyne advised the following:

- Find good-quality, complementary sites.
- Place a link to them on your site.
- Email the owner of the site a short, friendly note.
- Praise something on their site. If there is nothing to praise, delete them from your list.

- Tell the web site owner you have linked to their site and give them the URL of the page where you've placed their link.
- Ask for a link back to your site, suggesting a page where the link would be appropriate.
- Keep a record of sites you've linked to and requested links from.

COOPERATIVE EFFORTS, TEAMING UP, PROMOTIONAL TIE-INS

In addition to link swaps, there are other ways to forge partnerships with former competitors. You can arrange newsletter article exchanges with other newsletter publishers, perhaps reviewing their album in your newsletter and asking for the same consideration or swapping feature articles on each other. Be sure to include some in-context links in the article that will connect back to your web site. If the newsletters are archived online, the impact can be long lasting. Perhaps you want to team up to create a newsletter and then send it to the email lists for both artists. However, it is not advisable to just swap email lists. Your fans have entrusted you with their email addresses and granted you permission to use them to send them information on your artist. If they start receiving unsolicited correspondence from artists they are not familiar with, this could be considered spam, which is certainly unethical and, in some places, illegal.

By pooling efforts, several musicians can split the costs of CD manufacturing, advertising, web site development, and so on. In one instance, a local musician solicited one track each from 10 musicians to create a music sampler. For $300 per artist, the organizer raised $300 to cover manufacturing, printing, and marketing costs. Each participant got 100 records for their $300, and the organizer had an instant street team, each member eager to sell as many of their 100 CDs to friends and family as they could. The marketing materials had information on each of the artists and how to purchase more recordings from them and visit their individual web sites.

Artists can also group together to create themed marketing materials, such as a web site that brands a particular style of music and features the artists involved in the co-op. These centralized web sites take the place of the more disordered web rings of the 1990s and afford the co-op the chance to create a brand for the music of all participants by creating a brand name and a consistent image for the co-op. Consumers will quickly learn the meaning of the brand, and the marketing efforts will reach a wider audience.

VIRAL MARKETING

Publicityadvisor.com describes viral marketing as a new buzzword for the oldest form of marketing in the world: referral, or word of mouth, which the site has updated to "word of mouse."

Definition and Description

Viral marketing is any strategy that encourages individuals to pass on a marketing message to others, spreading exponentially as one group of people pass on the message to each of their friends. This creates an ever-expanding nexus of Internet users "spreading the word." Viral marketing capitalizes on social networking and the propensity for Internet users to pass along things they find interesting.

Ralph F. Wilson (2000), in his article "Six Simple Principles of Viral Marketing," identified six elements of viral marketing:

1. *Give away something of value.* Giveaways can attract attention. By giving something away up front, the marketer can hope to generate revenues from future transactions. For example, the tag can say, "Win free music at www.yourartist.com."
2. *Provide effortless transfer to others.* According to Wilson, "The medium that carries your marketing message must be easy to transfer and replicate."
3. *Expand exponentially.* Scalability must be built in so that one message transmitted to 10 people gets passed on to 10 each, for 100 new messages.
4. *Exploit common motivations and behaviors.* Success relies on the basic urge to communicate and share experiences and knowledge with others.
5. *Utilize existing communication networks.* Learn to place viral messages into existing communications and the message will "rapidly multiply in its dispersion."
6. *Take advantage of others' resources.* A news release reprinted elsewhere will include the viral message and perhaps the link.

Examples of Effective Viral Campaigns

The 2008 film *Dark Knight* used a fake web site to promote the political campaign of character Harvey Dent, complete with sharable political posters pitching "I Believe in Harvey Dent." The film grossed more than a billion dollars (McKee, 2011).

The viral marketing firm Fanscape was contracted to promote Clear Channel's music web site New! to unsigned bands to encourage them to upload their music to the site. The incentive was the opportunity for the band to expand its fan base through the site and increase its chance of being noticed by radio station program directors, because the music was to be made available publicly for streaming. The "most played" songs are prominently featured. Fanscape "reached out to unsigned bands by targeting them on MySpace, Facebook, Garage Band, etc., to personally invite them to check out New! and submit their music to the site, and to possibly be heard by radio program directors throughout the country" (www.womma.com). Fanscape generated more than 1,500 music submissions in two weeks.

Yahoo! Music "Get Your Freak On"

The growing popularity of online video-sharing and digital music created a perfect combination for Yahoo! Music to engage music enthusiasts with a one-of-a-kind promotion. Yahoo! Music created a program called "Get Your Freak On" where fans are invited to star in user-generated music videos for their favorite artists and songs.
This ongoing program has resulted in wildly popular videos for music stars like Jessica Simpson, Christina Aguilera, and 'Lil Jon, but it's the video for megastar Shakira that put this program on the map. Fans of the Colombian-born recording artist were asked to submit short, hip-shaking video clips to be included in a fans-only music video for the song "Hips Don't Lie."

After thousands of fans submitted their videos, Yahoo! and Epic took the best entries and created a video mashup that was exclusively available on Yahoo! Music. The fans-only video quickly raced to the number-one spot on Yahoo! Music, and even surpassed the popularity of the song's actual video. The user-generated music video was viewed more than one million times in the first few weeks of release, and has been seen more than 12 million times to date. At the time of the premiere of the fan-video, Shakira's actual video was #1 on Yahoo! Music, and the following week her record jumped 92 spots to become the 6th best selling album.

Yahoo! recognized the potential for this high-profile promotion to engage Yahoo! Music loyalists and spread virally throughout the Yahoo! network. The marketing team created a comprehensive campaign to kick-start the WOM nature of this promotion, including banner ads running across Yahoo, integration on Yahoo.com, and editorial placement in Yahoo! Music and within Yahoo!'s Buzz module.

Yahoo! also played close attention to Shakira's fan base and integrated the campaign on "Yahoo! En Espanol" and worked with Shakira Fan Clubs.

Credit Information

Client: Yahoo!
Agency: N/A
Budget: $15,000
Date of Campaign: Undisclosed

Figure 11.7
Yahoo! video viral marketing campaign

Ways to Incorporate Viral Marketing into your Communication Campaign

The five most commonly used viral marketing methods are (1) the email signature, (2) screensaver giveaways, (3) the tell-a-friend script, (4) the use of message boards, and (5) writing articles and allowing reprints.

- *Email.* The most common form of viral marketing is through email signature attachments. Hotmail.com and Yahoo! have successfully employed this technique by appending their message to the bottom of every email generated by its users. Hotmail offers free email service to users but also places a viral tag at the end of each message that its users generate. If that message is passed along, the advertising tag goes with it.
- *Customized screen saver.* The web site 2createawebsite.com suggests that an attractive screensaver that is distributed for free and is imbedded with links to your web site will promote your products while offering something of interest to the user. For example, photos of a band in concert could be

developed into a screensaver and include links to the artist's web site. The screensaver is then offered for free to fans through the email list and on the web site. There is always the possibility that fans will pass the screensaver program along to their friends. Other giveaway viral tools include skins for toolbars, email backgrounds, and wallpapers.

- *Tell-a-friend script.* These are usually found on the artist's web site and can encourage visitors to pass along your information to their friends. When a visitor clicks on a web site's "Tell A Friend" feature, a link to an artist's video or music clip is sent to their friend as well as a copy to their own email account. The friend then visits the link and perhaps forwards it on to others. The benefit to a tell-a-friend script, which is nothing more than a small piece of code you can paste into your web site, is that it can bypass spam filters because it arrives as an email from a friend rather than as an email from a business somewhere. (http://www.pctecmech.com/blog/Tell-A-Friend-Script.php)
- *Use of message boards.* Jeanne Jennings suggests that you participate in as many relevant forums, chat rooms, and message boards as possible. The more messages posted about your artist and the more often their name is associated with messages of value, the more likely people will seek out and visit the artist's web site. Some forums will allow a link in the signature file. For others that do not allow the use of links or the overt promotion of products, be sure to pick out a user name that can be easily traced to the web site. For example, the Sauce Boss would use that signature on all message board postings, expecting that interested readers would be able to easily find his web site at www.sauceboss.com.
- *Write articles of interest and allow reprints.* Because webmasters are always looking for good content, a well-written article on a topic of interest will be enthusiastically received and published. Links attached to the author name, at various points within the article, and in references at the end will create traffic to the web site. As the article is passed around electronically, anyone interested in the article will be exposed to the links. To find viral marketing articles, visit www.wilsonweb.com/cat/cat.cfm?page=1&subcat=mm_viral.

MANAGING STREET TEAMS ONLINE

The previous section dealing with grassroots marketing and viral marketing mentioned online street teams, but the Internet is also useful for managing on-the-ground street teams in multiple geographic locations. Unlike online street teams, traditional street teams are organized for geographic representation. In his article "Starting and running a marketing / street team," Vivek J. Tiwary defines a street team as "a group of people (the team members or 'marketing representatives') located in different areas who assist you in executing your marketing plan and expanding its reach to other territories." Street teams are members of your target market, or fans, who are willing to engage in grassroots marketing in their hometown. Street teams primarily evolved to promote urban music, however they have become a part of the marketing plan for many major

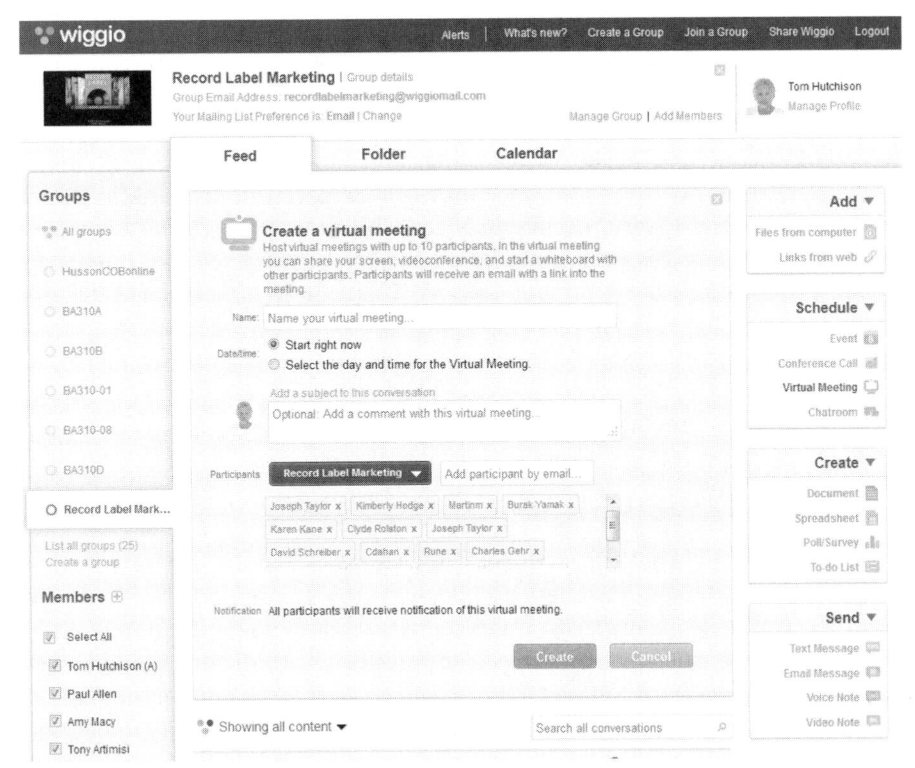

Figure 11.8
Wiggio used to organize street teams (courtesy of Wiggio)

label releases and are also used by indie labels and DIY musicians. The most successful team members are recruited from your fan base and may often be responsible for "sniping," which is the act of posting handbills in areas where the target market is known to hang out: record stores, clubs, college campuses, and so forth. Street team members are supplied with these and other marketing materials such as music samplers, swag, show tickets, etc. Street team members may also be called upon to "set up" an artist's live show in the area. Team members may be asked to visit local record stores, the press, and radio stations to set up meet-and-greets.

Maintaining a street team requires good communication and the Internet can assist with that function. It is important to communicate with the street team members, to motivate them and keep them informed of what the artist is doing and how their efforts are contributing to the success. Before the Web was in widespread use, street teams were organized using telephone conference calls and the postal service. Flyers were mailed out to members—or a master copy was mailed and the team members would have them printed locally. The particular needs and tasks to be performed were communicated via conference calls. Today, that function is covered via email and Skype.

Developing a sense of cooperation and competition among street team members can be achieved by using a closed-system social network platform such as Wiggio. This program was designed for teams to organize and communicate among their members. Setting up a Wiggio group for a street team is an easy task. Instructions, music samples, and camera-ready art (such as flyers) can be posted. The team manager can communicate with the group through conference calls, video conferencing, text messaging, and email blasts. Team members can communicate with one another through the social networking functions (message boards). Tiwary recommends that the street team organizer request verification for services performed by having team members take photos for documentation and file a progress report. This can all be managed through Wiggio.

Street team members should be compensated for their efforts. Free music and concert tickets can supplement a modest cash payment for services. Contests among street team members can also serve as an incentive.

FAN-BASED WEB SITES

Fan-based sites are unofficial web sites or web pages featuring an artist. Often they can be an extension of the fan's personal web site that also features other aspects of the fan's life. But many times they are dedicated fan sites honoring the creator's favorite artist and fostering a community of fans coming together. This ready-made street team represents some of the most motivated fans who can help propel an artist's career.

Importance of Fan-Based Sites

Often, the value of fan-based sites is overlooked in the grassroots marketing plan. They are obviously more common for well-known artists, but they exist for lesser-known artists in particular subcultures. At first, record labels were unsure how to deal with these sites, many of which had outdated or incorrect information about the artist. But labels quickly learned the value of this self-proclaimed street team—these are motivated fans who want to spread the word. Labels and webmasters now actively provide information to these sites in the form of updated photos, press releases, and RSS feeds (see the section that follows) so that they become part of the artist's online network.

For lesser-known artists, it might be necessary for the artist and management team to request that active fans create these sites. Some record labels have been known to have their interns set up and manage these sites or to create fan-based pages on social networks that appear to have no connection to the sponsoring label.

How to Find and Encourage Fan-Based Sites

To encourage and maximize fan-based web sites, marketers for established artists should scour the Internet using search engines to find these fan-based pages. Then, contact the web site owner of each site and ask them to join the "online network" to receive updates. Most fans will jump at the chance to receive official

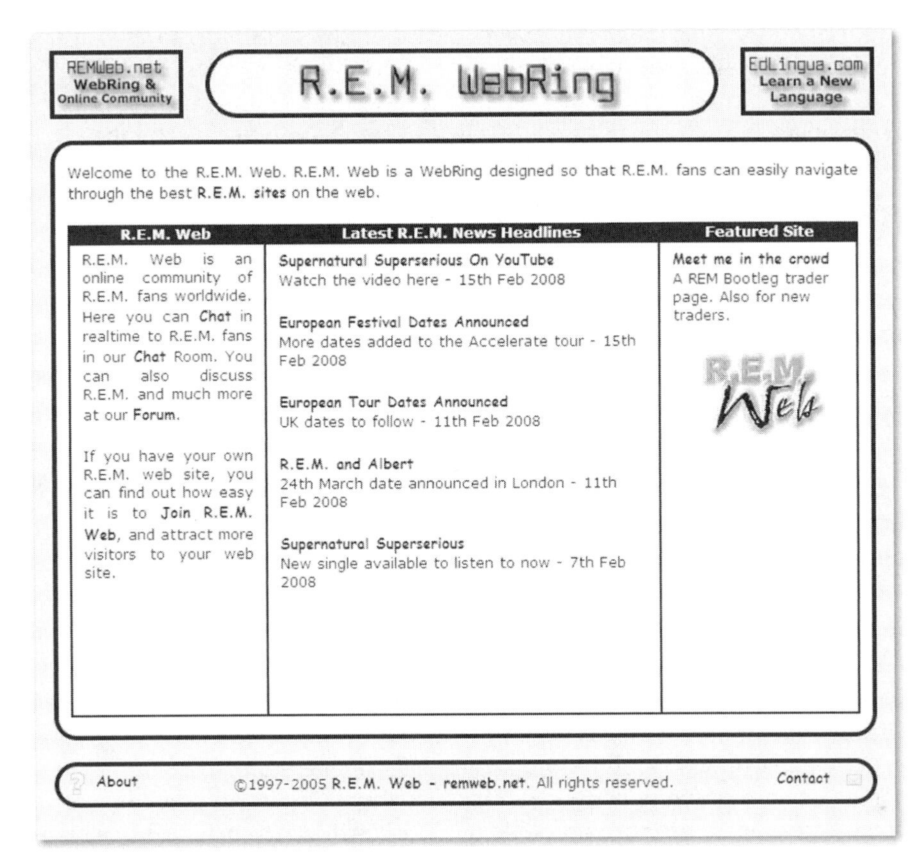

Figure 11.9
Example of a fan-based site (used with permission)

information and photos from the artist's management team before the general public has access to them.

Figure 11.9 presents an example of an R.E.M fan-based site, which shows how fans can successfully create and promote a site supporting their favorite artists. Care must be taken that no copyright infringement takes place and that all materials posted on the fan site are accurate and current. Notice how this site does not use officially licensed logos typically found on the official site, and no attempt has been made to deceive the visitor into thinking this might be the official artist site.

Tips on Helping Fans Maintain Their Fan-Based Sites

Social networking sites have recently facilitated fan-based music sites for users who may not have the savvy or initiative to create a standalone web site honoring their favorite music. Buzznet.com is one such social networking site that encourages its members to support and promote music on the site and rewards active trendsetters through the Buzzmob community. The Buzznet web site states, "The Buzzmob is the e-street team of Buzznet's most loyal fans who help promote Buzznet, bands and e-scene personalities across the Net. The Buzzmob is a place for Buzznet fanatics to get to know one another and help Buzznet grow" (www.buzznet.com/www/buzzmob/?t=footer|buzzmob).

RSS FEEDS TO FAN-BASED SITES

An RSS feed (really simple syndication) is frequently updated syndicated content published by a web site that appears on other web sites. It can also be used for distributing other types of digital content such as pictures, audio, or video. On Facebook, when a member changes out his or her primary photo, the updated photo appears in all the member's friends' "my friend spaces" and "bulletin spaces." Once information is in the RSS format, an RSS-aware program can check the feed for changes and react to the changes in the appropriate way.

Webmasters for recording artists have found that RSS feeds are an efficient way to update information on many "satellite" fan sites with little effort. This effort keeps the information on the artist current and accurate.

ADVERTISING

Online advertising revenue in the U.S. is predicted to surpass print advertising in 2012, after growing 23% in 2010 to $32 billion. In 2012 growth of 23.3% was expected, reaching $39.5 billion by year's end (eMarketer, 2012).

To generate traffic to a web site, it might be necessary to advertise on the Internet. Your return on investment (ROI) will help determine whether advertising is appropriate or not. For an emerging artist, most of your web site hits will come from word of mouth—announcing at gigs and posting your web address on all materials. Advertising would be more appropriate for a concert promoter or a record label, especially one that specializes in a niche form of music, such as Celtic. For example, when you type "Celtic music" into the Google search box, www.celticmusicusa.com pops up as a sponsored link. By visiting that web site, you learn of the vast catalog of music offered through this site. Because most advertisers pay based on the number of people who click on their ad, thus sending them to the advertiser's site (see the discussion on cost-per-click, presented later), those costs must be weighed against potential sales as a result of the advertising program. If, for example, only one in a hundred visitors to your site—generated through advertising—makes a purchase, and the average purchase is $10, how much would you be spending in advertising for each sale? If you pay 10 cents for each click, you have spent $10 in advertising for each sale—hardly worth the money. Sometimes companies will advertise anyway in hopes of building traffic, generating some repeat visits to the site, creating online word of mouth, and raising the site's profile in future search engine results.

Most Internet advertising takes the form of sponsored links, replacing the earlier style of choice—banner ads. Banner ads are basically a graphical

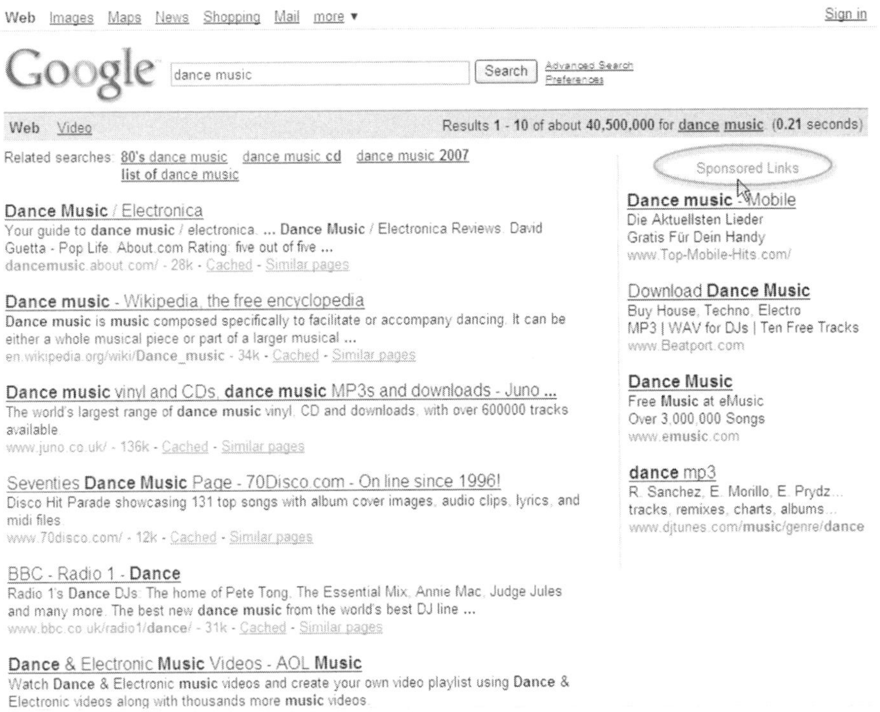

Figure 11.10
Example of keyword-driven ads on Google

advertisement placed across the top or down the side of a web page (also called a sidebar ad) that is linked to the advertiser's web site. Marshall Brain, in his article "How web advertising works," stated that "In the beginning [of the web] advertising on the Internet meant banner ads." This fueled a financial boom for the dot-com industry as advertisers switched more of their budgets from traditional media to the Internet. After this experimental phase, advertisers began to realize that banner ads were not as effective as magazines or "spot" commercials and scaled back on web advertising (Brain, 2002).

Advertisements would appear on popular content sites to pitch products available on e-commerce sites. Originally, these ad rates were set based on how it is done in the magazine industry—that is, they were based on the number of impressions. Thus, the advertiser paid based on the number of people who saw the ad. Now, the most popular form of charging for such advertising is "cost per click" (CPC), also known as pay per click (PPC). With CPC, the advertiser is charged a small fee each time a potential customer clicks on the ad and is taken to the advertiser's web site. This has proved more effective because the advertiser is paying only for those potential customers who respond to the ad by clicking on it and being directed to the advertiser's site. Google has popularized the use of sponsored links and offers web sites with similar content the ability to feature Google ads from sponsors who sign up for Google's AdWords.

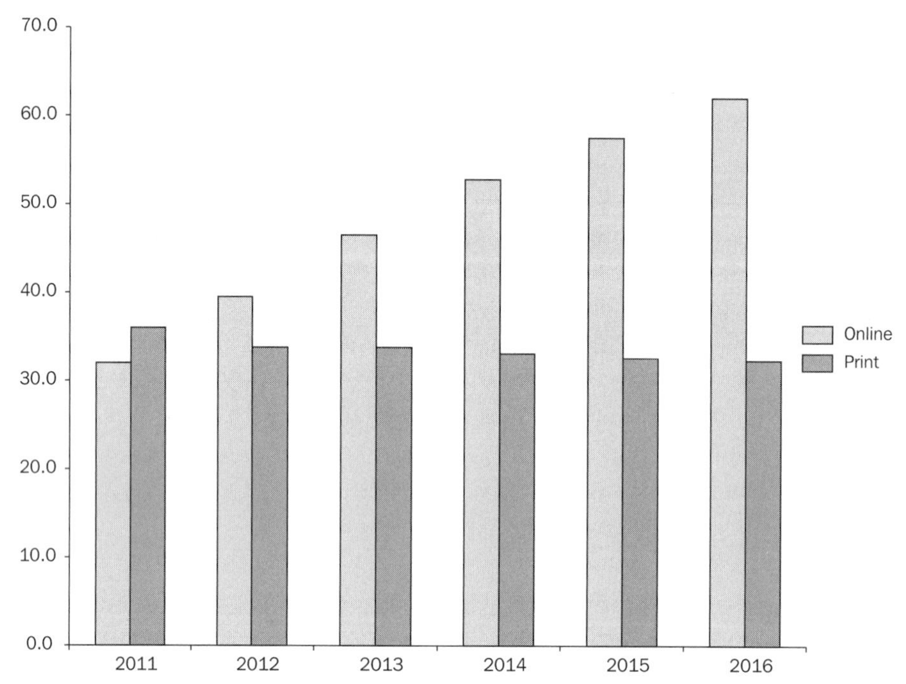

Figure 11.11
Internet advertising compared to print (source: eMarketer.com)

Here is how Google AdWords and Yahoo! Advertising Solutions work. As stated earlier, advertisers pay on a per-click basis; in other words, they pay a few cents for each time a web visitor clicks on their sponsored link. The advertiser enters in a series of keywords (search terms or words their customers are likely to use in a search engine when looking for a particular product or type of web site). Advertisers place bids to have their ad strategically located in the sponsored links category on search engine results. The advertiser can actually create a list of terms and bid independently on each one. The highest bid for that particular set of search terms has the top spot. You may not want to be the top spot for blues music if you depend on live shows for income because blues fans from all over would be likely to click to your site only to learn that you are not performing in their area. But if the term "blues music" was combined with "East Texas" and you are performing in that area, then perhaps you want one of the top advertiser spots. Finding the right keywords and combination of keywords may take a bit of trial and error at first.

Contextual Advertising

Contextual advertising is defined as advertising on a web site that is targeted to the specific individual who is visiting the web site based on the subject matter of the site and then featuring products that relate to that subject matter. For example, if the user is viewing a site about playing music and the site uses contextual advertising, the user might see ads for music-related companies such

Table 11.2 U.S. online advertising spending growth by format

	2010	2011	2012	2013	2014
Video	39.9%	52.1%	43.15%	35.9%	34.3%
Sponsorships	87.5%	26.4%	16.0$	12.3%	11.6%
Banner ads	23.1%	22.1%	17.6%	10.0%	10.4%
Search	12.2%	19.9%	18.4%	10.7%	7.1%
Classified and directories	15.2%	15.7%	11.4%	9.0%	8.9%
Rich media	2.2%	7.9%	4.3%	0.8%	–0.8%
Lead generation	–7.7%	6.1%	1.8%	1.7%	2.1%
Email	–33.2%	–16.5%	–0.5%	3.3%	3.4%
Total	14.9%	20.2%	17.6%	12.0%	10.4%

Source: www.eMarketer.com

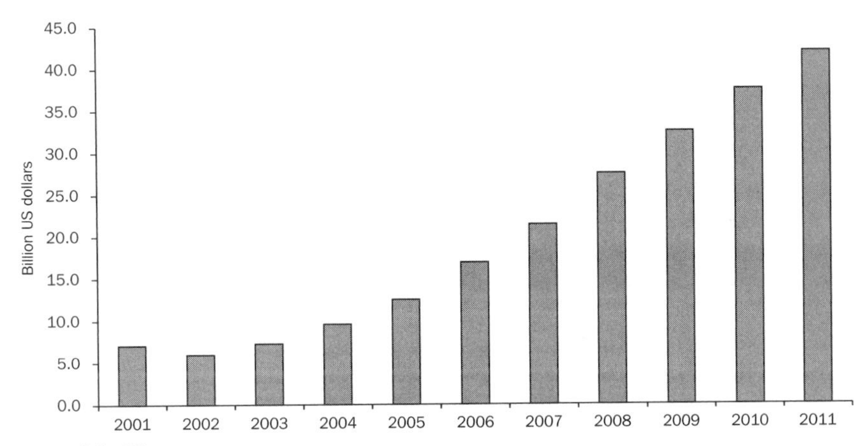

Figure 11.12
U.S. online advertising spending (source: www.eMarketer.com)

as music stores. Google has added AdSense, and more recently Google Content-Targeted Advertising as a way for web site owners to feature relevant advertising on their sites and share in the CPC revenue generated by sponsors. The source of these ads comes from the AdWords program, so that those who sign up for AdWords can specify if they want their ad to appear on these related web sites—the content network. Google's web site states, "A content network page might be a web site that discusses a product you sell, or a blog or news article on a topic related to your business."

In his article "10 facts and trends about contextual advertising," Larry Kim mentions the following:

- *Relevance.* The relevance of targeted, contextual advertising is constantly being improved through algorithms designed to match ads with site content.

- *Message customization.* Information gathered about the visitor to the site helps with a customized version of the advertisement beyond the context of the site where the ad appears.
- *A migration toward behavioral advertising.* In 2011, behavioral advertising started to appear within the contextual advertising realm. Behavioral advertising takes it a step further by customizing an advertisement not only related to the context of the web page in which it appears, but also customized to the particular visitor, based upon a profile of that visitor. This user profile is developed by tracking the user's web surfing behavior through the use of cookies. Click-through rates and conversion rates are much higher for behavioral advertising. Here is how it works: A web visitor has been to other sites, perhaps even shopping on Overstock.com and looking at particular products. Later, when that user visits a site with contextual advertising, such as a magazine's web site, any advertisements for Overstock.com will feature products similar to those the user had been viewing earlier.

CONCLUSION

This chapter has examined many of the promotional strategies available to musicians for web-based marketing campaigns. We have examined the importance of search engine optimization and how to submit web pages for inclusion in search engine results. Submitting articles and music to web-based publications such as e-zines and relevant bloggers is similar to techniques used by publicists for decades with traditional media.

The Web has made it possible for musicians to abandon direct mailings of postcards and newsletters to fans and concentrate on email newsletters at a fraction of the cost. Proper etiquette is important for developing email lists and formatting electronic newsletters. Bloggers represent new opportunities and blur the lines between journalists and fans.

The traditional efforts of getting radio airplay have their counterparts with Internet radio and streaming music services such as Pandora. Artists should make every effort to have their music available on these services and should register with SoundExchange to take advantage of any royalties generated from airplay. Grassroots marketing is important on the Web, and also includes creating cooperative efforts with other artists for the mutual benefit of all.

Viral marketing is a new concept made possible by the Internet—it's word of mouth on steroids. Various effective techniques have been developed to help a message spread through the virtual marketplace. Virtual street teams can play a role in viral marketing. And advertising has moved online. Expenditures on web advertising surpassed print advertising in 2012, because it is more targeted and has the potential to be more effective. All of these promotional efforts should be part of a larger plan that includes elements found in other chapters, such as social networking, setting up e-commerce, designing an effective web site, and much more. With web technology, "local" artists have a chance to compete on the world stage for a larger share of the global music business.

Glossary

AdSense An advertising program by which Google acts as the intermediary between content sites and web advertisers.

BCC (blind carbon copy) A way to send an email message without revealing the recipient's address.

Blogging or blog Short for weblog, it is a personal online journal that is frequently updated and intended for general public consumption.

Chat room Part of a web site that provides a venue for communities of users with a common interest to communicate to the group in real time.

Click through rate (CTR) The ratio/percentage of the number of times an ad is clicked divided by the number of times an ad is viewed.

Clipping service (See press clippings) A service that compiles relevant press clippings for a client.

Contextual advertising Advertising on a web site that is targeted to the specific individual who is visiting the web site.

Cost per click (CPC) The amount of money you pay to a search engine for one click on your ad.

Electronic press kit (EPK) A press kit equivalent in electronic form.

E-zines An electronic magazine, whether posted via a web site or sent as an email newsletter. Short for electronic magazine or fanzine, some are electronic versions of existing print magazines, whereas others exist only in the digital format.

Fan-based web sites Unofficial web sites or web pages featuring an artist, often an extension of the fan's personal web site.

Grassroots marketing Unconventional street-level marketing using word-of-mouth influence and that of opinion leaders to disseminate a marketing message among potential customers.

Guerrilla marketing Low budget, under-the-radar niche marketing, using both peer-to-peer (P2P) and any inexpensive top-down methods.

Keyword/keyword phrase The term or phrase that a user types in the search box of a search engine to receive more information that relates to that term/phrase.

Lurking Invisibly hanging out in chat rooms instead of actively participating in the discussion.

Mail merge The process of producing a personalized email letter for each person on a mailing list by combining a database of names and addresses with a form letter.

Message boards (online) A script on a web site with a submission form that allows visitors to post messages (called "threads" or "posts") on a web site for others to read. These messages are usually sorted within discussion categories, or topics, chosen by either the host or the visitor.

Online street team Similar to a record label street team, but its members engage in online word of mouth and other Internet marketing campaigns.

Peer to peer (P2P) A type of transient Internet network that allows a group of computer users with the same networking program to connect with each other and directly share files from one another's hard drives.

Podcast A digital media file that is distributed over the Internet for playback on portable media players and personal computers. Unlike real-time radio broadcasting, the podcast is usually played back at the convenience of the listener.

Press clippings Relevant excerpts cut from a newspaper or magazine.

Pureplay webcaster A company that provides webcasting or music streaming services as their core business. Examples include Pandora and Spotify.

Reciprocal link A mutual link between two objects, commonly between two web sites, in order to ensure mutual traffic.

Reprint A separately printed article that originally appeared in a larger or previous publication.

RSS feed (really simple syndication) A frequently updated and automated syndicated content feed provided by one web site to others for distributing digital content such as text, pictures, audio, and video.

Search engines Online directories of web sites and web pages that web visitors use to find a topic of interest or specific site.

Search engine optimization (SEO) A term that defines the sum of activities you run in order to promote your web site as high as possible among the organic results on a search results page.

Simulcast Simulcast is a contraction of "simultaneous broadcast" and refers to programs or events broadcast across more than one medium.

Sniping The act of putting up flyers or handbills in areas where the target market is known to hang out.

Social networking The practice of interacting with and expanding the number of one's business or social contacts by making connections typically through social networking web sites such as MySpace and Facebook.

SoundExchange An independent, nonprofit performance rights organization that is designated by the U.S. Copyright Office to collect and distribute digital performance royalties for featured recording artists and sound recording copyright owners when their sound recordings are performed on digital cable and satellite television, the Internet, and satellite radio.

Viral marketing A marketing phenomenon that uses online social networking to facilitate and encourage people to pass along a marketing message voluntarily and exponentially. Sometimes the marketing message is imbedded in, or attached to, a message that participants find interesting and are willing to forward to others.

Webcasting The concept of broadcasting much like a radio station, but via the Internet in lieu of a wireless signal.

Note

1 Levels of filtering in one's email program is usually consciously and proactively set by the user beginning with a default of no filtering on up to heavy filtering, then they work automatically.

Bibliography

American Media Services (2009). American media services survey shows popularity of Internet radio, even as regular radio continues holding its audience. http://www.americanmediaservices.com/press_releases/press_release_35.php

Arbitron (2010). The infinite dial 2010: Digital platforms and the future of radio. http://www.arbitron.com/downloads/infinite_dial_presentation_2010_reva.pdf

Arbitron (2012). Weekly online radio audience jumps more than 30 percent in past year says new Arbitron/Edison research study. Arbitron. http://arbitron.mediaroom.com/index.php?s=43&item=813

Atkinson, Robert (2007, July 12). The day the (web) music died. http://www.huffingtonpost.com/robert-d-atkinson-phd/the-day-the-web-music-d_b_56029.html

Borg, Bobby (2005). Web radio stations and getting played: A viable medium for new bands. www.indie-music.com/modules.php?name=News&file=article&sid=3604

Bracco, Chris (2011). How to really get your music on blogs. Tight Mix. www.TightmixBlog.com.

Brain, Marshall (2007). How web advertising works. April 8, 2002. http://computer.howstuffworks.com/web-advertising.htm

Brooks, Rich. Seven reasons to start an email newsletter today. Flyte New Media. www.flyte.biz/resources/articles/0303.php

Browne, Heather (2010). Getting heard by music blogs. ASCAP web site. http://www.ascap.com/playback/2010/05/features/gettingheard.aspx

Boyd, Danah, & Ellison, Nicole (2007). Social networking sites: Definition and conception. http://jcmc.indiana.edu/vol13/issue1/boyd.ellison.html.

CrunchBase (2012). Fanbridge company profile. http://www.crunchbase.com/company/fanbridge

Dawes, John, & Sweeney, Tim (2000). *The Complete Guide to Internet Promotion for Artists, Musicians and Songwriters*. Temecula, CA: Tim Sweeney and Associates.

Deitz, Corey (2005). Jimmy Buffet's Radio Margaritaville moves to SIRIUS Satellite Radio. http://radio.about.com/od/siriussatelliteprograms/qt/bljimmybuffet.htm

eMarketer, (2012). U.S. Online Ad Spend to Close in on $40 Billion.http://www.emarketer.com/Article.aspx?R=1008783

Experian Hitwise. Search Engine Analysis. http://www.hitwise.com/us/datacenter/main/dashboard-23984.html

Farrish, Bryan (2001). Radio Airplay 101: Traditional radio vs. the web. http://www.musicdish.com/mag/index.php3?id=3679

Gardyne, Allan (2005, December 20). How to get reciprocal links. http://www.associateprograms.com/articles/48/1/How-to-get-reciprocal-links

G-Lock EasyMail. Tips for successful Email marketing campaign. www.glocksoft.com

Hitwise (2007, March 14). IMeem and Bebo are the fastest movers. www.hitwise.com/press-center/hitwiseHS2004/socialnetworkingmarch07.php

Ives, Matt (2012). Online ad spending to pass print for the first time, forecast says. Ad Age. http://adage.com/article/mediaworks/emarketer-online-ad-spending-pass-print-time/232221/

Kim, Larry (2010). 10 Facts and trends about contextual advertising. The Search Engine Journal. http://www.searchenginejournal.com/10-facts-and-trends-about-contextual-advertising/24098/

MacPhearson, Michelle (2007, February 5). Viewing and posting new MySpace bulletins. www.articles3000.com/Internet/60055/Viewing-And-Posting-New-Myspace-Bulletins.html

Maloney, Paul (2012). Triton's Agovino wants radio to go for new revenue with online listening, not traditional on-air budgets. Radio and Internet Newsletter. http://www.kurthanson.com/news/triton%27s-agovino-wants-radio-go-new-revenue-online-listening-not-traditional-air-budgets

McKee, Ryan (2011, June 7). The 7 Best viral marketing campaigns in movie history. http://blog.moviefone.com/2011/06/07/super-8-best-viral-marketing-movies/

Mills, Elinor (2007). Advertisers look to grassroots marketing. news.com.com/2100-1024_3-6057300.html

Monash, Curt (2004). Search engine ranking algorithms. The Spider's Apprentice. www.monash.com/search_engine_ranking.html.

Nevue, David (2005). *How to Promote Your Music Successfully on the Internet*. Eugene, OR: Midnight Rain Productions..

PC Mechanic (2012) How to use a tell a friend script to drive traffic today. http://www.pctecmech.com/blog/Tell-A-Friend-Script.php

Pegoraro, Rob (2009). Web radio royalties resolved. *The Washington Post*. http://voices.washingtonpost.com/fasterforward/2009/07/web_radio_royalties_resolved_1.html

Search Engine Watch (2007). Submitting to directories: Yahoo! and the Open Directory. http://searchenginewatch.com/article/2065394/Submitting-To-Directories-Yahoo-The-Open-Directory

Search Web Services.com.

Skellie (2007). 10 tips for managing email effectively. Daily Blog Tips. http://www.dailyblogtips.com/10-tips-for-managing-email-effectively/

Sullivan, Danny. (2007a, March 14). Search engine placement tips. http://searchenginewatch.com/showPage.html?page=2168021#position.

Sullivan, Danny (2007b, March 15). How search engines rank web pages. http://searchenginewatch.com/showPage.html?page=2167961

Taglieri, John (2007, June 6). Internet radio—the new age of airplay. http://www.musesmuse.com/musicformats-4.html

Tiwary, Vivek J. (2008). Starting and running a marketing/street team. StarPolish. http://www.starpolish.com/advice/article.asp?id=31

Walker, Chris (2006). Search engine optimization versus pay-per-click. www.article99.com

Wilson, R. (2000, February 1). The six simple principles of viral marketing. *Web Marketing Today*, p. 70.

Wilson, Ralph (2000, June 1). Using banner ads to promote your website. *Web Marketing Today*.

www.adwordshowtos.com/learn/google-adwords-glossary.html

ZDnet.com (2008, February 20). 4.85 billion hours spent on online music in 2007. http://blogs.zdnet.com/ITFacts/?p=13961#

Chapter 12
Social Networking Sites

Social networking sites are defined as "web sites that allow members to construct a public or semipublic profile and formally articulate their relationship to other users in a way that is visible to anyone who can access their file" (Boyd and Ellison, 2007). Social networking sites allow anyone to have a personal web presence. Characteristics that have made social networking sites so popular include the following:

1. They allow for self-expression.
2. They require little or no knowledge of web design.
3. They allow for social interaction and networking.
4. They are free or inexpensive to use.

RISE AND POPULARITY OF SOCIAL NETWORKING SITES

The first significant social networking site was classmates.com, founded in 1995. Classmates.com allows for high school and college classmates and graduates to stay in touch. The site has been joined by, among others, Friendster.com, founded in 2002, MySpace, founded in 2003, and Facebook, founded in 2004. According to a press release from Hitwise (2004), a company that specializes in online web traffic and marketing, visits to social networking sites accounted for 6.5% of all Internet visits in February 2007. At that point in time, MySpace accounted for 80% of the market share, with Facebook holding 10% of the market. Other services such as Bebo, BlackPlanet, Classmates, Friendster, Orkut, and imeem accounted for less than 1% each. The same press release stated that "Buzznet and imeem are succeeding in building communities around music." By April 2007, Facebook had grown its market share to 11.5%, doubling its traffic since opening up its service to Internet users without school affiliation. The year 2007 was a watershed for music-oriented social networks as a host of upstarts were all vying to be the next MySpace. By the end of 2007, the other social networking services had begun to encroach on MySpace's commanding lead, which had declined slightly to 76% for the year and 72% for December 2007. Facebook continued its growth to 16% by December, with Bebo and BlackPlanet moving over the 1% mark. In early 2008, imeem acquired fledgling online retailer SNOCAP. imeem was the first network that revolved around music; relationships were formed and potential friends were recommended based on music preferences and the

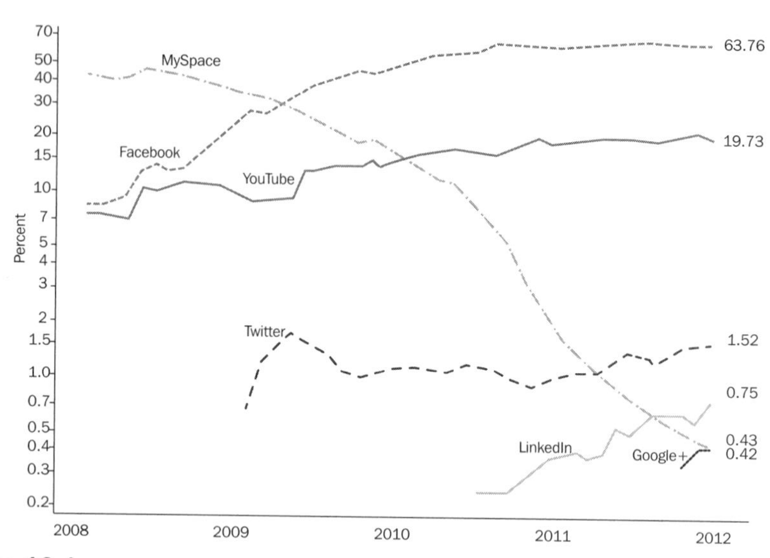

Figure 12.1
Social network market share by year

primary focus of interaction on the site was music. After struggling financially, imeem was acquired by MySpace in 2009 and shut down. By 2010, headlines such as "Twitter, Facebook soar as MySpace sags in U.S. market share" began to appear. Nielsen reported in 2010 that Facebook had increased its traffic by 70% in one year, for a total of 115 million users. Twitter was reaching 20 million U.S. users by mid-2010. Over the same time period, MySpace usage began to drop by 25% to 45 million page views. By January 2012, Facebook had almost 65% of the social network market, with YouTube in second place at almost 20% of the market. Twitter was a distant third with less than 2% of the market, and MySpace had fallen to less than 1% of traffic. At that time, the new upstart Google + also had less than 1% of social network traffic.

POPULARITY OF SOCIAL MEDIA

A 2011 study of Internet use by the Pew Research Center revealed that two-thirds of American adults use some type of social media (Smith, 2011). A 2011 Nielsen study found that in 2011, almost one-quarter of the time adults spent online was using social network sites. Staying connected to friends and family is the primary reason for usage. Middle age and older adults place high value on using social networking as a tool to connect with others around a given hobby or special interest, which includes music. In 2011, 65% of U.S. adults online were using social media sites, up 4% from the previous year, and compared to a mere 29% in 2008. Seventy percent of active social network users also shop online, 12% more likely than Internet users who do not use social media. Females are more likely than males to use social media, and people aged 18–34 are more likely to be social media users. When compared to the average adult Internet users, social network users are 75% more likely to be heavy spenders on music.

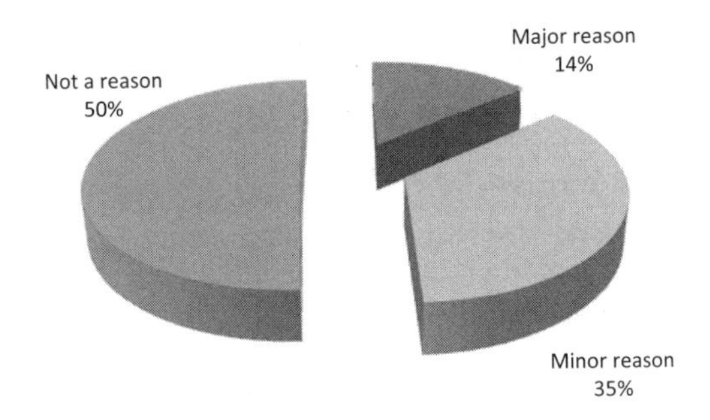

Figure 12.2
Use of social media to connect with others who share interests (source: Pew Internet Project)

GENERAL TIPS FOR MUSICIANS USING SOCIAL MEDIA

The article "Could social networking save the music industry," on www.cio-today.com (January 2007), professed that the new generation of social networking sites that emphasize music might be just what the industry needs to pull teens and young adults away from illegal file-sharing networks and back into the world of legal music consumers. These advertiser-supported sites offer what cio-today calls "a better form of free" for consumers.

Since that was written, the use of social networks has exploded and musicians today cannot ignore the importance of active participation. Several articles offer general tips for musicians on using social media to promote their music. Alex

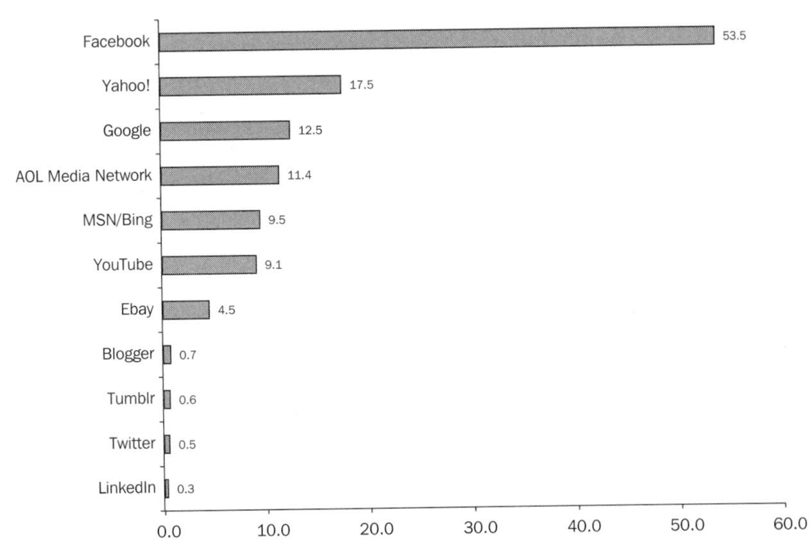

Figure 12.3
Minutes (in billions) spent on social networks by U.S. adults in May 2011 (source: Nielsen Social Media Report, 2011)

Pham of the *Los Angeles Times* wrote an article on "The five social media tips for indie musicians" in 2010. These are:

1. *Be real.* Don't focus solely on selling your product—connect with people.
2. *Stick with a couple services instead of hopping around.* Devote time to the few you have selected.
3. *Be unique.* Stand out from the crowd with something that gives you a competitive advantage.
4. *Share things that excite you.* Let the passion show.
5. *Embrace anarchy.* In other words, be flexible to the demands, expectations and limitations of socializing electronically.

Using social networking is a two-way street. Fans expect an interactive experience. Instead of posting information like a newsletter, a few efforts to create interactivity can vastly improve the effect of using social networks to sell your music. In an article titled "5 social media strategies to interact with your fans," Francis Bea has some suggestions to enhance the experience. A list based upon those five includes:

1. Recognize and reward loyalty with a "fan of the day" feature.
2. Music giveaways—a contest of sorts.
3. Set up a virtual scavenger hunt. You can enhance this by having fans search for something within your content, thus encouraging fans to comb through your content.
4. Make use of crowdsourcing. Crowdsourcing is described as outsourcing tasks to a loosely defined group of people to accomplish the task. It becomes an informal, temporary virtual street team.
5. Provide question and answer sessions and make every attempt to answer each inquiry.

MYSPACE

At the time of the first edition of this book, MySpace dominated the marketplace for social networks. Even then in 2008, Facebook was beginning to pull market share from MySpace. The introductory paragraph for MySpace was as follows.

> Because MySpace currently has the lion's share of social networking traffic at the time of this writing, it should be looked on as the most important social networking site worthy of marketing efforts. However, this could change as other sites vie to compete in the marketplace, hoping to offer new and exciting services to lure web visitors away from MySpace and Facebook.
>
> **Hutchison, 2008. *Web Marketing for the Music Business*, first edition**

Clearly, social network popularity is a moving target. In January 2012, the beleaguered MySpace launched a rebranding effort supported by investor Justin

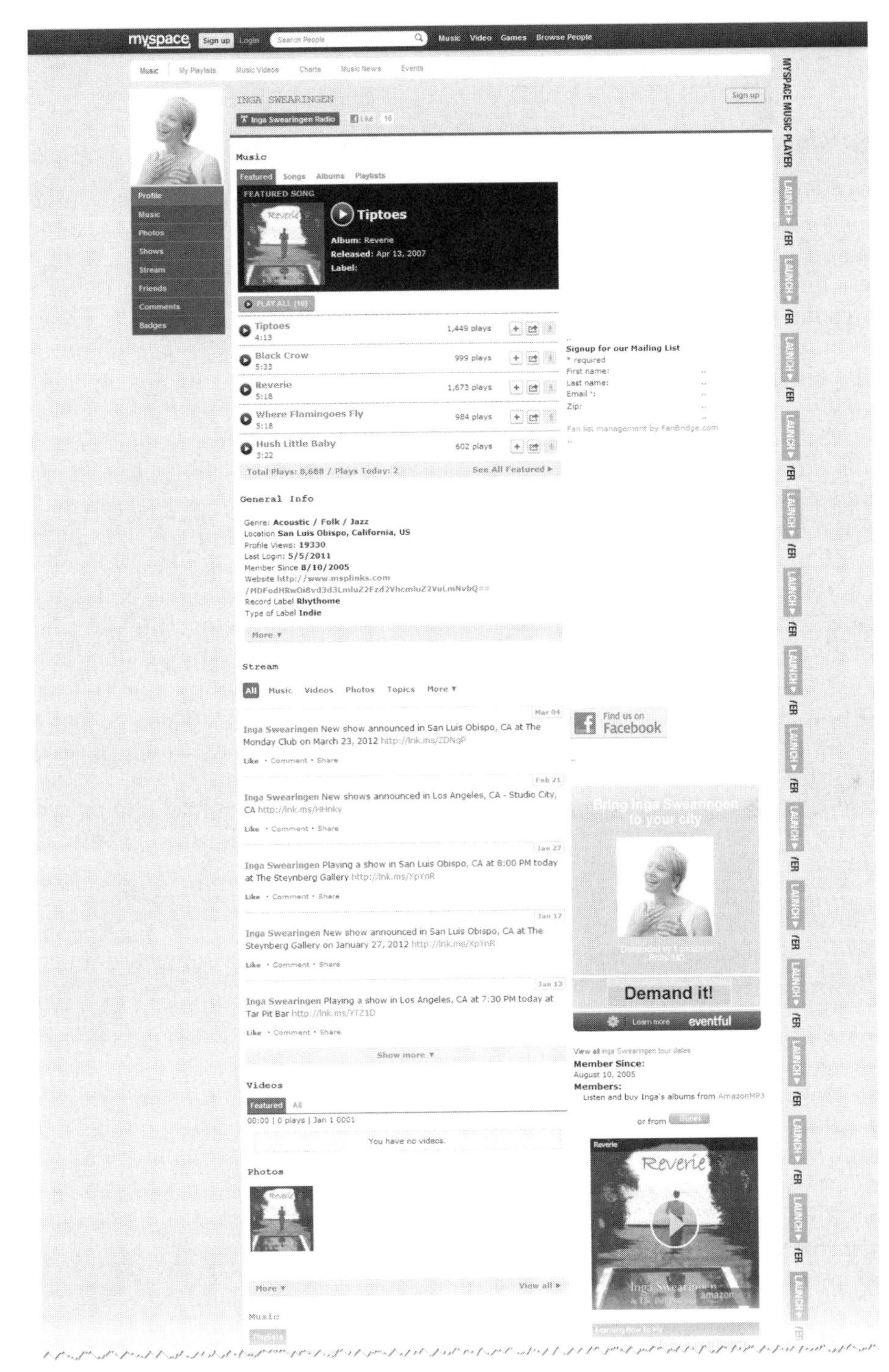

Figure 12.4
MySpace page for jazz artist Inga Swearingen (with permission from MySpace)

Timberlake after severing ties with News Corporation. The company planned to focus on music first (Greenwald, 2012).

SIGNING UP AND SETTING UP YOUR BAND SITE

For artists, MySpace's setup is ideal for promoting music. Some of the features for artists include a built-in music player, user ratings, reviews, artist rankings, featured bands, show listings by location, and music videos. Signup is easy, but you must select between a personal account or a musician account (they also offer accounts for comedians and filmmakers). By selecting "musician" under "Account Type" the signup form changes to reveal genre and label. Under the pull down "Label Type" menu, there is an option for unsigned artists. After signing up, there are a series of pages to set up the site, including uploading an image.

The band's name will become the artist's address at www.MySpace.com/artistname. To complete the registration, you get the chance to list the artist's web site address, the two genres under which you want the artist categorized, and the artist's label affiliation.

At that point, you have the option to upload music, videos and photos. On the "Artist-To-Do" menu, options are available anytime to customize the band profile, upload additional music, photos and videos, sync to Twitter and Facebook, install a Facebook App and to set up a mobile app. Through the "Edit Profile" button on the right side of the screen, you have the option of including crucial artist details in the artist profile, such as selecting and customizing a theme, posting a bio, and list of band members, listing upcoming shows, musical influences and comparisons, selecting three genres to describe the music. At this point, you get the chance to invite friends to sign up and to upload photos and audio files. You can specify whether listeners are allowed to download the song, stream it to their computers and share it with others.

To change the default photo on your site, scroll over the existing profile photo and an option box pops up.

To sell music on MySpace you need to have your music distributed through one of the MySpace partners listed below:

- www.ingrooves.com
- www.iodalalliance.com
- www.irisdistribution.com
- www.redeyeusa.com
- www.theorchard.com
- www.ada-music.com
- www.nettwerk.com
- www.redmusic.com
- www.carolinedist.com
- www.fontanadistribution.com
- www.dittomusic.com
- www.cdbaby.com
- www.tunecore.com

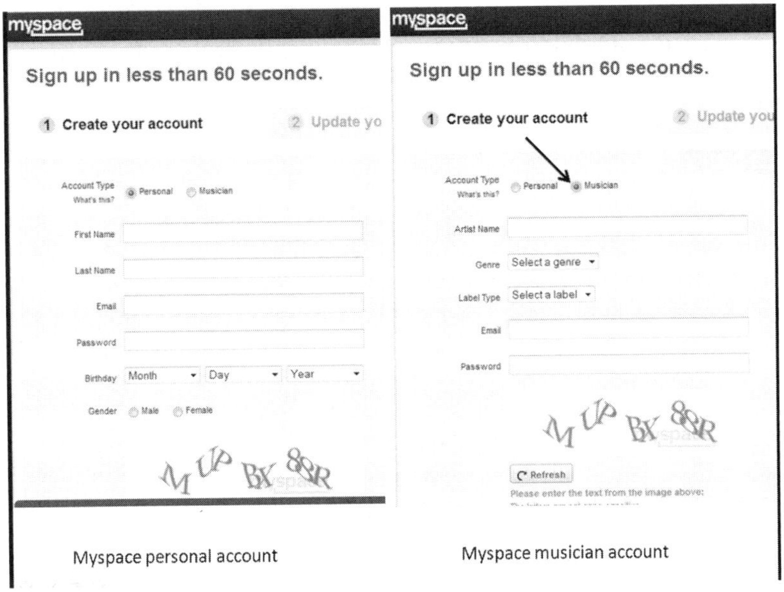

Myspace personal account

Myspace musician account

Figure 12.5
Registering a musician account on MySpace

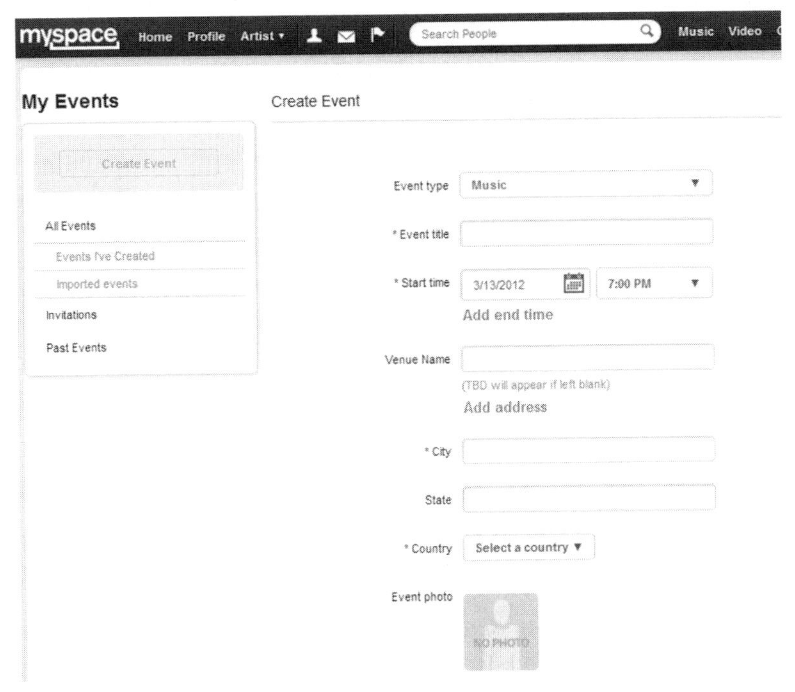

Figure 12.6
Add an event on MySpace (with permission from MySpace)

According to MySpace, "Once the distributor targets your content, your profile will have the official logo in searches, albums will show up under the Music tab on your profile, and the content within that album will have enabled buy links."

FACEBOOK

At the time of this edition, Facebook had the lion's share of social network users. Originally available only to students with a school affiliation, Facebook opened its ranks to the public in September 2006 and doubled its traffic in one year (Hitwise). In May 2007, the site opened its platform, allowing software developers the chance to create integrated programs. Many other sites, including MySpace, have developed widgets to feed content to your Facebook page. This has been accelerated by Open Graph Apps, described as an arrangement with third-party developers to allow for seamless integration with Facebook.

In early 2012, Facebook unveiled a new feature called the Timeline. Facebook describes it as "the new Facebook profile. Tell your life story through photos, friendships and personal milestones like graduating or traveling to new places" (web site). The Timeline feature allows users more control over featuring and hiding or burying certain items. One nice feature for bands is the ability to post a banner image that remains at the top of the page (Hyatt, 2012). This is good for branding. The feature allows the user to "pin" certain posts and items at the top of the Timeline, rather than having them disappear down the page as new items are entered. So, important information such as upcoming shows and new releases can retain a prominent spot near the top of the page. Another important feature mentioned by Ariel Hyatt in her post "Musician's Arsenal: Killer Apps, Tools & Sites – Facebook's Timeline," is highlight—the ability to highlight certain posts and stretch them across the entire width of the page. A third benefit of Timeline is the customization of content for each fan, incorporating items deemed to be more of interest to that visitor. Facebook has added an "Admin Panel" for viewing new notifications.

The only drawback initially was that the artist could not designate landing pages because of the removal of the tabs option for determining landing pages. Within three weeks of the Timeline launch, third party developers responded to complaints and analytics (that indicated reduced traffic) by offering tab applications via Open Graphs Apps. In the March 15, 2012 article "Tab applications for musicians look to innovate after Facebook removes default landing function," author Brittany Darwell stated

> ReverbNation, maker of Band Profile, announced a suite of new apps for specific functions. The icons for these apps take advantage of the larger thumbnails now available for tab applications. Instead of having a single app with videos, tour dates, merchandise and more, musicians can add individual apps with easy-to-identify icons that users are more likely to click on and use.
>
> **Brittany Darwell, 2012**

Figure 12.7
Facebook page with Timeline banner (courtesy of Facebook and Joseph Akins)

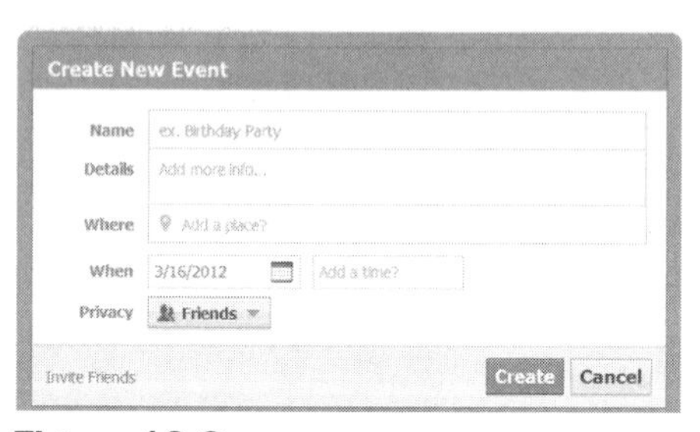

Figure 12.8
Create a new event on Facebook

Open Graph Apps were introduced by Facebook in early 2012, designed to seamlessly integrate third-party services into Facebook pages. Among the third-party companies participating in Open Graph Apps are ReverbNation and Vevo. The Open Graph App project was still in the early stages at printing time of this book.

Advice on how to maximize a Facebook presence should begin with determining what you want to accomplish. In his blog "10 tips to make your Facebook page work better" (anonymous, 2011), tip #1 is to determine your reasons for having a presence: to build a fan mailing list, to build stronger relationships with fans, to direct fans to your own web site, and so forth. Other noteworthy tips include committing to four posts per day, focus on quality and consistency rather than quantity of content, check your stats, and in addition to posting your thoughts, ask the fans for theirs. This tip is echoed by Brenna Ehrlich (2011) in her article "10 best practices for bands on Facebook." Ehrlich encourages artists to ask for input from the fans by making use of Facebook Questions. Facebook Questions is a feature that lets you get recommendations, conduct polls and learn from your friends and other people on Facebook. The poll feature is simple to use.

1. Click "Ask Question" in your sharing menu.
2. Enter a question and click "Add Poll Options." If you'd like to create a poll where people can add options, make sure the "Allow anyone to add options" box is checked.
3. Choose who can see your question by using the audience selector, then click Post.

Other best practices from Ehrlich include reaching out to other artists, creating the reciprocal relationship described in Chapter 11 under "Turning your competition into partners." She recommends going beyond the music by posting other things of interest, including plenty of visuals, and she suggests "making everything an event." Any time you post an event, it shows up in your fans' news feed.

TWITTER

Twitter is the popular short messaging social network designed for frequent but brief status updates. The service is popular among politicians who want to keep voters updated, celebrities who want to communicate to fans, and active people who want to be continuously in touch with their network of friends and followers. Twitter was introduced in 2006 as a way to send instant text messages to a group of followers, rather than the more common method of phone to phone. The company intentionally limited the content to 140 characters because the normal cell phone text message was limited to 160 characters (Picard, 2011). Messages are called "Tweets."

Twitter is a valuable tool for delivering "just in time" messages and has been influential in shaping world events such as the "Arab Spring" of pro-democracy movements in the Middle East, and for distributing communication after the Japanese earthquake and tsunami in 2011. College students have used it to create "flash mobs." Wikipedia defines a flash mob as "a group of people who assemble suddenly in a public place, perform an unusual and seemingly pointless act for a brief time, then disperse, often for the purposes of entertainment, satire, and artistic expression." In 2011, there were over 175 million Twitter accounts with 150 million messages sent daily.

All musicians who are actively recording and touring should have a Twitter account and build a fan base of followers. In mid-2011, the most popular Twitter accounts (those with the most followers) were Lady Gaga, Justin Bieber, and Britney Spears.

Twitter can be a valuable tool for music promotion, but should not be considered a replacement for Facebook or MySpace. The mission of a Twitter account is quite different. Twitter is more mobile; over half of Twitter users access their tweets via mobile devices. In the About.com article "Use Twitter for music promotion," Heather McDonald says that successful promotion for musicians depends on how you use it. She recommends that you begin by following others to get a feel for how it works, to actively jump into the conversation, tweet wisely, and use it to announce news events. She provides some ideas about things to tweet, including updates from the recording studio, updates on the manufacturing and distribution process (i.e. "arriving in stores two weeks from today"), updates from the road when on tour, and day-to-day work news.

Twitter allows music fans a window into the life and daily activities of the artist; it's the next best thing to being in the studio or backstage. Content should be a mixture of daily life, thoughts and reflections, and a soft sell for the music, shows and web visits. Tweets such as "check out my web site" are not compelling enough. Social interaction is the point behind tweeting, so creating a dialog with fans is more effective. Be regular with tweets, not a blast in one day and then silence for two weeks. But this needs to be balanced with relevance and avoid the "too much information" (TMI) syndrome. Tweets should be interesting: a news blurb, a reaction to a news event, a window into the artist's personal life, thoughts and feelings about life as a musician. Creating music is a process, and fans like to be engaged in that process.

You can also "Retweet" information of interest, passing it along using the RT in front of the username of the person whose tweet you are passing along. Set up a reciprocal "retweet" agreement with other local bands so that you are expanding the fan base for both artists. That will also help provide additional content of interest.

From a technical standpoint, the Twitter account should be linked to Facebook and other networks, so that updates are automatically cross posted. Hashtags are a method for putting your tweet into a thread of related tweets about a specific subject. This is likely to expand the reach of your tweets and provide searchability for Twitter users. This is done by using the symbol # in front of a keyword that relates to the subject. A list of common hashtag categories can be found at www.hashtag.org. Setting up a Twitter account is easy but is best done via a computer instead of a mobile device. It is important to create a profile to provide background information for potential followers. Then, you need to write a 160-character biography and include the band's web site address in the profile. The bio should include keywords that accurately describe the music and other information to help people find the Twitter account using a Google search. It is important to actively build up a list of followers and to provide a widget on your web site that says "follow us on Twitter."

YOUTUBE

YouTube is a video-sharing web site that opened in 2005; it has become very popular and created quite an impact in popular culture. Candid and bootleg videos have rocked political campaigns, spawned a new generation of videographers, launched new celebrities, and exposed others (i.e. the Michael Richards incident[1]). Artists have been exploiting YouTube to promote themselves and their videos, both professional and amateur. Much like video cameras produced a new genre of America's Funniest Videos, YouTube has become an outlet for everyone with a camera. Internet marketer Dan Ackerman Greenberg has described the strategy behind using promotional techniques to increase the popularity of videos on YouTube.

Both stars and emerging indie bands have found a home on YouTube for music videos. With the popularity of homemade videos, bands can create a music video or concert video at a fraction of what it would have cost just a few years ago. Postings on YouTube should be supported by a campaign to spread the word. Video "channels" can be created, and more traffic can be generated by joining forces with similar artists to set up a "channel" and promote it to the fans of all participating artists.

In 2010, YouTube launched a partner program for musicians (Ehrlich, 2010). The Musicians Wanted channel categorizes and features music videos by unsigned artists. The concept had not caught on by 2012 and only had 10,000 viewers signed up. YouTube created a Top Musician's Chart in 2011, and introduced the new YouTube music page that would feature recommendations in addition to the most popular music videos. In 2012, YouTube posted the following description of a musician account.

> A musician account is a type of YouTube account. In addition to performer information, Musician accounts can publish a schedule of show dates.
>
> **www.youtube.com**

A musician with an account can also sell merchandise through YouTube and their affiliates: Topspin, Songkick, iTunes and Amazon. According to YouTube, Topspin allows you to sell: t-shirts, posters, tickets, downloads, other physical/ digital goods; Songkick allows you to sell tickets, iTunes for music and Amazon for music and other products (http://support.google.com/youtube/bin/answer.py?hl=en&answer=1711134).

The Secret Strategies Behind Many "Viral" Videos: A Case Study

Dan Ackerman Greenberg

Have you ever watched a video with 100,000 views on YouTube and thought to yourself: "How did that video get so many views?" Chances are pretty good that this didn't happen naturally, but rather that some company worked hard to make it happen—some company like mine.

When most people talk about "viral videos," they're usually referring to videos like Miss Teen South Carolina, Smirnoff's Tea Partay music video, the Sony Bravia ads, Soulja Boy—videos that have traveled all around the Internet and been posted on YouTube, MySpace, Google Video, Facebook, Digg, blogs, etc. These are videos with millions and millions of views.

Here are some of the techniques to get at least 100,000 people to watch "viral" videos:

Secret #1: Not All Viral Videos Are What They Seem

There are tens of thousands of videos uploaded to YouTube each day. (I've heard estimates between 10–65,000 videos per day.) I don't care how "viral" you think your video is; no one is going to find it and no one is going to watch it.

Our clients give us videos and we make them go viral. Our rule of thumb is that if we don't get a video 100,000 views, we don't charge. So far, we've worked on 80–90 videos and we've seen overwhelming success. In the past 3 months, we've achieved over 20 million views for our clients, with videos ranging from 100,000 views to upwards of 1.5 million views each. In other words, not all videos go viral organically. There is a method to the madness.

We've worked with: two top Hollywood movie studios, a major record label, a variety of very well-known consumer brands, and a number of different startups, both domestic and international. This summer, we were approached by a Hollywood movie studio and asked to help market a series of viral clips they had created in advance of a blockbuster. The videos were 10–20 seconds each, were shot from what appeared to be a camera phone, and captured a series of unexpected and shocking events that required professional post-production and CGI [computer generated content]. Needless to say, the studio had invested a significant amount of money in creating the videos. But every time they put them online, they couldn't get more than a few thousand views.

We took six videos and achieved:

- 6 million views on YouTube
- ~30,000 ratings
- ~10,000 favorites
- ~10,000 comments
- 200+ blog posts linking back to the videos
- All six videos made it into the top 5 Most Viewed of the Day, and the two that went truly viral (1.5 million views each) were #1 and #2 Most Viewed of the Week.

The following principles were our secrets to our success.

2. Content Is NOT King

If you want a truly viral video that will get millions of people to watch and share it, then yes, content is key. But good content is not necessary to get 100,000 views if you follow these strategies.

Don't get me wrong: content is what will drive visitors back to a site. So a video must have a decent concept, but one shouldn't agonize over determining the best "viral" video possible. Generally, a concept should not be forced because it fits a brand. Rather, a brand should fit into a great concept. Here are some guidelines to follow:

- Make it short. 15–30 seconds is ideal; break down long stories into bite-sized clips
- Design it for remixing. Create a video that is simple enough to be remixed over and over again by others. Ex: Dramatic Hamster

- Don't make an outright ad. If a video feels like an ad, viewers won't share it unless it's really amazing. Ex: Sony Bravia
- Make it shocking. Give a viewer no choice but to investigate further. Ex: UFO Haiti
- Use fake headlines. Make the viewer say, "Holy crap, did that actually happen?!" Ex: Stolen NASCAR
- Appeal to sex. If all else fails, hire the most attractive women available to be in the video. Ex Yoga 4 Dudes

These recent videos would have been perfect had they been viral "ads" pointing people back to web sites:

- Model Falls in Hole on Runway
- Cheerleader Gets Run Over by Football Team
- PacMan: The Chase
- Dude
- Dog Drives Car
- Snowball—Dancing Cockatoo

3. Core Strategy: Getting onto the "Most Viewed" Page

Now that your video is ready to go, how is it going to attract 100,000 viewers?

The core concept of video marketing on YouTube is to harness the power of the site's traffic. Here's the idea. Something like 80 million videos are watched each day on YouTube, and a significant number of those views come from people clicking the "Videos" tab at the top. The goal is to get a video on that Videos page, which lists the Daily Most Viewed videos.

If you succeed, the video will no longer be a single needle in the haystack of 10,000 new videos per day. It will be one of the 20 videos on the Most Viewed page, which means that you can grab 1/20th of the clicks on that page! And the higher up on the page your video is, the more views you are going to get.

So how do you get the first 50,000 views you need to get your videos onto the Most Viewed list?

- Blogs. Reach out to individuals who run relevant blogs and actually pay them to post your embedded videos. Sounds a little bit like cheating/PayPerPost, but it's effective and it's not against any rules.
- Forums. Start new threads and embed your videos. Sometimes, this means kick- starting the conversations by setting up multiple accounts on each forum and posting back and forth between a few different users. Yes, it's tedious and time consuming, but if you get enough people working on it, it can have a tremendous effect.
- MySpace or Facebook. Plenty of users allow you to embed YouTube videos right in the comments section of their MySpace or Facebook pages. Take advantage of this.
- Facebook. Share, share, share. Take Dave McClure's advice and build a sizeable presence on Facebook, so that sharing a video with your entire friends list can have a real impact. Other ideas include creating an event that announces the video launch and inviting friends, writing a note and tagging friends, or posting the video on Facebook Video with a link back to the original YouTube video.
- Email lists. Send the video to an email list. Depending on the size of the list (and the recipients' willingness to receive links to YouTube videos), this can be a very effective strategy.
- Friends. Make sure everyone you know watches the video and try to get them to email it out to their friends, or at least share it on Facebook.

Each video has a shelf life of 48 hours before it's moved from the Daily Most Viewed list to the Weekly Most Viewed list, so it's important that this happens quickly. When done right, this is a tremendously successful strategy.

4. Title Optimization

Once a video is on the Most Viewed page, what can be done to maximize views?

4:07
Sunshine State
Grant Peeples sings his submission to the New State Song of Florida Contest. btw: he didn't win.
... dfusselman ... grant peeples sunshine state ...
by dfusselman | 4 years ago | 50,692 views

Figure 12.9
YouTube example of a video thumbnail

It seems obvious, but people see hundreds of videos on YouTube, and the title and thumbnail are an easy way for video publishers to actively persuade someone to click on a video. Titles can be changed a limitless number of times, so have a catchy (and somewhat misleading) title for the first few days, then later switch to something more relevant to the brand. Recently, I've noticed a trend towards titling videos with the phrases "exclusive," "behind the scenes," and "leaked video."

5. Thumbnail Optimization

If a video is sitting on the Most Viewed page with 19 other videos, a compelling video thumbnail is the single best strategy to maximize the number of clicks the video gets.

YouTube provides three choices for a video's thumbnail, one of which is grabbed from the exact middle of the video. As you edit your videos, make sure that the frame at the very middle is interesting. It's no surprise that videos with thumbnails of half-naked women get hundreds of thousands of views. Not to say that this is the best strategy, but you get the idea. Two rules of thumb: The thumbnail should be clear (suggesting high video quality) and ideally it should have a face or at least a person in it.

Also, when you feel particularly creative, optimize all three thumbnails, then change the thumbnail every few hours. This is definitely an underused strategy, but it's an interesting way to keep a video fresh once it's on the Most Viewed list.

6. Commenting: Having a Conversation with Yourself

Every power user on YouTube has a number of different accounts. So should you. A great way to maximize the number of people who watch your videos is to create some sort of controversy in the comments section below the video. Get a few people in your office to log in throughout the day and post heated comments back and forth (you can definitely have a lot of fun with this). Everyone loves a good, heated discussion in the comments section—especially if the comments are related to a brand/startup.

Also, don't be afraid to delete comments. If someone is saying your video (or your startup) sucks, just delete their comment. Don't let one user's negativity taint everyone else's opinions.

We usually get one comment for every 1,000 views, since most people watching YouTube videos aren't logged in. But a heated comment thread (done well) will engage viewers and will drive traffic back to your sites.

7. Releasing All Videos Simultaneously

Once people are watching a video, how do you keep them engaged and bring them back to a web site?

A lot of the time our clients say: "We've got five videos and we're going to release one every few days so that viewers look forward to each video."

This is the wrong way to think about YouTube marketing. If you have multiple videos, post all of them at once. If someone sees your first video and is so intrigued that they want to watch more, why would you make them wait until you post the next one? Give them everything up front. If a user wants to watch all five of your videos right now, there's a much better chance that you'll be able to persuade them to click through to your web site. Don't make them wait after seeing the first video, because they're never going to see the next four.

Once your first video is done, delete your second video, then re-upload it. Now you have another 48-hour window to push it to the Most Viewed page. Rinse and repeat. Using this strategy, you give your most interested viewers the chance to fully engage with a campaign without compromising the opportunity to individually release and market each consecutive video.

8. Strategic Tagging: Leading Viewers Down the Rabbit Hole

YouTube allows you to tag your videos with keywords that make your videos show up in relevant searches. For the first week that your video is online, don't use keyword tags to optimize the video for searches on YouTube. Instead, you can use tags to control the videos that show up in the Related Videos box. Why?

I like to think about it as leading viewers down the rabbit hole. The idea here is to make it as easy as possible for viewers to engage with all your content, rather than jumping away to "related" content that actually has nothing to do with your brand/startup.

So how do you strategically tag? Choose three or four unique tags and use only these tags for all of the videos you post. I'm not talking about obscure tags; I'm talking about unique tags—tags that are not used by any other YouTube videos. Done correctly, this will allow you to have full control over the videos that show up as "Related Videos."

When views start trailing off after a few days to a week, it's time to add some more generic tags, tags that draw out the long tail of a video as it starts to appear in search results on YouTube and Google.

9. Metrics/Tracking: How to Measure Effectiveness

The following is how to measure the success of your viral videos.

For one, tweak the links put up on YouTube (whether in a YouTube channel or in a video description) by adding "?video=1" to the end of each URL. This makes it much easier to track inbound links using Google Analytics or another metrics tool.

TubeMogul and VidMetrix also track views/comments/ratings on each individual video and draw out nice graphs that can be shared with the team. Additionally, these tools follow the viral spread of a video outside of YouTube and throughout other social media sites and blogs.

10. Conclusion

The Wild West days of Lonely Girl and Ask A Ninja are over. You simply can't expect to post great videos on YouTube and have them go viral on their own, even if you think you have the best videos ever. These days, achieving true virality takes serious creativity, some luck, and a lot of hard work. So, my advice: Fire your PR firm and do it yourself.

This was written by Dan Ackerman Greenberg, co-founder of viral video marketing company The Comotion Group and lead TA for the Stanford Facebook Class. Dan graduated from the Stanford Management Science & Engineering Masters program in June 2008.

http://techcrunch.com/2007/11/22/the-secret-strategies-behind-many-viral-videos/. Reprinted by permission of author

---

OTHER SOCIAL NETWORKING SERVICES

Vimeo is a video sharing service seen as an alternative to YouTube and has a free version that is similar to YouTube, but it also has a professional version that provides more flexibility in design and usage. While YouTube enjoys the lion's share of traffic and may provide more exposure, Vimeo is a good alternative as a professional service for small businesses.

LinkedIn is a professional network, designed to foster business relationships. While it's not the place to showcase your music, LinkedIn connections can be very beneficial for the business aspect of one's musical career. There are plenty of groups dedicated to the music business, including Music and Entertainment Professionals, Music Industry Forum, and MusicBiz.

Google+. Google launched this social network site in 2011. Unique features include integration with other Google products and the ability to create subgroups called "Circles." Content can be easily shared with one Circle while being withheld from other Circles. This makes targeting easier with specialized content. Another feature is "Hangouts." According to Google, "Instead of directly asking a friend to join a group chat, users instead click 'start a hangout' and they're instantly in a video chatroom alone. At the same time, a message goes out to their social circles, letting them know that their friend is 'hanging out.' Friends can then join the hangout as long as they have been placed in a Circle that was invited by the person who created the Hangout" (Google).

Shortly after the launch, Google launched the Google Music Service (GMS), designed to integrate with Google+ (CD Baby, 2011). Features include integration with Android for mobile music delivery, and the ability to share music with Google+ friends. For a one-time $25 fee, artists can set up a Google Music Account. Google keeps 30% of sales income.

Pinterest is an image-sharing social network site. According to an article by Lisa Irby (2012) of CreateAWebsite.com, it is a great way to generate traffic to your web site. Pinterest members create boards and "pin" images to them. It's the visual counterpart to Reddit and Digg. According to Irby, there are two ways to generate web traffic: (1) add your web site address to your Pinterest profile, and (2) pinned images are linked back to the source.

CONCLUSION

A social networking campaign is a necessary component for musicians in the twenty-first century. Their presence and popularity has been a boon for DIY and local musicians who lack the resources of the major and large indie labels. They have allowed local musicians to go global with their fan base. But social network marketing is an ongoing process, not a "campaign" with a beginning and end. Some larger businesses have been slow to discover that a static campaign does not succeed. Working the social network circle requires ongoing attention, fostering a dialog with fans, and providing fresh content regularly. Musicians should have a presence in all the major social network platforms and should provide material appropriate to the platform. Cross platform content updates have become more commonplace as social networks provide more opportunities for "one content feed." But specialized attention should be given to each platform, as each serves a different purpose and/or a different market.

Lists of Social Networking Sites

- The Social Network list: http://mashable.com/2007/10/23/social-networking-god
- The Wikipedia list: http://en.wikipedia.org/wiki/List_of_social_networking_websites
- Author's updated list: www.focalpress.com/cw/Hutchison
- The ultimate Web 2.0 reference site: www.go2web20.net

Glossary

Blogging or blog Short for weblog, it is a personal online journal that is frequently updated and intended for general public consumption.

Crowd sourcing Outsourcing tasks to a loosely defined group of volunteers to accomplish. Designed to tap the collective intelligence or skill sets of an undefined public group of people.

Flash mob A group of people who assemble suddenly in a public place, perform an unusual and seemingly pointless act for a brief time, then disperse, often for the purposes of entertainment, satire, and artistic expression.

Hypertext markup language (HTML) The predominant authoring language for the creation of web pages. HTML defines the structure and layout of a web document by using a variety of tags and attributes.

Open Graph Apps Introduced by Facebook in early 2012, designed to seamlessly integrate third-party services into Facebook pages.

Social networking The practice of interacting with and expanding the number of one's business or social contacts by making connections typically through social networking web sites such as MySpace and Facebook.

URL Uniform Resource Locator, the global address of documents and other resources on the Web. The first part of the address is called a protocol identifier, and it indicates what protocol to use; the second part is called a resource name, and it specifies the IP address or the domain name where the resource is located.

Viral marketing A marketing phenomenon that uses online social networking to facilitate and encourage people to pass along a marketing message voluntarily and exponentially. Sometimes the marketing message is imbedded in, or attached to, a message that participants find interesting and are willing to forward to others.

Web widgets A widget is a small application that can be ported to and run on different web pages by a simple modification of the web page's HTML.

WYSIWYG Pronounced WIZ-zee-wig, short for "what you see is what you get." A WYSIWYG application enables you to see on the display screen exactly what will appear when the document is printed or published to the Web.

Note

1 Michael Richards, who played Kramer on Seinfeld, was caught on video in a racially-tinged rant at a comedy club. One of the club's patrons used a cell phone to capture the moment and it was publicly disseminated, much to the dismay of Richards.

Bibliography

Bea, Francis (2011). 5 social media strategies to interact with your fans. Dotted Music. http://dottedmusic.com/2011/marketing/5-strategies-using-social-media-to-interact-with-your-fans/

Boyd, Danah, & Ellison, Nicole (2007). Social networking sites: Definition, history, and scholarship. *Journal of Computer-Mediated Communication* 13 (1) http://jcmc.indiana.edu/vol13/issue1/boyd.ellison.html

Cio-today (2007, January 30). Could social networking save the music industry?

Darwell, Brittany (2012). Tab applications for musicians look to innovate after Facebook removes default landing function. Inside Facebook. http://www.insidefacebook.com/2012/03/15/tab-applications-for-musicians-look-to-innovate-after-facebook-removes-default-landing-function/

Davies, Chris (2012, January 17). Facebook open graph apps due Wednesday tip sources. SlashGear. http://www.slashgear.com/facebook-open-graph-apps-due-wednesday-tip-sources-17209560/

Emerson, Ramona (2011). New study probes how we use social media. The Huffington Post. http://www.huffingtonpost.com/2011/09/28/social-media-study_n_985102.html?view=screen

Ehrlich, Brenna (2010). YouTube launches partner program for musicians. Mashable Social Media. http://mashable.com/2010/03/17/youtube-musicians-wanted/

Ehrlich, Brenna (2011) 10 best practices for bands on Facebook. July 11. http://mashable.com/2011/07/11/bands-facebook/

Friedman, Jacob (2010). Twitter, Facebook soar as MySpace sags in U.S. market. The Next Web. http://thenextweb.com/us/2010/05/04/twitter-facebook-soar-myspace-sags-market-share/

Greenberg, Dan Ackerman (2007). The secret strategies behind many "viral" videos: A case study. Reprinted from *Web Marketing for the Music Business*, 1st ed. http://techcrunch.com/2007/11/22/the-secret-strategies-behind-many-viral-videos/

Greenwald, Will (2012). MySpace is reborn at Panasonic press conference, unleashes Justin Timberlake. *PC Magazine*. www.pcmag.com/article2/0,2817,2398677,00.asp

Hyatt, Ariel (2012). Musician's arsenal: Killer apps, tools & sites—Facebook's timeline. Music Think Tank. http://www.musicthinktank.com/blog/musicians-arsenal-killer-apps-tools-sites-facebooks-timeline.html

Irby, Lisa (2012). Using pinterest to build traffic. 2CreateAWebsite.com. http://www.2createawebsite.com/traffic/pinterest-tips.html

Kalis, Priit (2012). Top 10 social networking sites by market share of visits [January 2012]. Dreamgrow. http://www.dreamgrow.com/top-10-social-networking-sites-by-market-share-of-visits-january-2012/

Make It In Music (2012). 10 tips to make your Facebook music page work better. www.makeitinmusic.com/facebook-musician-tips/

McDonald, Heather (2011). Use Twitter for music promotion. About.com http://musicians.about.com/od/musicpromotion/ht/twitterformusic.htm

Nevue, David (2005). *How to Promote Your Music Successfully on the Internet*. Eugene, OR: Midnight Rain Productions.

Pham, Alex (2010, January 29). Five social media tips for indie musicians. *The Los Angeles Times*. http://latimesblogs.latimes.com/technology/2010/01/five-social-media-tips-for-indie-musicians.html

Picard, André (2011). The history of Twitter, 140 characters at a time. *The Globe and Mail*. http://www.theglobeandmail.com/news/technology/tech-news/the-history-of-twitter-140-characters-at-a-time/article1949299/

Robley, Chris (2011). Google music store officially launches: How does this affect CD Baby artists? http://diymusician.cdbaby.com/2011/11/google-music-store-officially-launches-how-does-this-affect-cd-baby-artists/

Smith, Aaron (2011). Why Americans use social media. Pew Research Center. www.pewinternet.org.

Chapter 13

Professional, Legal and Ethical Issues

The worst mistake an Internet newbie can make is quitting his day job. The Web is filled with how-to books that promise a fortune of riches and success to those who buy and follow the book. While the Internet is full of opportunities to engage in commerce, competition is fierce. This chapter will address some of the Internet marketing no-no's, pitfalls, and outdated or useless strategies that may still be touted online. The Web is full of top ten lists of "things to avoid," mistakes, "no-no's," etc. Maybe the first item to mention should be "Leave it to Letterman and give us a break on the top ten lists." But here goes.

WEB SITE MISTAKES

1. *Launch when ready*–one of the worst mistakes a web designer can make is to prematurely launch a web site before it is ready and has undergone beta testing. Check for browser compatibility, spelling errors, broken links, missing components, etc., before telling the world about your site. This includes signing up with search engines. That should be done only after the site is in good working order. Ask friends and family to "test drive" the web site and report back any irregularities.[1]
2. *Oops! I did it again*—another mistake, one described by author Robyn Tippins in *What not to do on your web site* is "do not post things you will regret." Sometimes, words posted on a web site, blog or tweet can come back to haunt the author, just as public comments made by politicians can cause a backlash. And with the printed word on a web site, it's even more difficult to deny it, because ... there it is in writing. College graduates are finding that their shenanigans posted on Facebook are hindering in the job search. All comments, text, photos, email replies, etc., should be carefully critiqued as to their impact, before hitting that submit button.
3. *Go lightly on the ads*—too many ads clutter up a web site, compete for the viewer's attention, and drive people away. A professional site should concentrate on the product for which the site is created, instead of a billboard for selling ad space. Tippens suggests placing ads beneath a story or blog, in the left sidebar, or "wherever the eye naturally falls."

   With the introduction of Google AdSense, many web sites that were formerly ad-free have become littered with ads in an effort for the site owner to profit from maintaining the site. Many of these sites used to be a "labor of love" but they are now commercial enterprises. That is fine for the hobbyist, but professional music business entities, including artists, labels and affiliated music business web sites, should not have such advertising messages on the site except those that promote the products related to the purpose of the site.

In his book *How to Promote Your Music Successfully on the Internet,* author David Nevue states that banner ads can actually hurt your promotional efforts. There are many companies now on the Internet that foster banner exchanges, much like the link exchanges mentioned in Chapter 11. Among the reasons to stay away from such arrangements are (1) they lure visitors away from your site before they've had time to explore it, (2) they are unsightly and detract from the site, and (3) they can often be poorly targeted for products and services that have nothing to do with your site.

4. *Design for your audience, not yourself*—in his book *Don't make me think,* Steve Krug (2006) says to make web pages self-evident; the user is not familiar with your web site, so features should be intuitive and based on convention. Web site visitors don't want to spend a lot of time and energy trying to figure out what to do and where to find what they are looking for. He states that different web team members may have different perspectives of what is important for a good site; designers think design is most important, and developers may think lots of cool features and functions are important. In his article "Five big online marketing mistakes," author Ian Lurie states the mistake is in "thinking cool design equals good marketing." He states there is a difference between using the right design to appeal to customers and the cool design preferred by the company CEO. Don't let strong personal convictions cloud the issue—the site is not for you; it's for your customers.
5. *Put function before form*—don't let all the bells and whistles take priority over the purpose of the site. The visual aspect of the site should support the site's function, not detract from it. In the music industry, it is common to find music samples on web sites, but the visitor should be given the option to turn off the music if it is detracting from their mission or slowing down the process.[2]
6. *Relying too much on Flash*—one common theme among the lists is using slow-loading programs, too much Flash, making visitors wait while these programs load. This mistake was the first listed in several articles. An article on the AllBusiness site states that if you make viewers wait too long, "you can say goodbye to your potential customers." You need to build fast-loading web pages. A Flash-based logo is simply an ego stroke. No one cares as much about your animated logo as you do.

WEB MARKETING MISTAKES

In an attempt to recruit more business, commercial Internet marketing firms are generally more than willing to point out common marketing mistakes made by amateurs. Many have articles on the top five or top ten mistakes commonly made in Internet marketing. The following is a selection of those mistakes most

Table 13.1 Thirty-six web site mistakes that can kill your business

Poor load time	Multiple banners and buttons
Poor overall appearance	Use of frames
Spelling/grammar	Large fonts
No contact information	Pop up messages
Poor content	Over use of Java
Poor navigation	Poor use of tables
Broken links and graphics	Poor organization
Poor browser compatibility	Poor use of mouse over effects
Large slow loading graphics	Overpowering music set to autoplay
Too many graphics	Too much advertising
Pages scrolling to oblivion	Large welcome banners
Multiple use of animated graphics	Multiple colored text
Animated bullets	Text difficult to read
Too many graphic and/or line dividers	Multiple use of different fonts
Different backgrounds on each page	Confusing or broken links
Busy, distracting backgrounds	Under construction signs
Confusing	Large scrolling text across the page
No meta tags	Bad contrast between text and background

Source: Adapted from Shelley Lowery

Figure 13.1
This site boasts that it's the world's worst site: http://www.angelfire.com/super/badwebs/

likely to impact web marketing for musicians and their business entities (Pingel, 2004; Lurie, 2006; Gardner, 2000; Geisheker, 2006; Brauner, 2010; McGuinness, 2011). Many of these mistakes are merely omissions of important marketing recommendations found elsewhere in this book.

- *Depending too much on search engines:* A commonly mentioned mistake has to do with search engine optimization (see Chapter 7). They advise not to depend too heavily on SEO to draw traffic but also not to neglect submitting to search engines and optimizing. Lurie states "too many consultants are running around telling companies that they can achieve nirvana by simply changing meta-tags." An article on ShoestringBranding.com states

  > if we find ourselves spending too much time fiddling with page titles, changing the wording in our headlines or worrying about our keyword density, it is time to stop and ask ourselves if that time wouldn't be better utilized creating useful content and building loyal customers ...

- *Failing to collect information and act appropriately:* From email addresses to traffic information, consultants recommend that you get as much feedback and contact information from your visitors as is reasonable. You need to know as much about your visitors and your market as possible to be effective at marketing. However, moderation is the key in requiring visitors to divulge information about themselves (see following paragraph). And use that information with discretion.
- *Complexity:* Another commonly mentioned faux pas deals with how hard the visitor must work to find what they are looking for. In his article "Do you make these internet marketing mistakes?" Peter Geisheker says it's very annoying to try to navigate sites "where you have to click on generic graphics or parts of a flash picture to try and find the link to the information you are looking for." He says keep it simple. The AllBusiness article warns against "giving users the third degree" by making them jump through hoops or answer a battery of questions to move ahead with the transaction. They state that every question "beyond name and email address will cost you 10 to 15 percent of your potential customers."
- *Not knowing your competition*: Several articles mentioned the importance of "competitive intelligence" or competitive analysis. You can learn from observing what the competition is doing and analyzing their strengths and weaknesses.
- *Not focusing on your market or niche*: One advisor states "the more focused your message, the more it will influence your audience" (Brauner, 2010). Don't try to be all things to all people. Nothing discourages a potential customer more than messages that are clearly not targeted for them or what they are looking for.
- *Failing to have a coordinated plan:* All web marketing efforts should be designed to work together, creating synergy, to motivate the customer toward your call to action (making a purchase). Marketing efforts that work in a vacuum do not have a long-lasting effect, if any at all. This includes coordinating your online marketing with your offline efforts, which should not be neglected.

- *Social narcissism:* As described by Mark McGuinness (2011) in his article "The 10 biggest mistakes artists and creatives make at Internet marketing," the artist's fans are not interested in every detail of their day—what they had for breakfast. Name dropping or other navel-gazing activities are not the way to build long-lasting social network relationships. McGuinness says "It's not about you. It's about what you mean to your audience." Fans are interested in the work—the adventure, so keep social network postings focused on something meaningful for the fans.
- *Not using viral marketing:* Advertising and search engine listing are important but in the music business, the power of word of mouth is overwhelming. Viral marketing achieves this. AllBusiness suggests the "tell-a-friend" option online and the use of branded T-shirts that your customers can wear for offline WOM marketing. The web site e-consultancy offers several reasons why viral marketing campaigns fail. Among them are:

  1. *Neglecting seeding*—taking advantage of mailing lists, press releases, forums, etc., where the marketing message can be placed.
  2. *Failing to create an incentive for users to pass it along*—what good is a marketing message that is not passed along? It's not viral if it doesn't spread. Attach your message to something that is likely to be forwarded. Eighty-eight percent of web users say they have forwarded jokes.
  3. *Trying to copy a popular viral campaign*—what works in one situation may not translate to your specific needs.
  4. *Failing to integrate viral campaigns with other marketing efforts*—the integration of online and offline promotions is covered in another chapter in this book. The AllBusiness article states that "no one is online all the time. To successfully market an Internet site, you need to market offline, too."
  5. *Using a sledgehammer instead of a scalpel*—often, simple ideas—such as email signatures—produce better results than more complex campaigns.
  6. *Forgetting to ask the user to take action*—the success of a viral campaign depends upon the recipients actually doing something, such as visiting your web site, voting for your artist in some contest, or purchasing the product.

More good advice is available from Larry Brauner at http://online-social-networking.com/page/9.

INTERNET MARKETING ETHICS

Unethical business practices seem to be of major concern for the past decade, ever since Enron was first toppled. Since then, consumers and law enforcement have been more diligent when dealing with companies. We hear of various scams being perpetrated on the Internet, and warned about malicious software that can wreak havoc on your computer or Facebook account. It makes it hard to tell what is legitimate and what is fly-by-night. Anyone dealing with e-commerce needs to understand consumers' reluctance to click on suspicious links or give out personal information, especially credit card, banking or social security numbers. Predatory computer hackers roam the Internet causing problems and spreading malicious software.

Misrepresentation

Deliberate misrepresentation which causes another person to suffer damages, usually monetary losses, is fraud. There have been elements of the recording industry in the past that have crossed the line and been subjected to prosecution or lawsuits. *Payola* is a prime example, where the popularity of songs heard on the radio was not based upon decisions made by program directors, but instead based upon how much payment was exchanged in consideration of airplay. The payola law does not make it illegal to pay for airplay, but simply states that such payment must be disclosed to listeners. Why is it then so important to record labels that they not disclose that payment was made to secure airplay? It is because they want the public to believe the song is only being played because that is what the public requests to hear.

Other areas of marketing have existed in the gray area of ethics to where a certain amount of deception is now considered standard. When "nine out of ten doctors recommend" a certain product, you have to wonder who those nine doctors are and who they work for. In advertising, celebrity spokespeople are hired and paid to endorse products. Product features are enhanced. Photographs are manipulated to show products in the best light, and press releases are created to spin information in favor of the company sponsoring the release.

Having said that, a certain amount of perception is fabricated and manufactured by marketers to give the appearance of a naturally occurring rise in popularity for certain cultural products. Specifically, record labels and artist managers hire people to pose as fans to create a word-of-mouth buzz about the artist. Internet marketers are routinely charged with the task of creating multiple fictitious profiles on social networking sites for the purpose of posing as fans and promoting the artist. It is not that different from hiring celebrity spokespeople or actors to pose as consumers to tout products in advertisements. The ethical dilemma in both situations comes from the fact that there is no disclosure of who these spokespeople really are. The Internet makes it even more difficult for consumers to determine who they are getting information from since profiles can be easily falsified. Web 2.0 has created an atmosphere of user-generated content that is not subjected to the usual fact-checking, editing and accountability. Wikipedia and MySpace have been dealing with these issues for the past few years; other social networking sites may start employing more stringent methods of verification as these problems emerge. But for now, the Web is still the Wild West when it comes to integrity and accountability. A certain amount of consumer skepticism is to be expected.

Scamming, Phishing, and Pharming

There is no shortage of scam artists on the Internet, employing all sorts of techniques to separate honest people from their money. The most notorious is

DEAR SIR,

URGENT AND CONFIDENTIAL BUSINESS PROPOSAL

I AM MARIAM ABACHA, WIDOW OF THE LATE NIGERIAN HEAD OF STATE, GEN. SANI ABACHA. AFTER HE DEATH OF MY HUSBAND WHO DIED MYSTERIOUSLY AS A RESULT OF CARDIAC ARREST, I WAS INFORMED BY OUR LAWYER, BELLO GAMBARI THAT, MY HUSBAND WHO AT THAT TIME WAS THE PRESIDENT OF NIGERIA, CALLED HIM AND CONDUCTED HIM ROUND HIS APARTMENT AND SHOWED HIM FOUR METAL BOXES CONTAINING MONEY ALL IN FOREIGN EXCHANGE AND HE EQUALLY MADE HIM BELIEVE THAT THOSE BOXES ARE FOR ONWARD TRANSFER TO HIS OVERSEAS COUNTERPART FOR PERSONAL INVESTMENT.

ALONG THE LIME, MY HUSBAND DIED AND SINCE THEN THE NIGERIAN GOVERNMENT HAS BEEN AFTER US, MOLESTING, POLICING AND FREEZING OUR BANK ACCOUNTS AND EVEN MY ELDEST SON RIGHT NOW IS IN DETENTION. MY FAMILY ACCOUNT IN SWITZERLAND WORTH US$22,000,000.00 AND 120,000,000.00 DUTCH MARK HAS BEEN CONFISCATED BY THE GOVERNMENT. THE GOVERNMENT IS INTERROGATING HIM (MY SON MOHAMMED) ABOUT OUR ASSET AND SOME VITAL DOCUMENTS. IT WAS IN THE COURSE OF THESE, AFTER THE BURIAL RITE AND CUSTOMS, THAT OUR LAWYER SAW YOUR NAME AND ADDRESS FROM THE PUBLICATION OF THE NIGERIAN BUSINESS PROMOTION AGENCY. THIS IS WHY I AM USING THIS OPPORTUNITY TO SOLICIT FOR YOUR CO-OPERATION AND ASSISTANCE TO HELP ME AS A VERY SINCERE RESPONSIBLE PERSON. I HAVE ALL THE TRUST IN YOU AND I KNOW THAT YOU WILL NOT SIT ON THIS MONEY.

I HAVE SUCCEEDED IN CARRYING THE FOUR METAL BOXES OUT OF THE COUNTRY, WITH THE AID OF SOME TOP GOVERNMENT OFFICIAL, WHO STILL SHOW SYMPATHY TO MY FAMILY, TO A NEIGHBOURING COUNTRY (ACCRA-GHANA) TO BE PRECISE. I PRAY YOU WOULD HELP US IN GETTING THIS MONEY TRANSFERRED OVER TO YOUR COUNTRY. EACH OF THESE METAL BOXES CONTAINS US$5,000,000.00 (FIVE MILLION UNITED STATES DOLLARS ONLY) AND TOGETHER THESE FOUR BOXES CONTAIN US20,000,000.00(TWENTY MILLION UNITED STATESDOLLARS ONLY). THIS IS ACTUALLY WHAT WE HAVE MOVED TO GHANA.

THEREFORE, I NEED AN URGENT HELP FROM YOU AS A MAN OF GOD TO HELP GET THIS MONEY IN ACCRA GHANA TO YOUR COUNTRY. THIS MONEY, AFTER GETTING TO YOUR COUNTRY, WOULD BE SHARED ACCORDING TO THE PERCENTAGE AGREED BY BOTH OF US.PLEASE NOTE THAT THIS MATTER IS STRICTLY CONFIDENTIAL AS THE GOVERNMENT WHICH MY LATE HUSBAND WAS PART OF IS STILL UNDER SURVAILLANCE TO PROBE US.

YOU CAN CONTACT ME THROUGH MY FAMILY LAWYER AS INDICATED ABOVE AND ALSO TO LIAISE WITH HIM TOWARDS THE EFFECTIVE COMPLETION OF THIS TRANSACTION ON TEL/FAX NO:xxx-x-xxxxxxx AS HE HAS THE MANDATE OF THE FAMILY TO HANDLE THIS TRANSACTION.

THANKS AND BEST REGARD

MRS. MARIAM ABACHA

Figure 13.2
Nigerian email scam letter

the Nigerian 419 email scam. It preys upon people's sense of greed by purporting to have a large sum of inheritance money to give to the recipient, but only after they pay some transfer fees. These fees start out small, but once the scammer has hooked a victim, they keep going back for more fees. The unsuspecting victim then agrees to pay the other fees in an attempt to recoup the previous fees they have paid in an attempt to claim their "prize."

Other scams—called *phishing* and *pharming*—are not as notorious, more personal in nature and are designed to elicit private financial information that the scammers can use for identity fraud. Here is how phishing works: Internet fraudsters send spam email messages that direct the victim to a web site that looks just like a legitimate organization's site, typically a bank or PayPal. The site is not the official bank site but a bogus site that looks like it could be real.

It is designed to trick you into divulging your personal financial information by telling you that there is a problem with your bank account and you need to enter your password and other identification numbers to rectify the problem. The email will contain links to the bank's sites, with the legitimate bank domain name in the link, but the HTML code directs the victim to a different site. HTML code for links can be written as <a href="www.scam-operators-fake-site.com"> www.ReputableBank.com</a>. Note the web address is different from the text for the link.

Figure 13.3
Fraudulent email link redirects users to the scammer's web site

Pharming is the latest threat. In the article "Privacy and the Internet: traveling in cyberspace safely," pharming is described as "criminals' response to increasing awareness about phishing." Consumers have been alerted to the presence of phishing schemes and have been warned that, when receiving a message that purports to be from your bank, the safest option is to type in the web address of that site, rather than click on the embedded links in the email. Pharming works by redirecting the visitor after they have gone to the legitimate site. Presently, it is only possible for the scam artists to instigate on "http" sites, not on the more secure" https" (s for secure) pages.

Building Trust with Your Customers

Building customer trust and confidence when doing transactions on the Internet is an important aspect of e-commerce. In order to achieve your goal of converting web visitors into buyers, you must be aware of customers' concerns and take steps to overcome these. E-commerce Insights discusses seven top customer e-commerce fears and how to address each of those (Rodenborg, 2011).

1. *The top fear is credit card information being stolen.* Security concerns still cause online shoppers to abandon a purchase, if they sense a "red flag." Over 63% of shoppers who abandon shopping carts do so because of security concerns. Others are concerned about the information they are asked and how that will be used.

   To give your web customers a feeling of security, it is wise to become a member of one of the Internet security verification companies, such as VeriSign, BizRate or GoDaddy. Web customers need an easy way to see that their transactions are protected and that they are dealing with reputable firms. What has emerged is the online equivalent of the Better Business Bureau: companies that provide Internet security and verify the web site owner through a certificate authority. A certificate authority or certification authority (CA) is an entity which issues digital certificates for use by other parties. It is an example of a trusted third party. A June 2010 market share report from SecuritySpace.com determined that the company VeriSign and its acquisitions has a 44.2% market share of the certificate authority market, down from a 57.6% share in 2007. GoDaddy's aggressive pricing of under $100 per year has encroached on market share, closing the gap (and placing them second behind Verisign (Lewis, 2010). As a result, GeoSign has dropped their prices and offers a free trial.
2. *Fear that the web site is not a real store, but a phishing scam.* As mentioned in the phishing section above, scammers have gotten clever about mimicking a company's web site, often by stealing logos and design. This can be addressed by using clear, high-quality images when branding and a brand-specific URL. Also, provide contact information on the site.

   Another alternative is to make your products available with at least one of the reputable online retailers, for those customers who would otherwise hesitate to give out their credit card information to an unknown web site. It may cut in to your bottom line for profit on each unit, but will probably increase overall sales and exposure—keeping in mind that 63% of online shoppers are concerned about credit card theft.
3. *Fear that their information will be sold to marketers and email spammers.* This concern can be addressed with a clear privacy policy that states how the

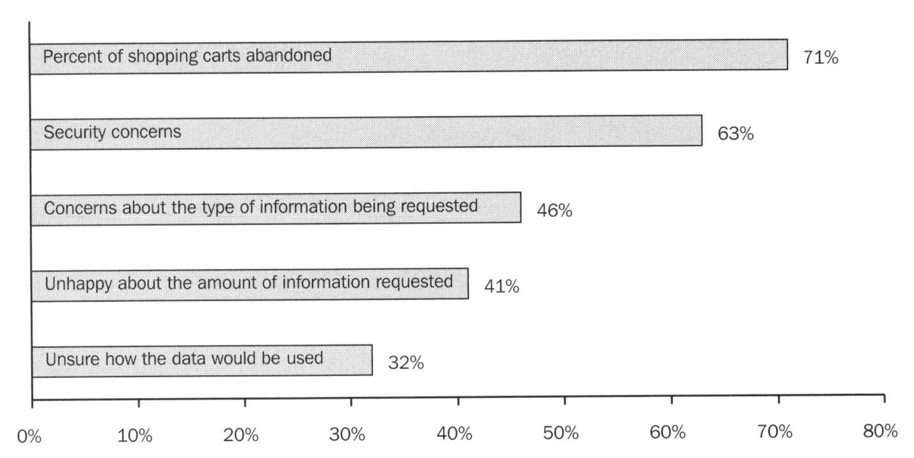

Figure 13.4
Reasons why online shoppers abandon shopping carts (source: BizReport.com)

information will and will not be used. Provide opportunities for customers to opt-out of receiving email specials and information from retail partners.

4. *Not being able to tell what a product is really like.* Customers like to touch or sample the items they are considering for purchase, like kicking the tires on a new car. For physical items for sale, provide a variety of images, product details, and video demonstrations. Customer reviews are also extremely valuable, with over half of online shoppers relying on them in 2011. For music products, extensive sampling is paramount. For reserved-seating concert tickets, a schematic of the venue with an indication of the exact seats the consumer is purchasing will alleviate concerns.
5. *Not being able to track orders.* One consumer fear is that their order didn't go through, and if they click again, they will be charged for multiple orders. This can be remedied with an immediate, automated email order confirmation to the customer. Additional emails should be sent when the order is processed and again when shipped, with a tracking number included. Provide tracking capabilities from your web site.
6. *Needing assistance from sales personnel.* The online shopping experience can be a solitary experience, with no opportunity to ask questions as a customer would normally do in a store. Remedy this by offering an instant chat function, so customers can ask questions. Several SaaS companies offer "Chat now" services. One company, Zopim, offers a free version, limited to two concurrent chats. The subscription versions start at less than $10 per month. The subscriber receives a widget for the web site, and dashboard access to the service features, which include online chat history, and rerouting to smartphones.
7. *Fear of being stuck with an item they don't like.* Again, customer reviews can assist with this problem. A simple return policy can also help.

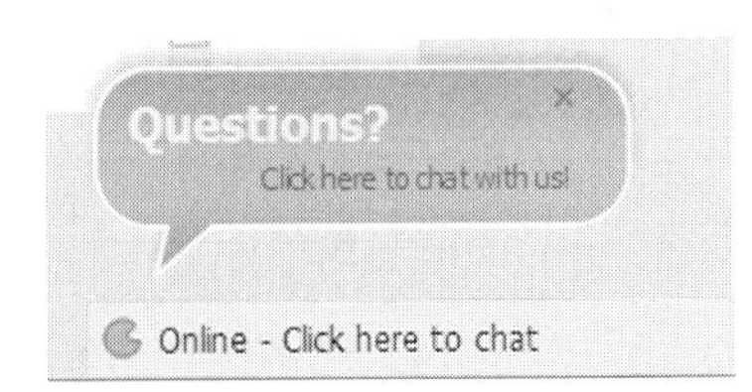

Figure 13.5
Example chat window from Zopim (courtesy of Zopim)

Copyright Protection

Copyright is a form of protection of "original works of authorship" so that the creator may benefit financially from their efforts. This includes literary works, music, dramatic performances, artistic creations and other intellectual works, referred to as *intellectual property* (IP). The original intent of copyright law was to provide this opportunity of exclusivity to the creator, but to also guarantee the public right to exploit these works in the public domain after the expiration of the copyright term. The U.S. Constitution states "to promote the progress of science and the useful arts, by securing for limited times to authors and inventors the exclusive right to their respective writings and discoveries …" (U.S. Constitution, Article 1, Section 8).

As owners of creative works, musicians and record labels are rightfully concerned about copyright infringement of their work. They should seek to protect their artistic works through copyright registration and diligent attention to how their works are being used. Record labels and publishers are known to hire interns and entry-level employees to search the Web for improper use of their protected works. Here are some of the basic copyright tenants and protections.

1. The copyright laws give the creator certain exclusive rights, including the right to create and distribute copies, to create derivative works, to perform the copyrighted work, and to assign copyright to a third party, such as a publishing company. More recently, changes have added the wording "to perform the work publicly by means of a digital audio transmission." Music publishers manage copyrighted works, collect and distribute income to the creator, and exploit the works for commercial gain.
2. For a recording, there are two copyrights: a copyright on the musical and lyrical composition (PA) and a copyright on the sound recording (SR). The musical composition is subject to different treatment from the sound recording. The composition may be bound by compulsory mechanical licensing, meaning that—after the song has been released the first time—the copyright holder must allow licensing of the song to others to "cover" subsequent renditions on recordings. For that, the copyright holder is paid a fee that is determined by a statutory rate. Currently that rate is 9.1 cents per copy for works less than 5 minutes. Arrangements for obtaining a license to cover a song can be made through the Harry Fox Agency or directly through the song's publishing company. More information is available at http://www.harryfox.com/public/ServicesPublisher.jsp.
3. Copyright holders of music compositions (not sound recordings) are currently entitled to royalty collection for public performance of the work, including terrestrial radio. The same royalty standard is not currently in place for the sound recording (SR) on terrestrial radio, but web streaming (Internet radio) and satellite radio are required to pay a license to both the SR copyright holder and the PA copyright holder.
4. Collection and distribution of SR licensing fees for web radio is generally handled through Sound Exchange and covers collection for the use of the sound recording—money that is paid to the label and artist. Owners should

register at http://www.soundexchange.com/performer-owner/performer-srco-home/.

Copyright Violation

Copyright violation occurs on two levels: consumer music piracy and unlawful use of copyrighted works by other artists. Consumer music piracy has become an issue of great concern since peer-to-peer file sharing became popular over the Internet. Copyright infringement by other artists is an ongoing battle with creative ownership being debated in courtrooms and through arbitration. The independent artist or label should take precautions to assure that they are not in violation of copyright law by accidentally or intentionally claiming ownership for the works of others. In cases of protecting one's works through litigation, the infringed party must be able to prove similarity of their original works with that of the infringement, and that the defendants had access to the original works. Some of the more famous cases include the Bright Tunes Music Corp. *v.* Harrisongs Music case of "My Sweet Lord" by George Harrison, which was determined to sound similar to a Chiffon's song "He's So Fine," and Coldplay being sued by Joe Satriani for similarities between his melody in an instrumental song "If I Could Fly" and Coldplay's hit "Viva La Vida."

MUSIC PIRACY

Illegal peer-to-peer file sharing first became big news in 1999 when Shawn Fanning introduced Napster to millions of music fans so that they could easily share recorded music files over the Internet. Since then, many steps have been taken by the record industry to protect copyrighted sound recordings, including seeking criminal and civil prosecution of offenders and lobbying governments to enact stricter laws to protect intellectual property.

The Recording Industry Association of America's (RIAA) web site states "If you make unauthorized copies of copyrighted music recordings, you're stealing. You're breaking the law, and you could be held legally liable for thousands of dollars in damages." Within Title 17, United States Code, Sections 501 and 506 is written the words "Federal law provides severe civil and criminal penalties for the unauthorized reproduction, distribution, rental or digital transmission of copyrighted sound recordings." Making a personal copy of a song that has been purchased is not against the law, but the copy must be for personal use and not distributed to another person. Any reproduction beyond this limited use is illegal—even if it is not for financial gain.

When creating a web site that will contain music for previewing, it is very important to include protection for the copyrighted materials. The various options of copy protection are discussed in Chapter 8.

THE UNAUTHORIZED USE OF COPYRIGHTED MATERIALS ON WEB SITES

These days, it's almost too easy to copy and paste items from other web sites on to your own. However, much of the material contained in web sites is subject to copyright protection and thus cannot be used without consent of the copyright holder. According to Virginia Montecino of George Mason University, the following elements of a web page are protected: links, original text (such as articles and blogs), audio and video, graphics, HTML and other markup and software code, lists that have been compiled by the web site creator or their organization. Among the elements that are legal to use are: original works including writings, images, recordings, videos for which you are the creator or hold the copyright. You can link to other web sites, although you cannot use any trademarked icons, wordmarks or other protected materials for the link. You can use anything that is specifically designated as free, unrestricted materials, or anything for which you have obtained permission or fulfilled obligations to purchase the rights. You can use limited portions of the works of others, as you would do in a research publication, provided appropriate credit is given to the original source (Montecino, 1996).

Abuse of Personal Information

The Web has given marketers the opportunity to learn more about customers than was ever possible before. As a result, consumer advocates are concerned about Web users' privacy rights and the abuse of data collected by web sites. There are several ways that Internet users give up personal information while surfing the web. Each computer connected to the Internet has a unique Internet Protocol Address (IP), in the form of four sets of numbers, each set separated by a period. The Internet Service Provider (ISP) that a person uses knows the IP address of each computer, and normally does not share that information. But marketers still have several ways to determine web usage patterns and purchase behavior online. Search engines track activity to help determine how they compile results for search terms. They can record IP addresses for comparison, even though they don't know about the individual conducting the searches. This helps them determine which sites, and which advertisements, are relevant.

The most widely recognized invasion of privacy on the Internet is the use of *cookies*. Cookies are pieces of information stored on a user's hard drive by various web sites to help those sites identify the user when they return to the site. They may contain information such as login or registration information, preferences, address information, and areas of interest. The site can then use those cookies to customize the display it sends to each user, based upon the information in the cookies. However, these cookies can be used by third party marketers in profiling each individual based upon what else they do on the Internet. Typically, this data is used in aggregate to determine behavior of groups rather than individuals, but has the potential to be abused by targeting individuals for unsolicited advertising messages.

Spyware, and the related programs adware and malware, are software programs that install themselves on a user's hard drive and gather information as the user travels the Internet. The original use was to collect information for marketing purposes, but it has more malicious uses and often installs itself on the user's computer without their knowledge. It can be used to track keystrokes, revealing sensitive passwords and financial account information. Most antivirus software programs now include anti-spyware features that will scour the user's computer, looking for and removing such programs.

Spamming

Spam is defined as unsolicited commercial email messages—the equivalent of junk mail or telemarketers. Because of the negativity and problems associated with spam, successful marketers have adopted a code of conduct. The U.S. government and several foreign countries have passed legislation regulating or outlawing the act of spamming. Many special interest sites (discussion groups) strongly discourage "harvesting" (collecting from posted messages) email addresses to be used for spamming. This is done by gathering up email addresses of people who have posted messages on the site and then using them for sending out email messages to them without their permission.

When sending out mass email messages, (which should be done sparingly and only to those who have given permission) do not make all the email addresses visible to the other recipients. They may have given you permission to send them emails, but that does not extend to one person on your list piggybacking by hitting the "reply all" button and sending out their own marketing message to your list.

Spamming can take on other forms besides just email. Bulletin boards are having trouble preventing the bot programs from finding them and bombarding them with spam messages. Spambots are automated programs designed to register on message boards or forums, disseminate the spam messages, and leave. Usually, they leave a fake name and email address and mask their true IP address. These annoying posts include links to commercial web sites with the dual purpose of generating traffic to the site and increasing search engine placement. The text may be unrelated to the forum topic, but the increase in incoming links may help improve the search ranking for the spammer's sites. In response to this, some message boards and forums now employ CAPTCHA programs requiring the user to visual identify a string of numbers and letters and type that in before the message is permitted to be posted (see Chapter 6).

Instant messaging (IM) has become another recent target for spammers, called "spim." The "spimmer" sends out an IM that includes a link in the message. Those who click on the link subject themselves to spyware that could be installed on their computer.

U.S. Congress, as well as many other governing bodies, has addressed this issue with legislation. A portion of the CAN-SPAM Act of 2003 states:

> Whoever, in or affecting interstate or foreign commerce, knowingly—
> '(1) accesses a protected computer without authorization, and intentionally initiates the transmission of multiple commercial electronic mail messages from or through such computer,

'(2) uses a protected computer to relay or retransmit multiple commercial electronic mail messages, with the intent to deceive or mislead recipients, or any Internet access service, as to the origin of such messages,

'(3) materially falsifies header information in multiple commercial electronic mail messages and intentionally initiates the transmission of such messages,

'(4) registers, using information that materially falsifies the identity of the actual registrant, for 5 or more electronic mail accounts or online user accounts or 2 or more domain names, and intentionally initiates the transmission of multiple commercial electronic mail messages from any combination of such accounts or domain names, or

'(5) falsely represents oneself to be the registrant or the legitimate successor in interest to the registrant of 5 or more Internet Protocol addresses, and intentionally initiates the transmission of multiple commercial electronic mail messages from such addresses,

or conspires to do so, shall be punished as provided in subsection (b).

'(b) PENALTIES—The punishment for an offense under subsection (a) is—

'(1) a fine under this title, imprisonment for not more than 5 years, or both...

CAN-SPAM Act of 2003

CONCLUSION

The Internet is a new format for marketing and communication and still has that pioneering spirit when it comes to entrepreneurship. Certain mistakes should be avoided and protocol followed in order to create an atmosphere of professionalism for your customers. The free-for-all aspect of the Internet brings with it the potential for abuse in many forms. Marketers should be aware of these pitfalls so that web sites can accommodate and address the concerns of its visitors.

Glossary

Adware A form of spyware that collects information about the user in order to display advertisements in the web browser based on the information it collects from the user's browsing patterns.*[3]

CAPTCHA Short for completely automated public Turing test to tell computers and humans apart, a technique used by a computer to tell if it is interacting with a human or another computer.*

Certificate authority or certification authority (CA) An entity that issues digital certificates to web site owners to verify the site owner and provide a measure of security to customers. It is an example of a trusted third party assuring customers that the site is legitimate.

Compulsory licensing A compulsory license, also known as statutory license or mandatory collective management, provides that the owner of a patent or copyright licenses the use of their rights against payment either set by law or determined through some form of arbitration.

Cookie The name for files stored on your hard drive by your web browser that hold information about your browsing habits, such as what sites you have visited, which newsgroups you have read, etc.

Flash Animated features typically found on high end web sites and generally created with Adobe software.

Harvesting Collecting email addresses by visiting user groups and copying email addresses from their message boards.

Instant messaging A type of communications service that enables you to create a kind of private chat room with another individual in order to communicate in real time over the Internet, analogous to a telephone conversation but using text-based, not voice-based, communication.*

Intellectual property "Intellectual property refers to creations of the mind: inventions, literary and artistic works, and symbols, names, images, and designs used in commerce. Intellectual property is divided into two categories: Industrial property, which includes inventions (patents), trademarks, industrial designs, and geographic indications of source; and Copyright, which includes literary and artistic works such as novels, poems and plays, films, musical works, artistic works such as drawings, paintings, photographs and sculptures, and architectural designs" (from the World Intellectual Property Organization).

Internet Protocol (IP) Address The particular identification number that each computer connected to the Internet possesses.[4]

Internet Service Provider (ISP) A company that provides access to the Internet, generally for a monthly fee.

Mechanical license Mechanical licensing is the licensing of copyrighted musical compositions for use on CDs, records, tapes, and certain digital configurations.

PA Copyright Copyright on performing arts. Works of the performing arts including: (1) musical works, including any accompanying words; (2) dramatic works, including any accompanying music; (3) pantomimes and choreographic works; and (4) motion pictures and other audiovisual works. (Note: the new CO form replaces the specific forms for SR, PA and others.)

Payola A bribe given to a disc jockey to induce him to promote a particular record.

Phishing The act of sending an email to a user falsely claiming to be an established legitimate enterprise in an attempt to scam the user into surrendering private information that will be used for identity theft.*

Search engine optimization (SEO) The process of increasing the amount of visitors to a web site by ranking high in the search results of a search engine.*

Secure Sockets Layer (SSL) A method of encrypting data as it is transferred between a browser and Internet server. Commonly used to encrypt credit card information for online payments.

Spamming The activity of sending out unsolicited commercial emails. The online equivalent of telemarketing or junk mail.

Spimming Like spamming only targeted using instant messaging instead of email.*

Spyware Any software that covertly gathers user information through the user's Internet connection without his or her knowledge, usually for advertising purposes.*

SR copyright Sound Recordings copyright. Covers the recorded musical, dramatic, or literary work as well as the sound recording.

Statutory rate The royalty rate (what gets paid to the music publisher and ultimately the songwriter) set by law for issuing a compulsory mechanical license. The current rate is 9.1 cents per copy for up to five minutes.

Viral marketing Marketing phenomenon that facilitates and encourages people to pass along a marketing message.

Wordmark A standardized graphic representation of the name of a company, institution, or product name used for purposes of identification and branding.

Notes

1 It's a good idea to also invest in a server plan that allows you to set up sub-domains (or have your developer do it on their site) and have the sub-domain to be the place that testing is done.

It's never good to test on a live domain, even if you don't announce the site to anyone. Using sub-domains also gives you the opportunity to test periodic upgrades to your site.

2 Or, better yet, make them turn on the music if they want to hear it. If they really want to hear the music, they'll be happy to click a button to play it. One of the problems with streaming music when a page loads is that, if the user is on dial-up or a low-speed DSL connection or any type of slow connection, loading and playing the stream is going to take away from the loading and operation of the visual aspects of the page and the user won't know why until the music starts playing, which may not be right away. They'll just bail out of the page before they even get to see all of it and never return.

3 Asterisked definitions are from Webopedia (www.webopedia.com)

4 Depending upon the circumstances of its connection, the IP address may not be the same every time a computer connects to the Internet. You'll more than likely get the same Class A and Class B numbers (the first two numbers in the IP address), but the other two may be different. It'll also be different if you connect in a different location such as a coffee house or a library.

Bibliography

AllBusiness (2012). Top 10 internet marketing mistakes. www.allbusiness.com/print/3969-1-22eeq.html

Brauner, L. (2010). 25 common social media and web marketing mistakes. *Online Social Networking*. http://online-social-networking.com/25-common-web-marketing-mistakes

eConsultancy (2006). Top ten viral marketing mistakes. http://econsultancy.com/us/blog/438-top-ten-viral-marketing-mistakes

Gardner, Jan (2000). Web site no-nos. http://www.inc.com/articles/2000/11/20883.html

Geisheker, Peter (2006). Do you make these Internet marketing mistakes? http://www.marketing-consulting-company.com/mistakes.htm

Krug, S. (2006). *Don't Make Me Think*. Berkeley, CA: New Riders Publishers.

Leggatt, H. (2009). NCSA: Security concerns drive shopping cart abandonment. BizReport. http://www.bizreport.com/2009/11/ncsa_security_concerns_drive_shopping_cart_abandonment.html

Lewis, H. (2010). DNS—VeriSign reveals SSL market share in June report. DNS Zone. http://dns-news.tmcnet.com/topics/internet-security/articles/88834-verisign-reveals-ssl-market-share-june-report.htm

Lowery, Shelly (April 4, 2006) 35 website mistakes that can kill your business. SWeCS Newsletter. http://www.southwestecommerce.com/newsletters/swecsnews030406.htm

Lurie, Ian (2006, April 24). Five big online marketing mistakes. iMedia. http://www.imediaconnection.com/content/9157.asp

Maven, Richard (2006, October). Top ten viral marketing mistakes. E-consultancy. http://econsultancy.com/uk/blog/438-top-ten-viral-marketing-mistakes

McGuinness, M. (2011). The 10 biggest mistakes artists and creatives make at internet marketing. Lateral Action. http://lateralaction.com/articles/artists-internet-marketing/

Montecino, Virginia (1996). Copyright and the internet. http://mason.gmu.edu/~montecin/copyright-internet.htm

Phang, Alvin (2007). Things you should not do when you market online. Ezine@articles. http://ezinearticles.com/?Things-You-Should-Not-Do-When-You-Market-Online&id=868179

Pingel, Cheryle (2004, May 27). Top 10 online marketing mistakes. http://www.imediaconnection.com/content/3547.asp

Privacy rights clearinghouse. Privacy and the Internet: traveling in cyberspace safely. www.privacyrights.org/fs/fs18-cyb.htm

SecuritySpace.com, (2007, October 1) Certificate authority market share report. http://www.securityspace.com/s_survey/data/man.200709/casurvey.html

Rodenborg, R. (2011). How to address your customer's top ecommerce fears. *Ecommerce Insites: Blog for Manufacturers, Distributors, and Retailers*. http://www.info.insitesoft.com/Insite-Software-Blog/bid/71345/How-to-Address-Your-Customer-s-Top-Ecommerce-Fears

Tippins, Robyn (2005). What not to do on your website. http://voices.yahoo.com/what-not-website-7602.html

Chapter 14
Mobile Media

The digital and mobile revolutions continue to present the entertainment industries with opportunities and challenges. Whereas the Internet and desktop computers have dominated the paradigm shift in the past, the future belongs to wireless mobile technology. The growth of cell phone adoption outstrips Internet adoption. The basic mobile phone, which has been around for nearly 30 years, has evolved from a voice-only form of communication to a multimedia device capable not only of communication but of delivering information and entertainment. Cell phone penetration has reached over 100% in many developed countries and has continued to increase in emerging countries.

FROM CELL PHONE TO MOBILE COMMUNICATION DEVICE: ADOPTION

The wireless handset, which has morphed into a multimedia personal communication device, continues to grow in popularity around the world. Beginning with the third-generation (3G) handsets a few years ago, mobile data networks and increasingly sophisticated handsets have been providing users a variety of offerings. The first generation of mobile phones was introduced in the 1970s, and the period lasted through the 1980s. These devices worked on an analog signal much like those used by two-way radios. The second generation began in the 1990s, used digital voice encoding, and included the geographically compatible Global System for Mobile (GSM). The third generation, or 3G

Table 14.1 Global penetration of cell phone adoption

Geographic area	Penetration level
Europe	125%
CIS	125%
Americas	90%
Arab States	77%
Asia Pacific	67%
Africa	40%

Source: The Mobile Divide Across Nations *Online Marketing Trends*, 2011

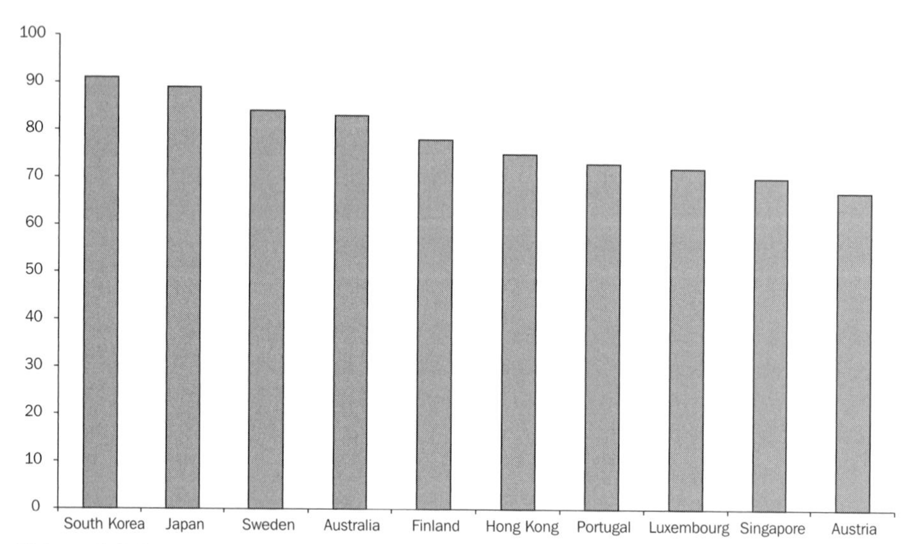

Figure 14.1
Top 10 countries by active mobile broadband subscriptions per 100 inhabitants

technology, allowed for enhanced multimedia, broad *bandwidth* and high speeds, and usability for a wide variety of communication tasks.

The 3G handsets, which are now widely in use in parts of the globe, were first to offer users a new data network with sufficient bandwidth compared to the original analog and first-generation digital cellular phone service. The wireless data services continue to evolve with 4G. As prices drop on handsets and service plans, adoption of web-based mobile handsets continues to outpace adoption of computer-based Internet devices. For developing countries, it is easier to install the infrastructure for wireless systems than to string coaxial cable from home to home. In 2012, Susan Huynh, Forecast Analyst for Forrester Research stated "In emerging markets, where the penetration of landline phone connections has been low, the adoption of mobile phones has soared over the past five years. Mobile handsets are able to provide a cheaper and more convenient means of telecommunications access." Also, the proliferation of Wi-Fi "hotspots" is fueling the growth of mobile-based browsing. Global Wi-Fi hotspots were predicted to increase from 1.3 million in 2011 to 5.8 million by 2015—an increase of 350% (Rasmussen, 2011).

As of the end of 2011, there were 5.9 billion mobile subscriptions—equivalent to 87% of the world population (The International Telecommunication Union, 2012). Growth is led by China and India, which now account for nearly 1 billion subscribers each (MobiThinking, 2012). The Pew Internet and American Life Project reported in 2011 that over 83% of Americans own a cell phone, and by early 2012, 46% of Americans had smartphones—an increase of 11% in less than one year.

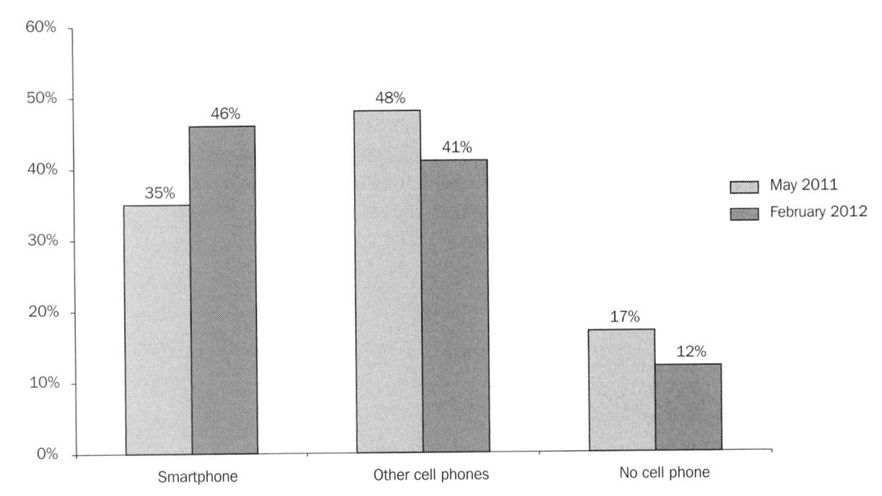

Figure 14.2
U.S. smartphone adoption over time (source: Pew Internet Project, 2012)

CONSUMER USAGE OF MOBILE COMMUNICATION DEVICES

In 2008, the Pew Internet and American Life Project reported that 62% of Americans had some experience with mobile access for activities other than voice communication, either through mobile phones or wireless laptops. In the first decade of this century, the most popular alternative use for cell phones was *text messaging*, with 58% of cell phone users stating they have used their device for text messaging. Equally, 58% had used their cell phones to take a picture, whereas only 17% had used their handheld device to play music and only 10% to watch a video. Of interest to the Pew researchers, nonwhites, specifically African Americans and English-speaking Hispanics, were more likely to engage in media activities other than voice phone service. That trend continued with their 2011 study that found 44% of Hispanics and African-Americans owning a smartphone and only 30% of Whites.

Mobile device usage also varies by age, with younger adults much more likely to use these devices for text messaging and all other non-voice activities. In all categories, usage decreases with age, thus underscoring the vast generational differences that exist between a generation that has grown up around computers and an older generation whose computer experience is limited to adulthood.

The results of the Pew Internet and American Life Project indicate that Americans have indeed become dependent on the use of mobile communication devices for activities other than voice phone, especially text messaging. That study and others have shown that mobile data and communication activities are especially important for user groups who do not also have landline Internet service, whether it's in developing countries who have bypassed landlines or demographic groups in the United States and other countries who had opted for

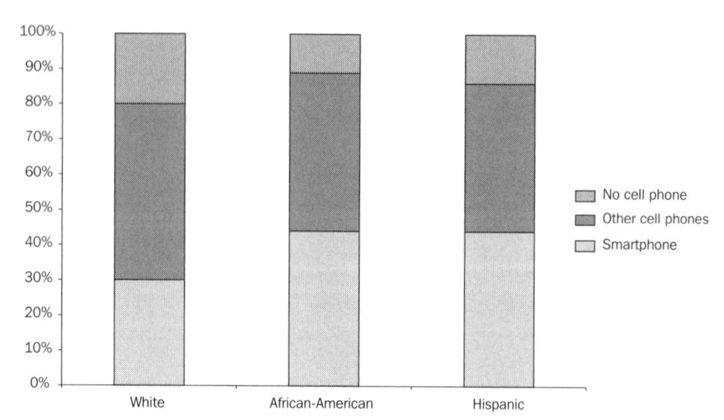

Figure 14.3
U.S. smartphone ownership by ethnicity (source: Pew Internet, 2011)

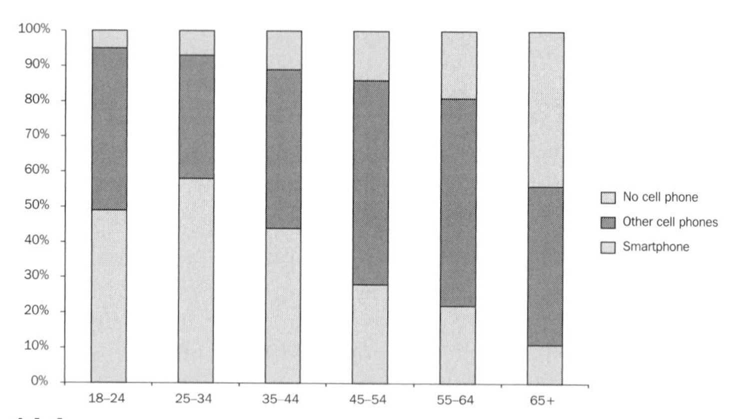

Figure 14.4
U.S. smartphone ownership by age (source: Pew Internet, 2011)

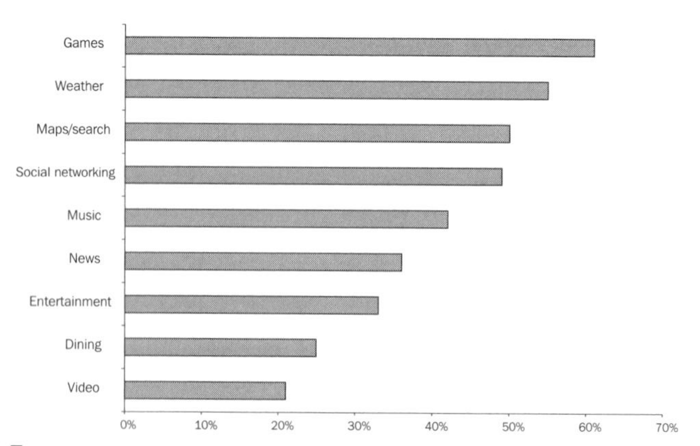

Figure 14.5
How people use mobile phones (source: Mobile Marketing Statistics)

wireless communication only and are less likely to have ever adopted landline communications.

The Pew study also indicates that mobile web adoption is still in the early stages among most demographic groups, with the 18 to 29 age group showing the most usage at 31% of cell phone users. This group is also the most inclined to be heavy users of social networking sites. With smartphone adoption reaching the tipping point, and with the proliferation of high-speed data networks and hotspots, consumers are turning away from desktop computer systems and relying more on mobile devices to conduct many online activities—including music purchase and consumption. And as mobile Internet becomes more popular, marketers are looking at ways to provide popular content to this age group, and social networking looks very attractive.

MOBILE E-COMMERCE

Mobile commerce is on the rise. Smartphone users have gotten comfortable using their mobile devices for shopping, comparing prices, looking up product reviews, banking, and making purchases. U.S. mobile commerce sales are expected to rise from $3 billion in 2010 to $31 billion in 2016—increasing to 7% of web sales (Mulpuru, 2011). In the United States, banks are beginning to offer banking services via cell phone, including checking on account balances, paying bills and transferring funds, and concert promoters are beginning to use cell phones for ticket sales and delivery (see Chapter 15).

In the article "5 Big trends in mobile commerce," author Lauren Indvik (2011) reports the results of a survey of smartphone owners, and presents these findings:

1. One-third of smartphone owners make purchases on their device.
2. 47% of smartphone owners and 56% of tablet owners expect to make more purchases on their devices in the future.
3. 30% of smartphone users use their device as an electronic wallet, and more intend to do so in the future.
4. The use of mobile devices to look up products and pricing, and use coupons is on the rise; 56% for product lookup and 54% for using coupons.
5. Data security and user inexperience are the two main barriers to expansion.

Forbes magazine declared 2012 as the year for mobile commerce: convergence and context (Caron, 2012). Caron makes the point that mobile e-commerce is not the same as mobile commerce. He states that mobile commerce is using your mobile device within a store to make a purchase, and thus bypassing the store's point of sale system. Mobile e-commerce involves using the device to shop at an online store such as Amazon. Caron describes the convergence as "When bricks and mobile finally integrate to provide a holistic shopping experience whereby each shopper has a unique experience with the physical store." Context involves personalizing couponing, features, discounts, and location-based information for each customer. Caron concludes by stating:

> Mobile is an incredibly personal channel by which retailers can directly engage their customers. The mobile experience must reflect this desire to be engaged directly, within the context of the shopper's needs, desires, preferences, timing and location. In 2012, those retailers who integrate mobile and the in-store experience with context will thrive while those who don't will become showrooms and warehouses for the mobile apps that have displaced them.
>
> ***Forbes*, 2012**

One technology that is making this possible is near field communication (NFC), defined as a short-range wireless transmitting technology that connects a smartphone with a receiver to transfer data. It works much like a toll booth EasyPass. NFC is cheap and easy, and while it is already available in other countries, neither Apple nor Android was offering NFC in early 2012. Despite security risks and the debate over who should control the transaction process (the phone service provider or the store), Lauren Brousell of CIO (2012) says adoption is inevitable.

MOBILE SOCIAL NETWORKING

As social networks rollout more sophisticated apps for mobile devices, mobile social networking continues to grow. As indicated in Figure 14.5, nearly half of all smartphone users engage in social networking via their mobile device. Growth in usage is up 41% from 2011. ComScore (2012) reports that more than half of smartphone owners who engage in mobile social networking do so on a daily basis.

> In the U.S., 64 million smartphone owners accessed social networking or blog destinations via their mobile device in December 2011, an increase of 77 percent from the previous year, while 48.4 million consumers accessed social media in Europe, an increase of 76 percent.

ComScore (2012) and Kharif (2006) believe that mobile customers are likely to use social networking services when they are stuck somewhere outside the home, for example, waiting for public transportation.

Mobile social networks are attempting to exploit the geographic benefits of mobile networking by providing location information of friends via global positioning information—called "location-based services" (LBS). This added dimension to social networking allows users to physically meet one another using mobile devices. In 2011, Facebook acquired the LBS Gowalla. Facebook had experimented with its own location-based service "Places" but had given up on the project. In 2012, FourSquare was the top location-based networking service with over 15 million users. Much controversy surrounds the use of proximity alerts, mostly concerning personal security. However, this feature could be beneficial to concert-goers who attempt to meet up with friends at a venue that may be filled with thousands of other fans (see Chapter 15 on cell phone usage at concerts).

As mobile phone users continue to find new ways to use phones for staying connected and for social networking functions, opportunities are developed to promote music, much the way computer-based social networking sites have found numerous ways to incorporate music into the social activities of their sites. Chapter 15 will explore ways to use mobile communication devices to promote music and music-related social events with the help of text-messaging, viral marketing, and mobile Internet features.

TABLETS: THE BEST OF MOBILE AND IN-HOME ENTERTAINMENT CONSUMPTION

The latest generation of mobile devices is the tablet, which combines the features of a laptop with those of a smartphone, in a mid-size device. Some tablets are set up for Wi-Fi only while others include a subscription service to a 3G or 4G mobile network. Advantages of tablets include the convenience and quick startup time of a smartphone combined with a larger screen and work area. The concept of a tablet-size computer is not new, but the iPad, first introduced in 2010, was the first to be successfully adopted by consumers on a large scale. Other tablets were introduced about the same time, although some, such as Amazon's Kindle and Barnes and Noble's "Nook" had limited functions. In late 2011, Amazon introduced the Kindle Fire, their response to the industry leader, the iPad. As of early 2012, Apple continued to dominate the market, but the Kindle Fire had quickly made a dent in Apple's market share.

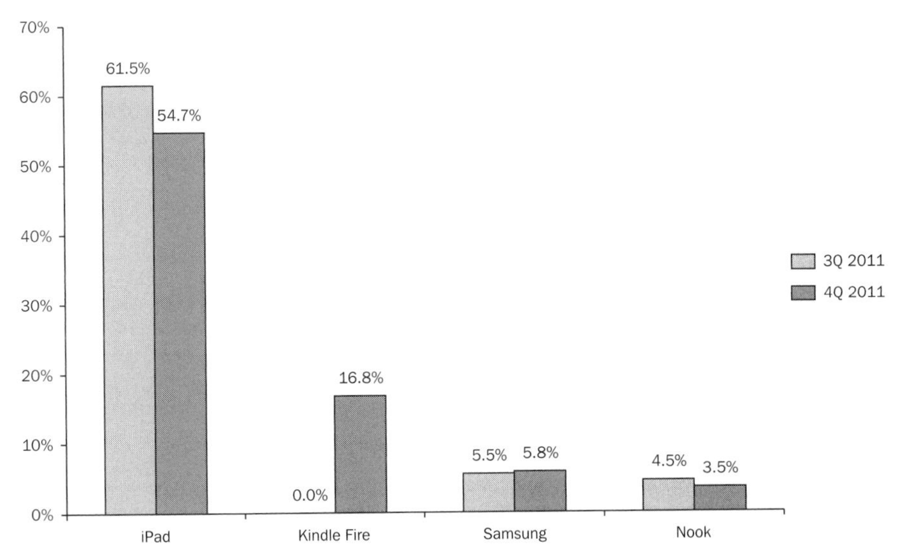

Figure 14.6
Tablet market share (source: Kahn, 2012)

OTHER NEW MEDIA DEVICES: IN-HOME STREAMING MEDIA DEVICES

One other class of new media devices that should be addressed because of their effect on music consumption is that of TV set-top streaming devices such as Roku and Apple TV. These units connect to the television set and use the Internet to stream entertainment content to the television. They are commonly used with Wi-Fi and can serve as a conduit for other computer devices in the home, streaming music and other content from the desktop computer, or from a mobile device. They are worth mentioning here because their presence allows music consumers to take full advantage of streaming music services like Pandora and Spotify. This has the potential to alter music consumption and thus music purchase behavior. In 2011, more than 200 indie labels pulled their songs from streaming services, believing that it was killing music sales. However, in early 2012, Sony Music's president of global digital business and U.S. sales Dennis Kooker did not believe the streaming services were having an effect on Sony's music sales (Chacksfield, 2012). And the RIAA reported in 2012:"Revenue generated by subscription services like Spotify, Rhapsody, Rdio, and others was up 13.5 percent (from $212.4 million to $241 million), according to the Recording Industry Association of America's (RIAA) year-end music shipment statistics for 2011"(Cheredar, 2012).

CONCLUSION

Mobile media continue to evolve, from simple cell phones, to texting, to smartphones, and now tablets. Each wave has forever altered the way people communicate and access entertainment. As this shift occurs, the music business must reinvent how it monetizes entertainment content in a way that will keep music consumers involved. Much of that shift is being facilitated by ubiquitous access to high-speed wireless Internet access, cloud computing that allows for storage of personal entertainment collections, and the plethora of interconnecting devices that allow for music access and consumption in a variety of settings. The MP3 player, or iPod, is no longer considered a new technology.

Glossary

3G technology The third generation of developments in wireless technology, especially mobile communications, that allows for the transmission of data and multimedia content.

Bandwidth The amount of data that can be carried from one point to another in a given time period (usually a second).[1]

Global positioning satellites (GPS) A group of well-spaced satellites that orbit the Earth and make it possible for people with ground receivers to pinpoint their geographic location.

GSM A digital cellular phone technology that is the predominant system in Europe but also used worldwide. GSM is the dominant second-generation digital mobile phone standard for most of the world. It determines the way in which mobile phones communicate with the land-based network of towers.

iPhone A smartphone made by Apple that combines an iPod, a touch screen, a digital camera, and a cellular phone. The device includes Internet browsing and networking capabilities. The iPhone supports both WiFi and cellular connectivity, depending on which signal is available.

Landline phone Refers to standard telephone and data communications systems that use in-ground and telephone pole cables in contrast to wireless cellular and satellite services.

Mobi Internet domain used for web sites that supply content to cell phones and other handheld devices with tiny screens.

Near field communication (NFC) A short-range wireless transmitting technology that connects a smartphone with a receiver to transfer data.

PDA Personal digital assistant; a handheld device that may combine computing, telephone/fax, Internet, and networking features.

smartphone A cellular telephone with built-in applications and Internet access.

Social networking sites Places on the Internet where people meet in cyberspace to chat, socialize, debate, and network.

Text messaging Sending short text messages from a mobile phone to other mobile phone users.

WAP Wireless application protocol is a specification for a set of communication protocols to standardize the way that wireless devices can be used for Internet access. A WAP browser provides all of the basic services of a computer-based web browser but is simplified to operate within the limitations of a mobile phone; for example, it has a smaller view screen.

Note

1 Bandwidth only accounts for the time that it takes to get into or out of your connection and is really just a potential speed. Each communication device has its own bandwidth based upon its transmission capabilities. The speed of any communication is limited by the device(s) that are communicating, sort of a lowest common denominator situation. Bandwidth also doesn't account for periodic bottlenecks anywhere in the system or extra "hops" that the signal may need to make through the Internet.

Bibliography

Brousell, Lauren (2012, March 28). Five things you need to know about near-field communications. CIO. http://www.cio.com/article/702631/Five_Things_You_Need_to_Know_About_Near_Field_Communications

Caron, John (2012). For mobile commerce: the year of convergence and context. *Forbes*. http://www.forbes.com/sites/ciocentral/2012/01/16/for-mobile-commerce-the-year-of-convergence-and-context/

Chacksfield, Marc (2012, January 30). Sony: music streaming isn't killing song sales: Spotify's business model working for labels. TechRadar http://www.techradar.com/news/internet/sony-music-streaming-isnt-killing-song-sales-1058615

Cheredar, Tom (2012, March 27). Streaming music services are giving the music industry a huge revenue boost. VB Media. http://venturebeat.com/2012/03/27/riaa-subscriptions-revenue-increase/

ComScore (2012). More than half of people that access social networks on their smartphone do so on a near daily basis. ComScore Data Mine. http://www.comscoredatamine.com/2012/02/more-than-half-of-people-that-access-social-networks-on-their-smartphone-do-so-on-a-near-daily-basis/

Huynh, Susan (2012). Mobile internet users will soon surpass PC internet users globally. Forrester Research blog. http://blogs.forrester.com/susan_huynh/12-02-21-mobile_internet_users_will_soon_surpass_pc_internet_users_globally

Indvik, Lauren (2011). 5 big trends in mobile commerce. *Mashable* blog. http://mashable.com/2011/06/21/mobile-commerce-trends/

International Telecommunication Union (2012). ICT facts and figures. http://www.itu.int/ITU-D/ict/facts/2011/material/ICTFactsFigures2011.pdf

Kahn, John (2012, March 13). Cheap Kindle Fire absorbed some tablet marketshare ahead of new iPad launch. 9TO5Mac. http://9to5mac.com/2012/03/13/cheap-kindle-fire-absorbed-some-tablet-marketshare-ahead-of-new-ipad-launch/ (general reference).

Kharif, Olga. (2006, May 31). Social networking goes mobile. *Businessweek*. www.businessweek.com/print/technology/content/may2006/tc20060530_170086.htm

Mi2n (Music Industry News Network). In 2010, half of all digital content will be delivered to mobile devices http://mi2n.com/press.ph3?press_nb=106654

MobiThinking (2012). Global mobile statistics 2012: all quality mobile marketing research, mobile Web stats, subscribers, ad revenue, usage, trends... MobiThinking. http://mobithinking.com/mobile-marketing-tools/latest-mobile-stats#subscribers

Mulpuru, Sucharita (2011, June 17). Mobile commerce forecast: 2011 to 2016. Forrester Research press release. http://www.forrester.com/Mobile+Commerce+Forecast+2011+To+2016/fulltext/-/E-RES58616?docid=58616

Pew Internet & American Life Project (2008, March). Mobile access to data and information www.pewinternet.org.

Rasmussen, Paul (2011, November 9). Global Wi-Fi hotspots forecast to increase 350% by 2015. Fierce Wireless. http://www.fiercewireless.com/europe/story/global-wi-fi-hotspots-forecast-increase-350-2015/2011-11-09

Chapter 15

Mobile Music Marketing

MUSIC GOES MOBILE

The global mobile music market is expected to rise to more than $18 billion by 2016 according to a report issued by IE Market Research Corporation in January 2012. This will be driven by subscription music services and *ringback tones*. The mobile music market is expected to be a boon for subscription services, unlike the PC-based network, which has seen disappointing numbers for music subscription services. Perhaps the difference is that consumers desire to own music that resides in the household, but they may be more willing to "rent" music on portable devices, much like a personal radio station.

The year 2007 was considered the tipping point for mobile music adoption (Holden, 2008). Windsor Holden, author of the Juniper report stated, "far more subscribers began downloading and subscribing to music content in developed markets, and it must be said that the publicity surrounding the iPhone launch undoubtedly contributed to consumer awareness of mobile music services per se." (http://hypebot.typepad.com). Until 2007, the majority of mobile music revenue took the form of *mastertones*—short excerpts from an original sound recording that plays when a phone rings. The market for ringtones, ringtunes, and the like developed throughout the first half of the decade but saw a decline in sales for the first time in 2007 as consumers moved away from phone personalization features and began to adopt full track downloads to mobile (International Federation of Phonographic Industries [IFPI]). The decline in ringtones may also be facilitated by consumers creating their own mastertones on their home computers and *sideloading* them into their mobile devices, and also because, in many cases, the ringtones cost more than the whole song. Ringtone sales have played an important role in the evolution of music to mobile devices. Most consumers' first experience with downloading music content directly to their mobile devices has been with ringtones, thus paving the way for consumer downloading of additional music content.

Ringtones accounted for 62% of the mobile music market in 2007 but only 35% in 2010. In 2010, Inform Telecoms and Media stated that ringback tone sales had surpassed ringtone sales in terms of the number of users, rising to 44% of mobile music users globally, with two-thirds of those coming from China and India. While ringtones and ringback tones remain the largest profit-generator for the mobile music market, sales vary widely from market to market, with much of

Phone Personalization Features

- *Polyphonic ringtones.* Cell phone polyphonic ringtones are ringtones that can produce more than one note at the same time. In music, groups of two or more notes played at the same time are called "chords." The more chords a polyphonic ringtone produces, the more exciting its texture.
- *Custom ringtones.* Some cell phones give you the capability of changing the built-in ringtones with others of your choice. Some cell phones have a built-in function that allows you to "compose" a custom ringtone, while with others you have to download ringtones composed on a PC or another phone.
- *Mastertones.* Short excerpts of original sound recordings generally sold to consumers.
- *Ringer profiles.* The ringer profiles feature allows a cell phone user to set up different ringer settings so that each "profile" can be activated for different situations. For instance, when you are at work you may want your ringer settings to be businesslike and quiet, while when you are on vacation you may allow yourself something more colorful.
- *Ringback tones.* Ringback is an intermittent audio tone or music snippet that a caller in a telephone system hears after dialing a number, when the distant end of the circuit is receiving a ringing signal. Ringback tones are the sound you hear when you call someone and are waiting to be connected.

Source: Wirefly.com

the continued popularity found in emerging, not mature markets. For example, in Indonesia, they account for 95% of mobile music revenue (Pandey, 2010).

While the market for ringtones appears to have peaked in the U.S. and is projected to decline elsewhere in the coming years, the future of ringtones may lie in their promotional value. Labels and artists are finding ringtones a useful contest giveaway or incentives for fans to interact and support the artist. Several online services provide ringtone sales platforms, but for promotional uses, ringtones can be created using the audio editing tools provided in Chapter 8 and a converter tool like Aukido (www. Aukido.net).

Ringtone vendors

- Myxer. http://www.myxer.com/make/
- Brinked. http://www.brinked.com/create-ringtone.html
- Aukido. www. Aukido.net
- Thumbplay OPEN. http://open.thumbplay.com/

MOBILE MUSIC SALES

After a half-dozen years of healthy growth in the mobile music market, 2010 saw a steep decline in mobile music income. *Billboard* reports that U.S. mobile sales suffered a 28% decline in 2010 (Peoples, 2011). Full track mobile downloads dropped 23% in 2010 and subscription services had mixed results with subscribership up and revenues down.

Despite the decline in mobile music revenues, revenue for full track mobile downloads is expected to grow. A 2011 Nielsen study found that "nearly one in four (24%) of the 20–24-year-old segment globally indicated they would be prepared to pay to download music videos on their mobile phone (Nielsen, 2011). However, the market for mobile revenue is evolving and now encompasses earnings generated from advertiser-supported streaming services, subscription-based services, and the download of full tracks and ringtones.

The emerging area of mobile music creates new challenges for marketing music to mobile devices. One of the first questions addressed in marketing is how to deliver the product to the consumer. What form should it take? How should it be distributed? And do the answers apply across the board to all segments of the market? Business models that work on the Internet do not translate seamlessly to mobile devices. The media industries are fraught with failed experiments into new technology where old paradigms were applied without regard to unique features of the new technology. For example, radio shows were adapted to the new medium of television, with drastic modifications as the medium developed. Consumers want their music with them at all times, and as a result, music has found its way into mobile devices beginning with car radios, through the development of the Walkman, to the portable CD player, the MP3 player, satellite radio, and now the mobile wireless handset.

Rent or Sell?

In the absence of experience in the area of music-to-mobile, questions arise as to whether the consumer will ultimately prefer to (1) purchase and download tracks, much like iTunes; (2) subscribe to a music service much like Spotify or Rhapsody; or (3) prefer to listen to streaming audio like a personal radio channel, much like Pandora or LastFM. In the article "Marketing Mobile Music," John Gauntt lamented,

> The most pressing question facing mobile carriers, music labels, and music service providers is how consumer behavior will evolve as mobile music transitions from a phase dominated by early-adopter, active music fans to one more influenced by mainstream, casual music fans.

He speculated that the variety of ways that digital music can be packaged and played on a variety of fixed and mobile devices such as dedicated music players, mobile phones, DVD players, stereos, TV, and automotive audio systems make it likely that mobile music will not be permanently fixed to a specific device. But even this varies from country to country.

> Despite all of the exciting online radio options, we are still seeing healthy growth in the market for digital-music downloads," NPD senior vice president of industry analysis Ross Crupnick said in a statement. "This growth is fueled by an increase in mobile devices and a core base of consumers who want to own the music they listen to, despite all of the emerging radio options ... As long as consumers want to own digital tracks and continue to have a passion for the physical format and a way to play their CDs, online radio and paid-to-own music will live in harmony.
>
> **NPD Press Release, 2012**

The subscription-based model that has not shown success for Internet music sales does show more potential for mobile music, with consumers more willing to "rent music" that is accessed remotely. Subscription services have been a hard sell to Internet users who prefer to possess their music collections on their computers. Subscription services require subscribers to sync their libraries each month to verify their subscription is still active. This limits consumers' ability to transfer the music to portable devices.

However, several sources, including Informa Media and Nielsen, believe subscription income will dominate future mobile music revenues.

Distribution is not a problem and financial transactions are not a problem because mobile users are billed for services monthly. The question becomes how to package and price music for the consumer. Until the industry sorts out what the consumer wants and is willing to pay for, companies that are currently in the digital music sales business, such as iTunes and Napster, wireless carriers such as Verizon and AT&T, and specialty upstarts such as Sony Music Unlimited will continue to experiment with their offerings to consumers until something hits.

Meanwhile, the mobile industry continues to test unique marketing ideas to get music consumers more involved with using mobile devices for music discovery and consumption. Informa's predictions for 2016 are that music streaming via subscription services will grow from 9% of mobile music users to 25%. They estimate that revenue from streaming will grow from US$650 million in 2010 to US$3 billion in 2016. The more recent music model has elements of both streaming radio and subscription-based on-demand services. Spotify allows customers to select and play music on-demand, much like Rhapsody, but the free version is advertiser supported.

MARKETING

Marketing music via mobile devices involves a host of activities and should take into consideration the difference between marketing music to be sold and delivered via mobile devices and mobile marketing that may direct a customer to a more traditional outlet for purchases and transactions. And as subscription services and ad-based services become more popular, marketing will shift toward encouraging consumers to add songs to their playlist rather than purchasing them for download, since the artists and labels will be paid based upon total number of plays.

Music ID

One of the most challenging problems for record labels since the 1950s has been finding ways to help music lovers connect with the music they hear and are then motivated to purchase. A gap existed between listening to the song and determining where and how to find it for sale. FM radio, with its long blocks of music, has done a poor job of helping listeners identify the music they are hearing, and therefore, the label loses out on a sale. A campaign in the late 1980s attempted to encourage radio stations to back-announce songs

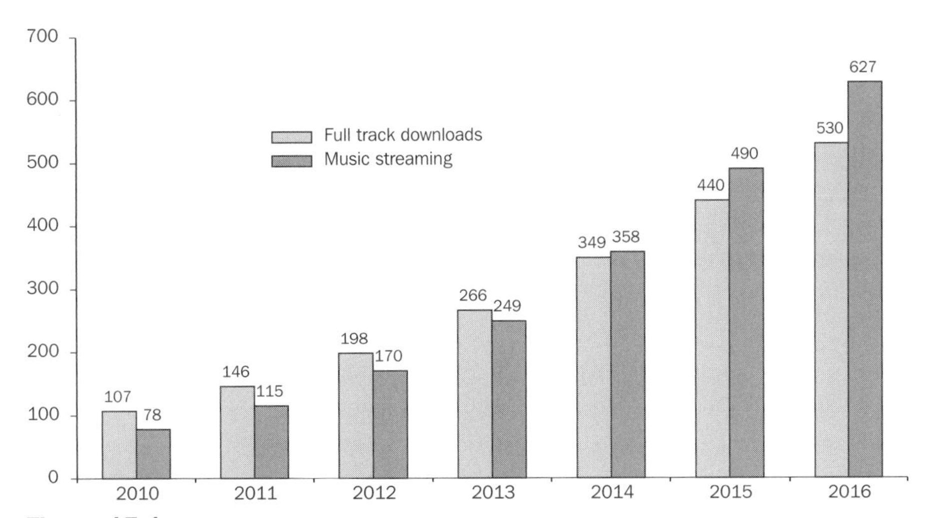

Figure 15.1
Global mobile music users (millions) (source: Informa Media)

they played with a reminder to disk jockeys to "when you play it, say it." In the 1990s, early versions of retail kiosks sought to provide reference information that would help consumers quickly locate music they were searching for. Some kiosks provided music samples, and others had a keyword index. The Internet introduced other methods for identifying music, the most important being the music sample, which allowed consumers to recognize and confirm they had found the titles they intended to purchase, either online or in the store.

With the introduction of advanced services for the cell phone, wireless providers introduced simple song identification services. The customer would hold the phone up to the music source, and the service provider would send back a text message with the song identification information. Smartphones have taken this experience to a new level with the introduction of applications designed for music identification. Currently, the most popular music identification application is Shazam. Unfortunately, getting your music included in their database is not a readily-available feature.

Advertising via Cell Phones

Mobile advertising is poised to explode as advertisers move their budgets from traditional media to mobile media in an attempt to reach more targeted markets. Mobile advertising is predicted to grow from $3.4 billion in 2010 to $22.4 billion in 2016 (Mobile Marketer, 2012).

A 2012 Nielsen Consumer Usage Report found that 51% of consumers said they were OK with advertising on their mobile devices if they were able to access content for free (NielsenWire, 2012). Advantages of mobile advertising include:

- Potential consumers are on their devices day and night.
- Consumers will research a brand because they are about to make a purchase decision.

- One-click purchasing is possible.
- Advertising can take advantage of context—when the consumer is most receptive to the product's pitch.

Mobile advertising can direct music fans to a mobile web site or directly to iTunes or Amazon to purchase music. For small indie labels, artists, and other entrepreneurs, perhaps Google AdWords is the easiest way to place ads on mobile devices, using device platform targeting. This allows you to select which mobile platforms, or a combination with computer-based platforms, your ads will run. Using a standard AdWords account, this option is selected on the "campaign settings" page. There, you can select geographic locations and "networks and devices." The Google ads show up as text ads on smartphones and text and/or image ads for tablets. Choose your landing pages wisely. Google advises that if

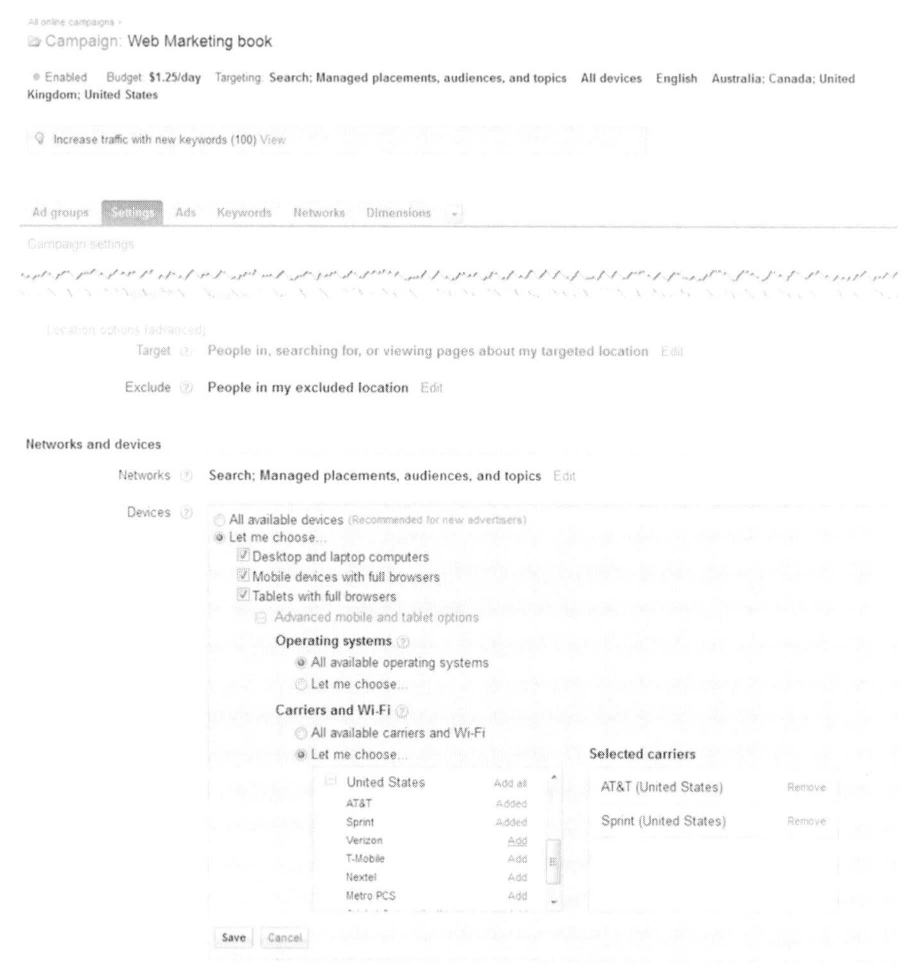

Figure 15.2
Google AdWords device selection for mobile ads (with permission from Google)

customers are taken to a page with Flash for a device that does not support it, Google will limit your ads from running on those devices.

QR Codes and Pull Marketing

QR codes (quick response codes) are bar codes designed to be read by the camera of a mobile device and then direct the device to perform a specific task. Smartphone applications have been developed to read and respond to QR and the Microsoft counterpart: MS Tags.

QR codes can contain a phone number, text message, plain text or a URL. They are mainly placed in print materials as a convenience, allowing the reader to quickly access the online content as instructed via the code. QR code generators are available for free on the Web, and codes can be developed quickly and easily. From a QR code generator site, select the type of communication (URL, text, SMS, phone) and then type in the information (URL, phone number, etc.). An image will be generated immediately. Some code generators will allow for formatting,

Figure 15.3
Example of QR code generator (http://goqr.me)

Free QR Code Generators

- http://goqr.me/
- http://qrcode.kaywa.com
- http://www.qrstuff.com/

such as color, size, background color, etc. The image can be downloaded. (Go ahead and try out the code in Figure 15.3.)

When a QR code is developed using the text communication, textual information is sent to the user's smartphone. This can be used to provide additional product information to the potential customer, such as details about an upcoming event. The text is delivered within the scanner application. When creating a QR code for SMS communication, an SMS message is created on the smartphone to send a specified third party phone number. This can be used if you want people to communicate with you immediately. When setting up the code, you can specify your cell number, and the default information in the message. If you use several QR codes, strategically setup for different situations (such as location-based), the default information in the SMS field can provide you with contextual information.

Musicians should incorporate QR codes into all physical printed materials, including CD covers and even T-shirts. You can use QR codes to send someone to a *squeeze page*: a landing page created to solicit opt-in email addresses from prospective subscribers. Squeeze pages provide very little information other than the "opt-in" features. For more elaborate email or mobile campaigns, the opt-in page can allow users to select from a list of categories to opt in, or opt out. For example, the customer may want to receive information about upcoming tours in the area, but not weekly news updates on the artist's career.

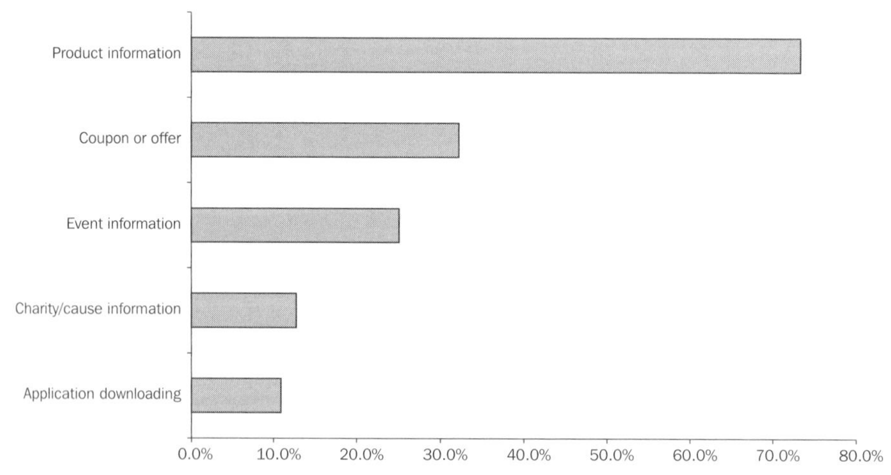

Figure 15.4
Results of scanning QR codes into smartphones (source: calculated from data from ComScore)

Microsoft Tags	QR Codes
Controlled by Microsoft	Open source code
Multi-colored	Two-tone
Can be artistically altered	Standard image
Claims to be faster	Seems fast enough
Being actively marketed by Microsoft	Advantage of first to market
Must have a Microsoft account	No signup needed
Get the free mobile app at http://gettag.mobi	

Figure 15.5
Comparison of QR codes and Microsoft Tags

Microsoft has a competing product called Microsoft Tags. The tags are also free to generate, from the Microsoft site. A comparison of the two codes is presented in Figure 15.5. The true value of tag features is in the ability to collect metrics on usage. Some of the QR code generators offer a premium account that will provide such analytics.

Push/Pull and Viral Marketing via Text Messaging

In 2010, a Pew Internet & American Life Project report found that 95% of cell phones in the U.S. had **short message service** (SMS) capabilities; and a 2011 Pew study found that that three-quarters of them regularly send and receive text messages. Usage is much higher among young people, who are traditionally the most difficult to reach with marketing messages. Ninety-five percent of 18- to 29-year-olds and 82% of 30 to 49 year olds regularly use their mobile phones for texting. *Pull* marketing campaigns are those that use some other form of communication to target the customer to take some form of action on their smartphone, such as visiting a web site to learn more about a product. Nester and Lyall used the example of a "text to win" campaign, or special offers that encourage users to log on using their cell phones to receive free wallpaper or ringtones. The TV show *American Idol* uses a pull strategy on the live program to encourage viewers to text in their vote using their AT&T phones. Pull campaigns are being enhanced by QR codes in print materials (including billboards) and Shazam audio signals embedded in commercials and television programming. The user activates their mobile app, which upon receiving the signal or scanning the code, directs the mobile device to load up a particular web page or perform a function. Advertisers can use this to engage potential customers in interactive relationships such as contests, coupons, and so forth.

Cell phone marketers have begun using push messages to consumers' cell phones to market products. A *push message* is a specially formatted SMS text message that provides the recipient with marketing information they have requested or agreed to receive. SMS messages are limited to text only and to 160 characters. Despite the limitations of the media, it has proven to be highly effective generating high brand recall and response rates.

Mobile marketing campaigns using SMS or text messaging will often use a service to manage the campaign. Mobile marketing messages are strictly regulated because often the recipient is charged for the message. Therefore, messages should be sent only to those potential customers who have agreed to receive the messages. Often, this requires some type of enticement, such as exclusives, coupons, and privileged information. There are two levels of SMS marketing: (1) Sending out one-way text messages with a traditional call to action (such as come to the show tonight) or with no call to action, (2) campaigns that encourage the recipient to take some action on their mobile device, such as "text 'FREEMUSIC' to 55432 now to win."

Viral marketing campaigns are new to cell phones and incorporate the notion of recipients passing along the message to others. In the blog "How to make viral marketing appealing for phones," JayVee mentioned three factors to successful viral campaigns on cell phones:

1. Offer exclusive content: something that is not available on the Internet but only via cell phones. It could include ringtones, free music files, wallpaper, graphics, and so on.
2. Make it useful and timely: for musicians, something centered around a live concert date would be effective, offering something of value related to the concert or the artist.
3. Clearly define objectives: the goals should be simple and straightforward; create awareness for the product or service; encourage recipients to pass along the message; elicit a call to action (log on, download, etc.)

One-way SMS campaigns should be managed with one of the inexpensive group SMS services such as SendHub. This is one of many services offering a simple messaging solution. You sign up, add your list of numbers from customers or fans who have "opted-in" and agreed to receive your messages. Then you send out a blast to the group. With SendHub, you can schedule messages to be sent at a future time. The free version allows you to accommodate up to 50 people. For a list of 1,000 members, the cost is $50 per month. Messages can be sent out providing fans with "just in time" information, such as "gig start time changed to 8:30" or "early show sold out—second show added for 10PM."

GupShup is a site that allows users to easily start an SMS group. Musicians can start a group, invite friends and fans, and have a convenient method to send out blast messages in support of their career. The GupShup site outlines three simple steps to start a GupShup from your mobile phone:

1. Create your group by sending an SMS to 567678. Example: CREATE mybandname.
2. Invite friends to your group. SMS the following to 567678: INVITE phone1 phone2 phone3. For example: INVITE 9876543210, 9876543211, 9876543212.
3. SMS your message to 567678 to post a message to your group. Example: "Hi, we're the Rock Dudes and this is our group of friends and fans. Please check out our new recordings at ..."

Interactive SMS campaigns require that the marketer (that would be the musician or manager) set up an account with a service that will provide the interactive components, including keywords and short codes. The *short code* is similar to a phone number in that it identifies a company, but a short code contains only five or six digits, making it easier for the recipient to type into a smartphone (SMS Marketing Services, 2012). This must be coupled with the keyword: a word or words that you choose to associate with the SMS marketing campaign. For example, you might send out a text message that says "For advance tickets, text TICKETNOW to 58546." Then, an auto-responder message set up by the service will send back the instructions, phone number or link to purchase the tickets. This can be used with coupons, contests, polls, and other interactive campaigns.

SMS Marketing Services has some suggestions of what to look for in a mobile text marketing service. They include:

"Under the CAN-SPAM law regulated by the Federal Communications Commission (FCC), businesses engaging in mobile marketing must obtain permission from recipients before sending them text messages. Opt-in and opt-out guidelines and tools are typically included in plans offered by sms marketing services. According to federal laws, opt-in consent must be obtained from the mobile phone owner via a text message. This applies even if a business captures opt-in leads by phone, in writing at their establishment, or on the Internet. Once permission is received, businesses can manage their contact lists using the account management tools supplied by their sms marketing service."

"Regulations also require businesses to provide a means for consumers to opt-out of their sms marketing campaign at any time. To meet this requirement, businesses typically provide a link, an opt-out keyword and short code, or other opt-out instructions as part of their text message."

SMS Marketing Services, 2012

Figure 15.6
SMS marketing services (http://www.smsmarketingservices.net)

1. Is the dashboard easy to use? You will be managing user lists, uploading content, scheduling text blasts, managing incoming responses, and so forth.
2. Are all mobile carriers supported? You want a service that will reach everyone on your list. What about in other countries?
3. Is two-way messaging available? This is how you make your marketing campaigns truly interactive.
4. What is the cost structure and what is included?
5. Does the service offer the level of support and security that you need?

BoomText is an interactive SMS service provider for those who want to expand their marketing reach to the mobile devices of customers and prospects that have smartphones or cell phones with text messaging capabilities. The service has call to action capabilities. Recipients text back the code number to receive the offer.

Pushr is a reasonably priced full-service SMS service. Services include two-way messaging, on-demand and scheduled messages, message subscriptions on a regular basis, and a dedicated local phone number instead of a short code that would drive up costs. RockStartTexting is a service that focuses on the music business (Raleigh, 2011). According to their web site,

> every person that texts your BANDNAME to 411247 is automatically added to your text message database. They are sent an automatic text message (which you set up). Send them a free mp3, coupon code, Facebook url, venue address whatever! Rock Star Texting is the text messaging solution for local, indie, and major bands.
>
> **http://www.rockstartexting.com/**

Zlango is a platform that transforms SMS into a colorful icon-based experience; a combination of text and icons can be sent to mobile phones. Their plug-ins and widgets allow for some creative communication.

The best known service for SMS campaigns, and one used by major labels and artists, is *Mozes*. Mozes boasts,

> Take your fan club mobile … Broadcast text messages to your various groups and get quick replies, or poll them and see instant results on the web too. You can ask your mobile mob trivia questions, send coupons, run contests and more.

In the case study shown in Figure 15.7, Mozes developed a marketing program for the Linkin Park tour where fans used their mobile phones to text the keyword LPSUN to the short code 50505 to receive a free digital download that was emailed to their account. More about this campaign is mentioned in the section on concert music downloads. Mozes has a basic plan for $67.50 per month that includes 3,000 messages per month.

PROXIMITY MARKETING

Proximity marketing is the localized wireless distribution of advertising content associated with a particular location, generally a retail establishment,

PRODUCT CUSTOMERS COMPANY

« Back to Customer Case Studies

CASE STUDY

Warner Music Group

Linkin Park

41,000
EMAILS COLLECTED

23%
REDEMPTION RATE

OBJECTIVES

Warner Bros. Records wanted to reinvent the official concert bootleg with mobile on Linkin Park's 2011 'A Thousand Suns' North American Tour.

SOLUTION

Warner Bros. Records used Mozes to deliver fans a free, professionally-mixed, recording of the show they attended on the Linkin Park 'A Thousand Suns' North American Tour. Concertgoers texted LPSUN to 50505 to receive their digital download, which was emailed to them following the concert. Linkin Park displayed in-venue calls-to-action to provide fans with mobile instruction.

RESULTS

- Nearly 41,000 fan email addresses collected to-date (average of 1,950 email addresses per show)
- 23% redemption rate per show of download

Figure 15.7
Mozes case study

SMS Messaging Services

- GupShup: www.smsgupshup.com
- Pushr: http://pushr.net/
- Rock Star Texting: http://www.rockstartexting.com/
- Mozes: www.mozes.com
- Zlango: http://www.zlango.com/

More links at www.go2web20.net

transportation hub, or venue (Wikipedia). The electronic marketing messages are delivered to willing customers' handsets and may include store coupons and discounts. A press release from Ace Marketing and Promotions states:

> What if you could tap every potential customer on the shoulder within 100m of your location and send them a business card, advertisement or special offer directly to their mobile phone? What if you could do this constantly, without worry about the cost?

Proximity marketing allows businesses to do this as long as the customer has the equipment to receive and has agreed to accept marketing messages. Messages should not be sent to customers who have not opted-in to receive them. Bluetooth-enabled cell phones and PDAs, as well as cell phones equipped with a GPS system, are among the current generation of mobile devices that are capable of being targeted by proximity marketing. Proximity marketing services are capable of sending text messages, video clips, ringtones, audio files, and promotional or discount coupons complete with a graphic representation of bar codes. An article in South Africa's BizCommunity.com touts eight advantages of proximity marketing over other forms of marketing (from Proximity Marketing Revealed, Imajinn, 2006):

- *Speed.* Content is delivered to mobile devices in seconds.
- *Cost.* Significantly cheaper than direct mail or other media.
- *Relevance.* Content to users in relevant places at relevant times.
- *Interactivity.* Interact with or engage consumers nearby.
- *Proximity.* The mobile device is always in the consumer's pocket.
- *Reach.* 100% of current mobile phone market, for example.
- *Personalization.* Bespoke content per individual consumer.
- *Viral.* Content shareability factor allows for peer-to-peer transfer.

Proximity marketing is also capable of responding to pull marketing, where consumers seek out nearby retail establishments by logging on to the Internet using their mobile devices. The system would note the location and then send a customized list of appropriate retailers in the vicinity of the consumer's device. For example, someone traveling who would like to know the names and locations of nearby pizza parlors could simply log on, type in "pizza," and get a local list.

One of the most relevant applications of proximity marketing in the music business is proximity marketing at concerts and venues. One of the tenants of proximity marketing is having members of the target market in the area who will be responsive to the marketing messages. In all applications of proximity marketing, these messages need to have some appeal to the recipient in the form of (1) information they may desire to have, (2) special offers, (3) privileged access or opportunities, or (4) "advertainment," something that entertains as well as advertises. Proximity marketing may be facilitated by the adoption of *near field communications*: a set of short-range wireless technologies, typically requiring a distance of 4cm or less to initiate a connection.

CONCERTS

Marketers have been finding new ways to incorporate cell phone marketing into live performances since the mass emergence of cell phones at concerts began to influence shows in the 1990s. Although recording devices and traditional cameras are not allowed into concert venues, some cell phone use is often encouraged. Journalist Neil Strauss of the *New York Times* wrote in 1998 of the emergence of the use of cell phones at concerts for something other than talking: fans were holding up cell phones, allowing someone not in attendance to hear the music. This first display of cell phone interactivity at concerts was just the beginning. A few years later, bloggers started to write of their experiences at concerts where the crowd was prompted by the artist to hold up their opened clam-shell phones to bask the audience in the blue glow from cell phone screens. At that point, the cell phone had replaced the old (and hazardous) tradition of holding up lit cigarette lighters at concerts (Alderman, 2004). Now, some acts will provide specific wallpaper for fans to download that they can use when they participate in the "screen glow" events.

In an effort to engage music fans in interactive marketing endeavors, many new forms of interaction via the cell phone have emerged to enhance the concert experience that maximize marketing potential for the artist. The advantage for artists is that they generate a list of phone numbers to be used for future marketing projects. Generally fans are given an incentive to register their cell phone numbers, even if it's just a screen saver, wallpaper, or a music track.

TEXT TO SCREEN @ CONCERTS

Text to screen is a concept that incorporates mobile phone texting into concerts and clubs by providing attendees the chance to post a "shout out" on a screen on stage or television monitors in a club. Major concerts are now routinely offering display screens above the stage that allow for fans to use their cell phones to post a text message on the screen, for a fee or in exchange for registering their phone number and agreeing to accept push messages in the future (Leeds, 2007). For smaller venues such as clubs and parties, there are several services that provide entry level packages. In the club setting, these services allow attendees to post their text messages on video monitors to request songs, answer trivia questions, vote on drink specials, submit song requests (thus displacing the request written in lipstick on a napkin), or ask someone to dance. The event coordinator only needs to hook up a computer with Internet access to a video projection system. Some systems are service driven, charging a monthly or per-message charge. They often use a short code number to which fans text their message. Others are software packages that connect your existing cell phone account and merely project your incoming text messages to television screens. Premium features include filtering mechanisms to censor unsuitable content, either automatically, or via a delayed posting that gives the operator time to review each message.

This technology is still relatively new to the entertainment industry, especially for indie artists working in smaller venues. New services are still being launched.

Some of the more notable, affordable services available in 2012 include FireText, Text-to-screen.com, iWall, iVision, and Text Live.

Top 10 Reasons to Try FireText

FireText's patented Entertainment System is widely being used by sporting events, nightclubs, live events, tradeshows and retailers to directly target and communicate with their clients. Our system works by enabling your clients to send text messages to a designated phone number which would then be projected onto a large screen at your venue.

- Entertains your audience like never before.
- Powerful filter keeps your video chats clean.
- Automatically builds a database of phone numbers.
- New advertising platform for sponsors.
- Instant way to notify customers about upcoming events.
- Your customers can interact with the venue.
- Compatible with Excel for data management.
- Advanced logic exclusive to FireText, makes text games fun.
- Transparency templates to overlay live video, DVDs, PowerPoint presentations.
- User friendly and easy to use.

www.firetext.com

The benefits of these services include:

1. Engaging fans in a more interactive experience.
2. Providing fans with an additional form of entertainment during the show.
3. The ability to capture mobile numbers and have participants agree to accept future push messages from the band in exchange for access.

Text to Screen Services

- FireText: www.firetext.com
- Text-to-screen: www.text-to-screen.com.
- iVision: www.ivisionmobile.com
- Text Live: www.textlive.com/
- iWall: www.iwall.com.ar/

Some major acts have incorporated text messaging into other aspects of their concerts. The band Korn encouraged fans at the concert to vote by phone on which song should end their show. Others have featured contests in which concert-goers enter via their cell phone for a chance to win better seats, VIP passes, or other prizes. Fans who are encouraged to participate in these concert-related activities submit their phone numbers and grant the artist permission to send them future marketing messages about the artist. As fans leave the show, the artist has the ability to send them each a message thanking them for coming to the show and offering discounts on merchandise (LaPlant, 2008).

Figure 15.8
FireText text-to-screen features (courtesy of Firetext)

MOBILE APPS

The iPhone introduced the world to mobile apps (short for application). These small software programs are customized to allow the user to perform certain software functions and/or access specific content. A 2011 Nielsen study found that over 30% of mobile users in North America, Europe and nearly 40% of respondents in the Middle East expressed an interest in artist-based smartphone apps.

ReverbNation provides its members with a way to create an app specific for the artist. Artists are able to create an app for the iPhone, Android, and Blackberry platforms. Some suggested uses for such an app include contacting your fan base and quickly signing up new fans at an event. This app is not designed to be distributed to fans, but for use by the artist on their personal smartphone. http://www.reverbnation.com/main/overview_artist?feature=mobile_control_room

MobBase is a service that will assist in the creation of a mobile app that can be given away or sold to fans. While the service supports both the Apple and

Figure 15.9
ReverbNation smartphone app

Android platforms, a recent change in Apple's policy requires each app designer to have their own developer account, thus adding to the cost of implementation. http://www.mobbase.com/.

MOBILE TICKETING

Cell phones are beginning to replace paper tickets in a variety of settings and are expected to soon replace the paper tickets for events and air travel. In addition to reducing the environmental impact and costs, it provides a convenience to both the concert promoter and the ticket holder. Fans can purchase digital tickets on their cell phones, and then a graphic with the bar code is sent to the ticket holder. Fans simply have to scan the onscreen bar code at the entry gate to gain access to the concert. The concert promoter then has access to the phone numbers of all ticket holders, many of whom may have opted in to receive future marketing messages.

Live Nation now includes this feature on their smartphone applications, allowing the ticket holder to scan in the bar code to enter the show. Showclix is a company dedicated to mobile ticketing to events. Concert fans purchase the tickets either online or via their phone, and the QR code image is sent to their phone via SMS. The scanners at the entrance to the venue can verify each ticket, preventing fraud and duplicates.

CONCERT MUSIC DOWNLOADS

As music in the United States moves toward mobile handsets, another opportunity is created to sell recorded music to fans—impulse purchases of concert recordings. Still in its infancy at the time of this writing, there is great

Figure 15.10
Showclix iPhone image (courtesy of Showclix)

potential for artists to "sell back" the concert experience to attendees, even before they've reached their cars in the parking lot. Portions of the event can be audio recorded, mixed on the fly, and offered to attendees for sale via cell phone as they leave the venue. Many times concert attendees have lamented that they wish they had a recording of a concert they attended, and in the future, they will have the opportunity to purchase one on the spot.

Warner Brothers teamed up with Mozes on a campaign to provide concert attendees with a "bootleg" recording of the show they attended in exchange for texting LPSUN to 50505 to receive their digital download, which was emailed to them following the concert. The campaign resulted in the collection of 41,000 fan email addresses for future marketing purposes.

PREPARING YOUR WEB SITE FOR MOBI

Cell phone access to the Internet is bringing new challenges to web design. In the article "Web sites adapt to mobile access," Amanda Kooser described some of those challenges: the small screens, the different formatting of the various mobile devices, limited maneuverability, and so on. Mobile handsets have been relying on the *wireless application protocol* (WAP) to create web sites specifically formatted for mobile devices. The domain system used for wireless devices is the *.mobi* (top level) domain extension on the end of the URL address. The implementation of HTML5 will allow for some standardization that was lacking with WAP (Spooner, 2011).

Just as with an Internet web site, design for a mobile site must be preceded with an understanding of goals for the site. What are you trying to accomplish and provide with a mobile web site? The goals may not be the same as for the computer-based site. An Internet site might focus on more detail, more long-term information, more entertainment, and creating branding and customer awareness. A mobile site will most likely be used by fans on the go who need quick access to information that they can use immediately. This may include information on live shows, such as directions to venues and start times. A tour schedule may be important if groups of fans are together and want to attend an event at a future date. A quick glance at your tour schedule may be what they are looking for. Contests and coupons may be important for fans as well as access to music to download and listen to. Pull messages that appear on billboards, bus boards, or in newspapers, perhaps as QR codes, may create an immediate impulse among viewers to respond to the message via cell phone, such as signing up to win something, downloading new content or visiting the mobile web site. With these goals in mind, a separate web site must be developed that allows mobile fans to quickly access what they are looking for without scrolling through a lot of pages and screens. And the site must be updated often to reflect the changing needs of its visitors.

GENERAL MOBILE WEB SITE FORMATTING RULES

Creating mobile web sites may be as simple as stripping out all CSS formatting to reveal a text-only site similar to how a search engine spider renders a site, although much of the content must also be reduced for easier browsing. Or better yet, professional "mobile" developers can create a site that dynamically produces formatting from a variety of options depending on the type of device accessing the site. In other words, the formatting of the site can vary, and for each visitor, the presentation is customized after the server detects what type of handset the visitor is currently using. HTML5 adoption may be driven by its mobile framework abilities. HTML5 can be used across almost all smartphone platforms; only the Blackberry platform is lacking compatibility. With HTML5, mobile developers can store more content on the device, making it less dependent on connectivity. Canvas and video make it easy for developers to add graphics and video without additional plugins (no need for Flash). Geo-location and advanced forms are other benefits.

Great mobile sites start with emphasizing function over form. The purpose of the site is to address a need for information or access (Henderson, 2011). Performance is important, as is simplicity. Don't build an app when a mobile site will suffice. Apps should be left for things that a mobile site can't do. Mobile apps restrict design and limit access to only those who have installed the app.

When designing a mobile web site, here are some basic rules to consider:

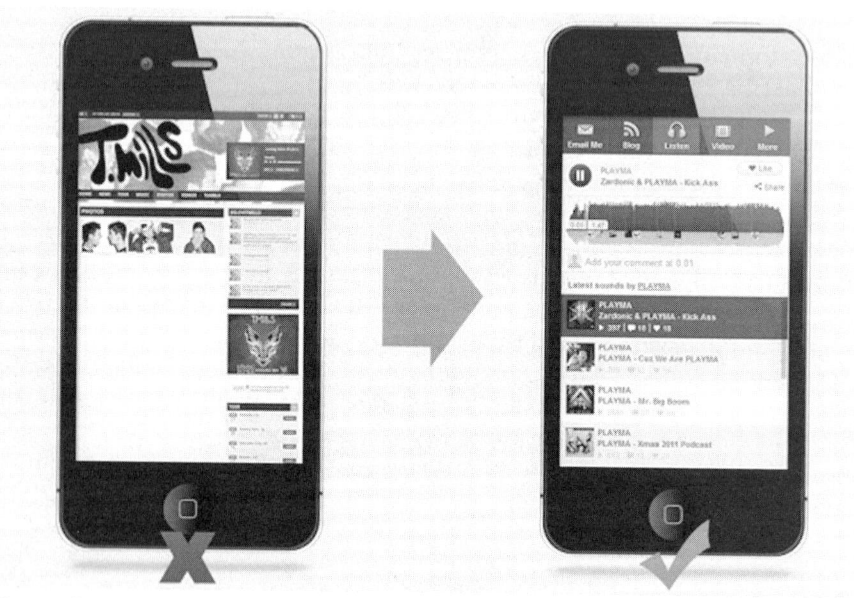

Figure 15.11
Converting computer-based web design to mobile-based web design

- Design for the small screen. Be realistic about what will fit on the screen of a mobile device. Paring down your information for the screen size is first and foremost.
- Design for touch screens. Links should be spaced far enough apart to be maneuvered by large fingers.
- Eliminate all features and information that is not important to the mobile user. Less is more on a mobile site. Visitors are usually looking for specific information or completing a specific task.
- Avoid vertical scrolling. Web users don't like it, and it's even less popular with mobile users.
- But use horizontal swiping. smartphone users have become accustomed to swiping right to left to move on to more content, but make sure swiping features are unambiguous. Have cues when swiping is possible and be consistent.
- Reduce the number of clicks. Don't go deep in page numbers; it's slow going for the mobile user.
- Keep it clean. Use readable text on a readable background.
- Test your design on a variety of handheld devices to see if it holds up on all of them.
- Use abbreviations and succinct wording wherever possible.
- Have clearly visible links from the mobile site to the full site and back again.
- Access keys are helpful for speeding up navigation. (Access keys are those just below the screen on the left and right that can take on a variety of functions, usually with the name of that function on the bottom of the screen just above the key.)

MOBILE SITE DESIGN PROGRAMS

Mobile web site development is still not commonplace among web designers. Adobe 5.5 has some new mobile development features including templates (Henderson, 2011). ShareSquare offers an easy conversion process whereby you enter in your site name and the system automatically pulls content from the site for your mobile site. Then you tweak the mobile site to offer the features most needed by visitors. ShareSquare offers a free version, or a premium version for less than $10 per month.

SiteSpinner Pro (mentioned in Chapter 5) offers mobile site design software. The cost is about $100. Its development software is powered by mobiSiteGalore. The mobiSiteGalore offers a free WYSIWYG mobile web site builder including multiple pages, a link manager, and an image editor. Google launched a free mobile web site building tool in 2011. They offer free templates, e-commerce, analytics, and some customization. Mofuse offers both DIY and custom designs. Plans start at less than $10 per month. Wirenode's simple editor lets users personalize their mobile site with colors and personal images, and create, edit, rearrange and delete individual pages. Zinadoo is a quick, easy-to-use service that offers a variety of subscription plans and a 14-day free trial period. MobileMo

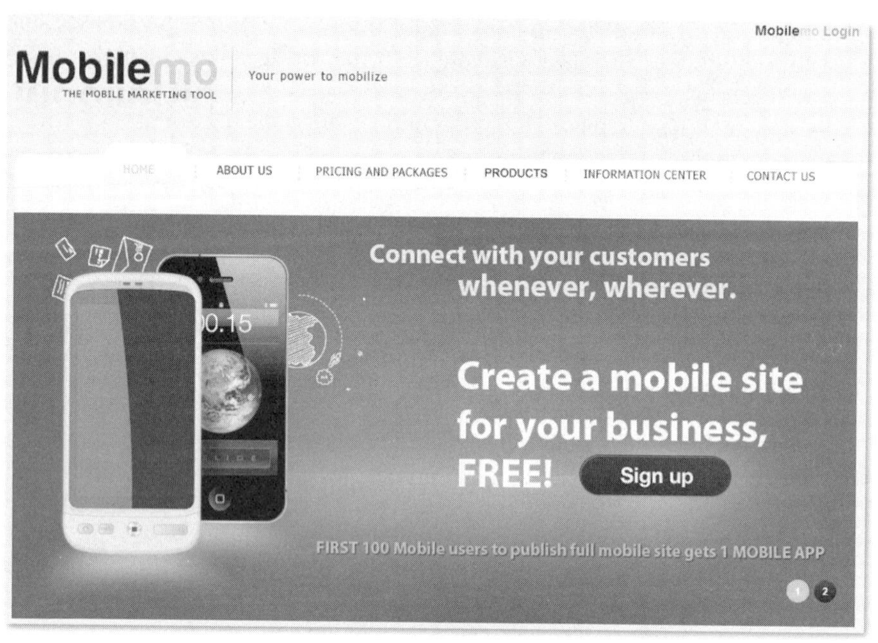

Figure 15.12
Mockup of mobi site from mobilemo (courtesy of www.mobilemo.com)

offers a free basic version and several premium levels. In addition to site creation, MobileMo offers integration with Facebook and mobile apps.

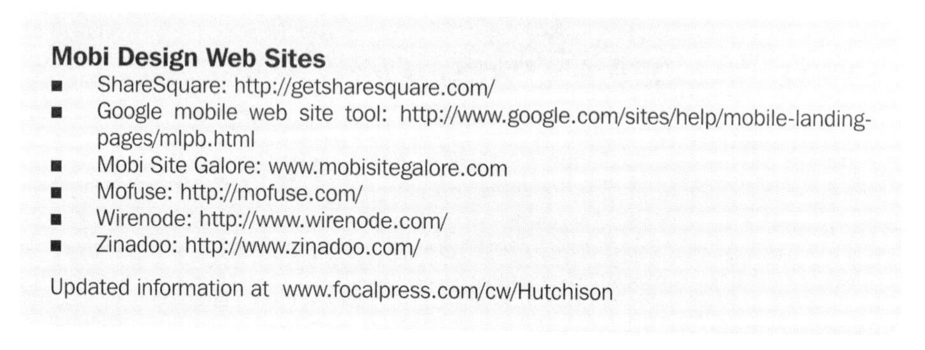

Mobi Design Web Sites

- ShareSquare: http://getsharesquare.com/
- Google mobile web site tool: http://www.google.com/sites/help/mobile-landing-pages/mlpb.html
- Mobi Site Galore: www.mobisitegalore.com
- Mofuse: http://mofuse.com/
- Wirenode: http://www.wirenode.com/
- Zinadoo: http://www.zinadoo.com/

Updated information at www.focalpress.com/cw/Hutchison

CONCLUSION

Mobile marketing has come to music in a big way. With over 14 million Americans using QR codes, and with 91% of all U.S. citizens having their mobile device within reach at all times (Bahaijoub, 2011), having a mobile presence is no longer a luxury. For artists who want to stay up to date, creating a "mobile" site now may increase marketing success with a small investment and may generate incremental sales and broaden the fan base. The combination of viral/push text messaging and having a mobile web presence allows for two-stage marketing campaigns, where a push message is sent to fans, who then can respond to the call for action by visiting the mobile site to buy music, check the tour schedule, or enter contests. The incorporation of mobile devices into social networking, live concerts, ticketing, and QR codes has made mobile marketing a necessary part of any artist's marketing campaign.

Glossary

CSS Short for cascading style sheets, a feature added to HTML that gives both web site developers and users more control over how pages are displayed. With CSS, designers and users can create style sheets that define how different elements, such as headers and links, appear. These style sheets can then be applied to any web page.

Mastertones Short excerpts from an original sound recording that plays when a phone rings.

Mobi Top-level Internet domain used for web sites that supply content to cell phones and other handheld devices with tiny screens. The .mobi suffix was introduced in 2005.

Mobile music Music that is downloaded to mobile phones and played by mobile phones. Although any phones play music as ringtones, true "music phones" generally allow users to import audio files from their PCs or download them wirelessly from a content provider.

Nearfield communication Short-range wireless technology in use by stores for communications and transactions with customers' mobile devices.

Opt in/Opt out A consumer can specify whether they want to receive marketing messages or remove their email address from the list.

Phone personalization features Customized phone features such as ringtones, wallpaper, and skins that distinguish a person's cell phone as unique and allows consumers to express themselves.

Proximity marketing Localized wireless distribution of advertising content associated with a particular location, retail establishment, transportation hub, or venue.

Push message A specially formatted cell-phone text message that gives you the option of connecting directly to a particular mobile Internet site, using your phone's browser.

QR codes Quick response codes are square codes similar to bar codes and are used to communicate information to smartphones, such as to call a certain phone number or visit a particular web page.

Short code A short version of a phone number that cell phone owners type in to send a text message to vote or participate in a marketing activity.

Sideload A term used in Internet culture, similar to upload and download; the process of moving data between two electronic devices without involving the local computer in the process.

Short message service (SMS) Usually refers to wireless alphanumeric text messages sent to a cell phone.

Squeeze page A landing page created to solicit registration of consumers to collect email addresses and/or cell phone numbers for marketing purposes.

Text-to-screen A phenomenon where event attendees can post messages on large public screens using their mobile devices and a text messaging service.

Text messaging Also called SMS (short message service); allows short text messages to be sent and received on a mobile phone. Messages can be sent from one phone to another by addressing the message to the recipient's phone number.

Text to screen A system allowing event participants to send text messages that are displayed on a screen at the event.

Top level domain The last part of an Internet domain name—that is, the letters that follow the final dot of any domain name.

XML (eXtensible Markup Language) A standard that forms the basis for most modern markup languages. XML is an extremely flexible format that only defines "ground rules" for other languages that define a format for structured data designed to be interpreted by software on devices. XML by itself is not a data format (phonescoop.com).

WAP Wireless application protocol; a protocol designed for advanced wireless devices; allows the easy transmission of data signals, particularly Internet content, to microbrowsers built into the device's software.

Bibliography

Alderman, John. (2004, March 11). Mobile phones are the new lighters. Textually.org. www.textually.org/textually/archives/2004/03/003233.htm

Bahaijoub, Jem (2011). Mobile marketing for independent artists. http://www.musicthinktank.com/blog/mobile-marketing-for-independent-artists-strategy.html

Comscore (2011) http://www.comscoredatamine.com/2011/12/20-million-americans--scanned-a-qr-code-in-October/

Digital Strategy Consulting (2011). Mobile music sales to reach $5.5bn by 2015. http://www.digitalstrategyconsulting.com/intelligence/2011/02/mobile_music_sales_to_reach_55.php

Gauntt, John. (2006, May 26). Marketing mobile music. EMarketer. www.imediaconnection.com/content/9747.asp

Henderson, Andrew (2011). Ten best practices for designing mobile websites. Inspire. http://www.adobe.com/newsletters/inspire/november2011/articles/article6/

Holden, Windsor (2008) Mobile music adoption passes tipping point with revenues set to reach $17.5bn by 2012, February 26, 2008 http://juniperresearch.com/viewpressrelease.php?pr=80

Imajinn (2006, August 25). Proximity marketing revealed. BizCommunity.com. www.bizcommunity.com/Article/196/16/11399.html (link no longer active).

International Federation of Phonographic Industries. Digital Music Reports 2009–2011. http://www.ifpi.org/content/section_statistics/index.html

Jayvee (2006, November 28). How to make viral marketing appealing for phones, www.cellphone9.com/how-to-make-viral-marketing-appealing-for-phones

LaPlant, Alice (2008). MP3 unplugged: rethinking the digital music future. Information Week. www.informationweek.com/shared/pritableArticleSrc.jhml:jsessionid=DTLSCSY...

Leeds, Jeff (2007, August 15). They've just got to get a message to you. *New York Times*. http://www.nytimes.com/2007/08/15/arts/music/15conc.html

Mobile Marketer (2012). Classic guide to mobile advertising. www.mobilemarketer.com/cms/lib/10465.pdf

Nester, Katharine, & Kurt, Lyall (2003). Mobile marketing: a primer report. FirstPartner Research and Marketing. pdfsource.org/pdf/mobile-marketing-a-primer-report.html

Nielsen (2011). Music, money & mobile: A global music outlook. http://blog.nielsen.com/nielsenwire/global/music-money-mobile-a-global-music-outlook/

Nielsen (2011). The hyper fragmented world of music. Report prepared for the MIDEM conference.

NielsenWire (2012, January 9). Consumers OK with ads… if the apps are Free. NielsenWire. http://blog.nielsen.com/nielsenwire/consumer/consumers-ok-with-ads-if-the-apps-are-free/

Pandey, Shailendra (2010) Effective Monetization of Music on Mobile. October 25. http://blog.midem.com/2010/10/effective-monetisation-of-music-on-mobile-midemnet-2010-presentation/

Peoples, Glenn (2011). Mobile sales plummet, subscription revenues slip in RIAA 2010 year-end stats. *Billboard Magazine*. http://www.billboard.biz/bbbiz/industry/digital-and-mobile/mobile-sales-plummet-subscription-revenues-1005160832.story

Perez, Sarah (2011, June 29). Google launches free mobile website builder. ReadWriteWeb. http://www.readwriteweb.com/archives/google_launches_free_mobile_website_builder.php

Pew Internet and American Life Project (2011). Americans and text messaging. http://pewinternet.org/Reports/2011/Cell-Phone-Texting-2011/Summary-of-Findings/Summary-of-Findings.aspx

Raleigh, Bobby (2011, April 20). SMS text-messaging tools for bands. eHow. http://www.ehow.com/info_8263704_sms-textmessaging-tools-bands.html

Smith, Clyde (2012, February 16). ShareSquare offers quick & easy HTML5 mobile sites for musicians, bloggers. Hyperbot.com. http://www.hypebot.com/hypebot/2012/02/sharesquare-offers-quick-easy-html5-mobile-sites.html

SMS Marketing Services (2012). Exploring SMS marketing for your business. http://www.smsmarketingservices.net/exploring-sms-marketing-for-your-business/

SMS Marketing Services (2012). The key to keywords. http://www.smsmarketingservices.net/the-key-to-keywords/

SMS Marketing Services (2012). All about short codes. http://www.smsmarketingservices.net/all-about-short-codes/

SMS Marketing Services (2012). Boost customer relations with automated messaging campaigns. http://www.smsmarketingservices.net/boost-customer-relations-with-automated-messaging-campaigns/

Spooner (June 14, 2011). WAP v. HTML5 v. native: WAP, a language from the past. www.ubelly.com/2011/06/wap-v-html5-v-native/

Strauss, Neil (1998, December 9). A Concert communion with cell phones; press 1 to share song, 2 for encore, 3 for diversion, 4 to schmooze. *New York Times*, http://query.nytimes.com/gst/fullpage.html?res=9C01E2DF153AF93AA35751C1A96E958260

Wikipedia (2008). Proximity marketing.

Index

Figures are shown by *italic* page numbers and tables are shown with **bold** page numbers